Modern Power Transformer Practice

MODERN POWER TRANSFORMER PRACTICE

Edited by

R. FEINBERG, Dr.-Ing., M.Sc., F.I.E.E.

A HALSTED PRESS BOOK

JOHN WILEY & SONS
New York

First published in Great Britain 1979 by The Macmillan Press Ltd

Published in the U.S.A. by
Halsted Press, a Division of
John Wiley & Sons, Inc.
New York

Printed in Great Britain

Library of Congress Cataloging in Publication Data

Main entry under title:

Modern power transformer practice.

   Includes bibliographies.
   1. Electric transformers. I. Feinberg, R.
TK2551.M54   621.31′4   78–5608
ISBN 0 470–26344–X

Contents

The Contributors

Preface

Generators and transformers are two major cornerstones in the fabric of any electric power supply system. In 1975 the installed generator capacity was in the region of about 650 GW in Europe and about 585 GW in the USA and Canada. By taking a ratio of about 7:1, the associated installed transformer capacity is about seven times the generator capacity which gives an idea of the magnitude of transformer capacity in service in those parts of the world alone. All over the world the total transformer capacity in service is substantially larger; this signifies the importance and vital duty of power transformers.

The book is intended essentially as a statement on the current state of the art of design, manufacture and operation of power transformers. It arose from a well-attended course of lectures given to practising engineers of the industries of electric power supply and of transformer manufacture. An editorial effort was made to integrate the entire material into a book approached and presented at a standard level.

Standard specifications play an important part in the choice, design, manufacture and operation of power transformers. A selection of specifications of the International Electrotechnical Commission (IEC), the British Standards Institution (BSI) and the USA Standards is given in section 1.14. Of necessity the list is far from complete. Its primary purpose is to stimulate interest and to provide a pattern of sources for authoritative information also in other countries.

The terminology used is uniform and in line with the current revision of the *International Electrotechnical Vocabulary* of the IEC; the classification of power transformers within the context of this book is explained in section 1.1. The letter symbols for quantities and for the SI units of measurement are in accordance with the IEC publication *Letter Symbols to be Used in Electrical Technology*, Part 1: General, 27-1 (1971).

A systematic guide to the contents of the book is given in chapter 1. The book is addressed to a wide range of practising engineers, and students may use it as a factual reference at the initial stages of project or research work.

Cheadle, Cheshire, 1978   R.F.

List of Quantity Symbols

These quantity symbols are common to chapters 1 to 11.

Note. For electrical and thermal quantities the subscript 1 refers to the low-voltage and the subscript 2 to the high-voltage winding.

Symbol	Meaning	Unit
A_C	external cooling surface of a coil	m^2
A_{Cu}	total copper cross-section per phase	m^2
A_{Fe}	core leg net-cross-section area	m^2
$A_{Fe,g}$	core leg gross cross-section area	m^2
A_T	external effective tank cooling surface	m^2
A_t	$= A_{Cu} + A_{Fe}$	m^2
A_w	nett core window area between core circles	m^2
A_θ	cross-section area for thermal power flow	m^2
$A_{\theta l}$	heat transfer surface between winding layers	m^2
B_m	peak value of magnetic flux density in core leg	T
B_{lm}	peak value of leakage flux density in axial duct	T
C_t	total cost of active copper and iron	U_m
I, I_1, I_2	winding currents	A
I_k	winding current under fault condition	A
I_m	normal full-load winding current	A
IN	$= I_1 N_1 = I_2 N_2$ winding ampere-turns	A
J, J_1, J_2	winding current density	$A\,mm^{-2}$
J_k	short-circuit winding current density	$A\,mm^{-2}$
K	see equation 3.15	
K_A	$= A_{Fe} \times A_{Cu}$	
K_{AS}	output coefficient, see equation 2.55	
K_C	cost coefficient	
K_c	empirical heat transfer coefficient	
K_{et}	empirical coefficient for total heat transfer from a tank surface	
K_T	tank heat transfer coefficient, see equation 2.70	
K_t	$= (1/K_{et})^{0.8}$	
K_{VS}	output coefficient, see equation 2.59	
M_e	total thermal power transfer from unit surface area	$W\,m^{-2}$

Symbol	Meaning	Unit
M_{ec}	thermal power transfer by convection per unit area	$W\,m^{-2}$
M_{er}	thermal power transfer by radiation per unit area	$W\,m^{-2}$
M_{et}	thermal power transfer from tank surface per unit area	$W\,m^{-2}$
M_{eT}	average of M_{et} over effective tank cooling surface	$W\,m^{-2}$
N, N_1, N_2	winding turns	
P_{Cu}	$= P_{Cu1} + P_{Cu2}$	kW
P_{Cu1}, P_{Cu2}	copper loss per phase at 75 °C	kW
$P_{Cu,t}$	total copper loss	kW
P_{eh}	vertical thermal power flow	W
P_{Fe}	total iron loss	kW
P_1	total load loss at 75 °C	W
P_R	I^2R loss per phase	kW
P_{tot}	total transformer loss	kW
$\%P_{Cu}$	percentage copper loss	
$\%P_i, \%P_{i1}, \%P_i$	percentage conductor eddy current loss	
R_1, R_2	winding resistance per phase at 75 °C	
$\%R$	percentage resistance	
$R_{\theta b}$	horizontal equivalent thermal resistance in winding	$K\,W^{-1}$
$R_{\theta h}$	vertical equivalent thermal resistance in winding	$K\,W^{-1}$
S	rating per phase	MVA
V, V_1, V_2	winding voltage per phase	V
$\%X$	percentage reactance	
$\%Z$	percentage impedance	
a, a_1, a_2	cross-section area of a conductor strand	mm^2
a_θ	see equation 9.1	
b_0	radial clearance between low- and high-voltage windings	mm
b_1, b_2	radial widths of windings	mm
b_{01}	radial clearance between core leg and low-voltage winding	mm
b_{02}	radial clearance between high-voltage windings	mm
b_C	breadth of rectangular conductor	mm
b_c	breadth of conductor strand	mm
b_{cen}	distance between centres of core legs	mm
b_d	width of vertical duct between coil sections	mm
b_{Fe}	half the width of widest core leg plates	mm
b_w	width of core window	mm

Symbol	Meaning	Unit
b_X	reactive width of windings, see equation 2.60	mm
c	distance between top and bottom entry pipes	mm
c_{Cu}	cost per kilogram of copper	$U_m\,kg^{-1}$
c_{Fe}	cost per kilogram of core steel	$U_m\,kg^{-1}$
d	diameter of circle circumscribing core leg	mm
d_c	diameter of round conductors	mm
f	frequency	Hz
h	assumed equal height of low- and high-voltage windings	mm
h_0	assumed equal total axial clearance between windings and core yoke	mm
h_1, h_2	height of low and high-voltage windings, respectively	mm
h_{01}, h_{02}	axial clearance between low- and high-voltage windings and core yoke	mm
h_c	height of bare conductor strand	mm
h_d	height of horizontal ducts between coil sections	mm
h_{Fe}	height of core	mm
h_w	height of core window	mm
h_x	meaning h_1 or h_2 whichever is the greater	mm
h_θ	height of centre of transformer heating	mm
k, k_1, k_2	winding space factors	
k_C	cost ratio of copper to total of active material	
k_c	core circle space factor	
k_e	$= k_{i2}/k_{i1}$	
k_f	$= 1/k_1 + 1/k_2$	
k_F	see equation 3.7	
k_{Fe}	space factor of core laminations	
k_h	$= h_2/h_1$	
k_{hX}	see equation 3.4	
k_i, k_{i1}, k_{i2}	conductor eddy current loss factor, see equation 3.6	
k_J	$= J_2/J_1$	
k_{Js}	see equation 3.5	
k_S	fractional load for maximum efficiency	
k_s	$= 2b_{Fe}/d$	
k_{sX}	see equation 3.3	
k_{t1}, k_{t2}	correction to N_1, N_2 to allow for tappings	
k_v	$= \Delta\theta_v/\Delta\theta_o$	
k_w	window space factor, see equation 2.35	

Symbol	Meaning	Unit
l_{Fe}	total length of core legs and yokes	m
m_{Cu}	total mass of active copper	kg
m_{Fe}	total mass of active iron	kg
n	N_2/N_1; number of copper layers in a coil cooling only from vertical surfaces	
n_0	number of conductor strands in parallel	
n_b	number of horizontal conductor strands	
n_h	number of vertical conductor strands	
n_{sb}	number of horizontal sections in a winding	
n_{sh}	number of vertical sections in a winding	
p_{Cu}	specific copper loss	$W\,kg^{-1}$
p_{Fe}	specific iron loss	$W\,kg^{-1}$
p_i	specific conductor eddy current loss	$W\,m^{-3}$
p_R	specific I^2R loss	$W\,kg^{-1}$
s	$= \frac{1}{2}(s_1 + s_2)$	mm
s_0	length of mean turn of interwinding axial ducts	mm
s_1, s_2	length of mean turn of winding	mm
s_t	mean winding circumference on each leg multiplied by number of legs	m
t_k	duration of short circuit	s
ϕ_m	peak value of magnetic flux in core leg	Wb
δ_b	horizontal insulation thickness between conductor strands	mm
δ_{b1}	insulation thickness at sides of a coil section	mm
δ_{be}	external insulation thickness at coil cooling surfaces	mm
δ_{eb}	equivalent horizontal insulation thickness between conductor strands	mm
δ_{eh}	equivalent vertical insulation thickness between conductor strands	mm
δ_h	vertical insulation thickness between conductor strands	mm
δ_{h1}	insulation thickness between upper and lower surfaces of coil section	mm
ε	thermal emissivity of a surface	$W\,m^{-2}\,K^{-4}$
$\Delta\theta$	temperature difference between surface and cooling medium	°C
θ_a	ambient temperature	°C
θ_H	hot-spot temperature	°C
$\Delta\theta_H$	hot-spot temperature rise	°C
$\Delta\theta_o$	top oil temperature rise	°C
$\Delta\theta_{om}$	average oil temperature rise	°C
$\Delta\theta_R$	mean winding temperature rise	°C

Symbol	Meaning	Unit
$\Delta\theta_{so}$	temperature drop at a coil surface	°C
$\Delta\theta_{v}$	oil temperature difference between top and bottom of tank	°C
θ_{w}	winding temperature measured by resistance	°C
$\Delta\theta_{wo}$	mean winding temperature rise above oil	°C
$\Delta\theta_{wom}$	maximum winding temperature rise above oil	°C
λ	thermal conductivity	$Wm^{-1}K^{-1}$
ρ	resistivity of copper strip, taken as $21.4 \times 10^{-3}\,\Omega\,m$	
$\rho_{d,Cu}$	mass density of copper, taken as $8890\,kg\,m^{-3}$	
$\rho_{d,Fe}$	mass density of core steel, taken as $7650\,kgm^{-3}$	

1

General Information

R. Feinberg*

1.1 POWER TRANSFORMER CLASSIFICATION

The types of transformers considered in the following chapters are commonly employed in the chain of electric power supply from generating stations to consumers of electric energy. They include some special transformers designed for particular industrial purposes.

Within the context of this book, transformers have been classified as power system and as distribution transformers, where the phrase power system refers to the transformer chain from generating station to first distribution points and the phrase distribution system means the subsequent chain to final distribution points. Power system transformers and inductors are discussed in chapter 10, and distribution transformers in chapters 9 and 12 with special types in sections 11.1, 11.2 and 11.5. The remainder of chapter 11 deals with other special types of transformers.

1.2 REVIEW OF BASIC THEORY

1.2.1 Electromagnetic induction

The phenomenon of electromagnetic induction, discovered by Michael Faraday in 1831, manifests itself in two ways: (1) as self-induction produced in inductors and underlying the principle of an auto-transformer and (2) as mutual induction which is the basis of transformers with two or more windings.

Assume a coil with N turns in a magnetic field and that each of the turns is linked with the same magnetic flux ϕ which varies with time. The law of

---

* Consultant, formerly Ferranti Limited.

electromagnetic induction gives for the instantaneous voltage v produced across the coil the expression

$$v = -N\,d\phi/dt \tag{1.1}$$

Let ϕ be a sinusoidally varying flux of amplitude ϕ_m and frequency f

$$\phi = \phi_m \cos(2\,\pi ft) \tag{1.2}$$

then, from equation 1.1,

$$v = 2\pi fN\phi_m \sin(2\pi ft) = V2^{1/2}\sin(2\pi ft) \tag{1.3}$$

which means

$$V/N = 2^{1/2}\pi f\phi_m = 4.44f\phi_m = 4.44fB_m A_{Fe} \tag{1.4}$$

where B_m is the amplitude of sinusoidal flux density and A_{Fe} the nett cross-section of the iron core carrying the sinusoidal flux (see equation 1.2). The quotient V/N in equation 1.4 denotes the voltage per turn of the given coil.

In the two-winding transformer depicted in figure 1.1(a) with N_1 and N_2

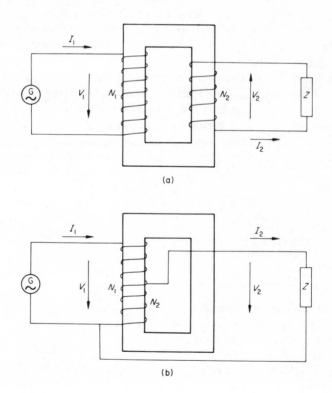

(a)

(b)

Figure 1.1   Transformer principle: (a) two-winding transformer; (b) auto-transformer

denoting the turns of primary and secondary winding, respectively, the turns ratio n of secondary to primary winding is

$$n = N_2/N_1 \qquad (1.5)$$

Assume an ideal transformer, that is a transformer without any leakage flux or without any losses, then in accordance with equations 1.4 and 1.5

$$V_2/V_1 = n \qquad (1.6)$$

where V_2 and V_1 are the respective rms voltages across the secondary and the primary winding. A transformer with $n > 1$ is a step-up transformer, and that with $n < 1$ a step-down transformer.

The auto-transformer shown in figure 1.1(b) is in principle a tapped inductor with N_1 turns for the whole coil and N_2 turns for the tapped part of the coil. As shown in figure 1.1(b) the whole coil functions as the primary winding, it is connected to the source of power supply, and the tapped part is the secondary winding; the turns ratio is $n < 1$, the transformer operates as a step-down transformer. By reversing the input and output connections, the turns ratio is changed to $n > 1$, and the transformer becomes a step-up transformer.

The two transformers in figure 1.1 are of the single-phase type. In either case, a three-phase transformer represents a logical extension of the single-phase version.

1.2.2 Ampere–turn balance

Assume idealised transformers in figure 1.1, that is a magnetic flux without a magnetising current and the absence of leakage flux and of losses. With the respective currents I_1 and I_2 in the primary and secondary circuits (see figure 1.1) the law of ampere–turn balance in the transformer windings requires the relationship

$$I_2 N_2 = I_1 N_1 \qquad (1.7)$$

which means, with equation 1.5,

$$I_2/I_1 = 1/n \qquad (1.8)$$

1.2.3 Change of turns ratio

According to equation 1.6 the voltage ratio of a transformer is altered by changing the turns ratio. This is achieved by means of tappings on either the primary or secondary winding of the transformer. The principle of tapping on the secondary winding is indicated in figure 1.2(a) for a two-winding transformer and in figure 1.2(b) for an auto-transformer.

1.2.4 Magnetising current

The relative permeability of the core steel of a power transformer varies with the magnitude of magnetic flux density in the core and the effect of magnetic

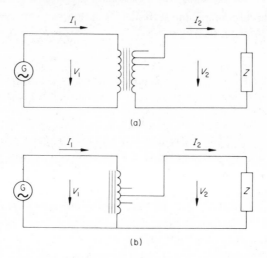

Figure 1.2   Transformer with tappings on secondary winding: (a) two-winding transformer; (b) auto-transformer

hysteresis. Power transformers are normally operated with a sinusoidal primary voltage; this means that the magnetic flux density in their cores is sinusoidal and that the magnetic field strength and hence the magnetising current are non-sinusoidal.

The magnetising current contains third and higher-order odd-number harmonics which substantially increase in magnitude as the amplitude of magnetic flux density exceeds the knee of the magnetisation characteristic of the core steel. There are circumstances where the harmonic content of the magnetising current could affect in an undesirable manner the operation of a transformer and of the power supply system. Remedial measures in transformer design are explained in sub-section 10.6.2.

1.2.5   Equivalent circuit

If we assume a sinusoidal magnetising current and the absence of leakage flux and losses, figure 1.3 demonstrates the equivalent circuit for the transformers in figure 1.1 in accordance with equations 1.6 and 1.8. With the notations of figures 1.3 and 1.1 it is

$$I'_2 = nI_2 \tag{1.9}$$

and

$$Z' = Z/n^2 \tag{1.10}$$

where Z is the impedance of the load in figure 1.1 and Z' the equivalent impedance in figure 1.3.

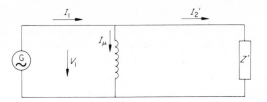

Figure 1.3   Equivalent circuit for the transformers in figure 1.1(a) and (b)

From figure 1.3, the current supplied from the source is

$$I_1 = I_\mu + I_2'$$

(1.11)

where I_μ is the phasor of the magnetising current supplied to the transformer primary winding. Figure 1.4 gives the phasor diagram for the equivalent circuit figure 1.3.

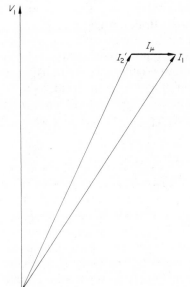

Figure 1.4   Phasor diagram for the equivalent transformer circuit

1.3   TRANSFORMER REACTANCE

A transformer core has unavoidable magnetic reluctance just as an electric conductor has resistance (apart from the special case of a superconductor). Because of the existence of reluctance, some of the magnetic flux is channelled outside the core of a transformer; it is described as leakage flux and sometimes also as stray flux. The magnitude of leakage flux is proportional to the current in the windings.

The effect of leakage flux in respect of the performance of a transformer is symbolised in the equivalent circuit diagram of the transformer by appropriate inductances, known as leakage inductances. The corresponding total leakage reactance between the input and output terminals of the transformer denotes the transformer reactance. This quantity is of basic importance in the operation and thus for the design of a transformer.

1.4   TRANSFORMER LOSSES

1.4.1   Classification

Thermal power generated in a transformer is an undesirable but unavoidable by-product of normal transformer operation. Transformer losses are the electric power, supplied by the source, that is converted to thermal power in the transformer and which has to be dissipated.

A distinction is made between the no-load loss, produced when the transformer is connected to the power supply system without being connected to a load, and the load loss that arises additionally when the transformer is connected to a load. In the design and operation of a transformer the two aspects of loss are considered separately, but they are considered jointly in respect of the dissipation of the thermal power generated in the transformer.

1.4.2   No-load loss

The no-load loss, also called iron loss, in a transformer arises in the core from the effects of magnetic hysteresis and of eddy currents. The loss component due to magnetic hysteresis of the core steel is proportional, on the one hand, to the power supply frequency and, on the other hand, to the peak value of magnetic flux density raised to the power of between 1.6 and 2.5 depending on the core material used and the range of flux density. The loss component due to eddy currents in the steel laminations of which the core is assembled is proportional to the squares of frequency, peak value of magnetic flux density and thickness of the steel laminations; to keep the eddy currents produced of a low value, the steel laminations are thin and electrically insulated from one another (see sub-section 4.2.1).

1.4.3   Load loss

The load loss in a transformer results from the load current carried in the windings. The bulk of load loss, usually described as copper loss, arises in the conductors of the windings, and the remainder of load loss, usually called stray loss, is caused outside the windings, by magnetic leakage that traverses metal parts such as clamps and tank walls. The copper loss consists of two components:

(1) one known as the $I^2 R$ loss and (2) the other designated the conductor eddy current loss.

The $I^2 R$ loss is due to the Joule effect in the conductors. It arises in each winding and is, as the term indicates, equal to the product of the winding resistance and the square of current carried in it. To keep its magnitude low, the winding resistance is made as low as possible which means that the winding is designed to have in its conductor a current density as high as permissible at full-load current.

The conductor eddy current loss is produced by circulating currents induced in those parts of the conductor which are traversed by magnetic leakage flux. Low conductor eddy current loss is achieved by using thin conductors with the thin side perpendicular to the direction of the leakage magnetic flux density, that is perpendicular to the axis of the winding, or otherwise, where necessary, by employing conductor strands in parallel and insulated from one another combined with transposition, that is cross-overs, of the strands as described in sub-section 5.6.2 and also considered in a worked example in sub-section 9.10.1.

The magnitude of stray loss is discussed in sub-sections 5.6.3 and 5.6.4.

1.4.4 Cooling

The thermal power generated in the core and in the windings is removed from the places of origin by means of a cooling medium. In the core the thermal power is conducted through the laminations to the core surface and is then transferred to the cooling medium. In the windings the thermal power passes through the conductor body to the conductor surface, from where it passes through the insulation to the insulation surface for transfer to the cooling medium.

The cooling medium is air or gas in a dry-type transformer (see sections 11.1 and 11.5) or mineral oil in the commonly used oil-immersed transformers. The cooling medium transports the thermal power to the places for dissipation, either to ambient air or to running water.

The cooling arrangement of a transformer has to be designed in such a manner that the areas of maximum temperature in the core and in the windings, the so-called hot spots, always remain below specified maximum values. Overheating of the transformer core can lead to damage as outlined in sub-section 4.5.3, and overheating of the windings means accelerated ageing of the insulation and thus reduction of the life of the transformer.

1.5 TRANSFORMER IMPEDANCE, RESISTANCE AND REACTANCE VOLTAGES

A short-circuit test as described in sub-section 7.4.8 provides measured information on the load loss and the reactance of a given transformer. In such a test the terminals of the transformer secondary winding are short circuited which means that the transformer represents at its input terminals a complex network configuration with an impedance of which the resistive part is a measure of the load loss and the reactive part a measure of the transformer reactance.

The voltage that is required at the transformer input terminals to produce a short-circuit current equal to the rated value of load current is called the impedance voltage of the transformer. Its active component denotes the resistance voltage and its reactive component the reactance voltage of the transformer.

1.6   PERCENTAGE RESISTANCE, REACTANCE AND IMPEDANCE OF A TRANSFORMER

The resistance voltage V_R of a transformer is an indicator of the magnitude of load loss of the transformer, and the reactance voltage V_X an indicator of the transformer reactance. For design and operational purposes it is convenient to express both V_R and V_X as percentages of the corresponding rated voltage V of the transformer, all voltages being expressed per phase.

The percentage values of the ratios V_R/V and V_X/V are expressed by the symbols $\%R$ and $\%X$, respectively. Thus

$$\%R = 100V_R/V \tag{1.12}$$

and

$$\%X = 100V_X/V \tag{1.13}$$

Similarly, the percentage impedance of the transformer is

$$\%Z = 100V_Z/V = 100(V_R^2 + V_X^2)^{1/2}/V = \{(\%R)^2 + (\%X)^2\}^{1/2} \tag{1.14}$$

From equations 1.12 and 1.13 for a three-phase transformer with load loss P_1, reactance X, voltage V and current I, all per phase, the percentage values are

$$\%R = 100P_1/VI \tag{1.15}$$

and

$$\%X = 100X/VI \tag{1.16}$$

1.7   TRANSFORMER DESIGNING PROCEDURE

Skilfully designing a power transformer, particularly in the range of larger sizes, provides considerable scope for creative and imaginative activity based on a profound understanding of the underlying principles of utilising the characteristic properties of the materials employed for transformer construction so as to achieve a satisfactory compromise between conflicting demands of a technical and an economic nature.

The basic information from which a transformer designer starts to create his design is outlined in chapter 3, where section 3.2 deals with the customer's specification and section 3.3 with the designer's specification. Chapter 2 explains the general principle of design procedure without using a computer. The necessary calculations are shown for the design of a 750 kVA distribution transformer in chapter 2 and in sections 9.9 to 9.12. A design procedure with the aid of an electronic digital computer is described in chapter 3.

The various aspects in preparing a decision for a customer's specification on distribution transformers are discussed in sections 12.1 to 12.6. A designer's specification is controlled by the particular manufacturer's code of practice or by standards for the relevant design. In addition, there are standard specifications of the International Electrotechnical Commission and of national standardisation authorities; examples are listed in section 1.14. Examples of customer's specifications are given as references in chapters 9 to 12.

1.8 TRANSFORMER MAIN PARTS

1.8.1 General

The main parts of a power transformer are the core, the windings with means for tap changing either off-load or on-load, the tank containing the core and windings, the terminal arrangements for entry for the external electric circuit connections and the cooling arrangement to remove the heat generated in the core and in the windings and to dissipate it. In addition, there is protective gear which operates when a fault arises in the transformer.

1.8.2 Core construction

The various practical aspects concerning the magnetic material and the techniques employed to construct a core are described in chapter 4 and in section 9.3. Aspects for gapped-core power inductors are discussed in sub-section 10.4.2.

1.8.3 Winding construction

Common practical aspects concerning the construction of various types of transformer winding, together with questions of relevant insulation and conductor material, are treated in chapter 5 and in section 9.4. Chapter 5 also deals with measures in the electric design of high-voltage windings to meet the conditions of surge-voltage testing described in chapter 7.

Additional points of interest are set forth in sub-sections 10.2.2, 10.3.2, 10.3.4 and 10.3.5, in section 10.6, in sub-sections 11.1.2, 11.1.3 and 11.3.2 and in sections

11.5 and 11.6. Windings of power system inductors are represented in sub-section 10.4.3 and section 10.5.

1.8.4 On-load tap changers

Chapter 6 states the conditions of operation of on-load tap changers and the principles of their practical implementation in those transformers which are normally employed in the chain of electric power supply systems from generator transformer to distribution sub-station transformer.

1.8.5 Tanks and terminal arrangements

Tank construction is referred to in section 9.5 and in sub-section 10.7.2. Except for very large transformer sizes, tanks are made from mild steel. The tanks for very large transformers are often constructed from aluminium in order to obtain a reduction of their mass.

Terminal arrangements are treated in section 9.5 and sub-section 10.7.4.

1.8.6 Arrangements for cooling and heat dissipation

There are various methods to remove the heat generated in the core and windings of a transformer in operation. These methods are indicated by the use of appropriate letter symbols, for example as follows.

(1) AN indicates cooling by natural circulation of air.
(2) AF indicates cooling by forced circulation of air.
(3) ON indicates cooling by natural circulation of oil.
(4) OF indicates cooling by forced circulation of oil.
(5) OB indicates cooling by a combination of natural circulation of oil and forced circulation of air. - - - *
(6) OW indicates cooling by natural circulation of oil through an oil – water heat exchanger.
(7) OFAF indicates cooling by forced circulation of oil through an oil – air heat exchanger with forced circulation of air.
(8) OFW indicates cooling by forced circulation of oil through an oil – water heat exchanger.

The cooling methods (1) and (2) apply to dry-type transformers and the methods (3) to (8) to oil-immersed transformers.

Provisions to facilitate cooling of a transformer core, where necessary, are considered in sub-section 4.5.3. The practical problem of cooling in windings is treated in section 5.7.

Transformer cooling and heat dissipation in general are described in section 9.5, in sub-section 10.7.1, in sub-sections 11.2.2 and 11.2.3 and in section 12.7.

Preservation of oil to maintain its insulation quality is an important problem concomitant with its use as a cooling medium in a transformer. Relevant practical information is given in sub-section 10.7.3.

1.8.7 Transformer protection

Numerous faults may arise in a transformer in service, and various external causes may result in an internal fault. Internal and external faults and protective measures are listed in section 12.8.

1.9 TRANSFORMER NOISE

The main source of transformer noise is the core which vibrates in a complex manner. The vibrations are caused by the action of magnetostriction and of magnetic flux carried between adjacent steel laminations of the transformer core. Their frequency is twice that of the power supply system, and their magnitude increases with increasing peak value of magnetic flux density. Noise is also generated by cooling fans when in operation to produce the air blast for heat exchangers attached to a transformer.

The subject of noise from transformers and from any cooling fans is treated in chapter 8, together with practical remedial measures to reduce the magnitude of noise at source by appropriate considerations in the design and manufacture of transformers, on the one hand, and by attention being given to site conditions, on the other hand (see also sub-section 4.5.4).

1.10 TRANSFORMER TESTING

A power transformer, before leaving the factory, undergoes a test procedure which may be quite elaborate depending on the type of transformer that is being tested. A complete test procedure for larger power transformers is described in chapter 7.

1.11 TRANSFORMER TRANSPORT AND SITE ASSEMBLY

Apart from the smaller types, all transformers have a substantial mass and require careful attention in their transport from factory to site. The problem of transport is accentuated for very large transformers. The practical approach to various aspects of transformer transport is outlined in sub-section 10.8.2.

In particular cases, and for a variety of reasons, it may be impossible to transport a large three-phase transformer in one piece from the factory to the site. Such a transformer, with appropriate design, is partially dismantled in the factory after construction and complete testing, the parts are separately transported to the site and reassembled there, and the reassembled transformer subjected to a simplified test procedure, as referred to in sub-section 10.8.3.

1.12   TRANSFORMER MAINTENANCE

Transformers in service require regular maintenance. A pattern for the maintenance schedule of transformers in a distribution system is presented in section 12.9.

1.13   MONITORING OF GAS-IN-OIL IN LARGE POWER TRANSFORMERS

There are faults in a power transformer which are very slow in developing. Their incipience, none the less, produces in the oil of the tank minute quantities of gas which can be quantitatively analysed by gas chromatography. The result of analysis can be used to diagnose the nature of an incipient fault.

Such diagnosis of incipient faults in a transformer may be employed as a means to monitor the state of health of a large power transformer and to indicate, where necessary, that prophylactic action has to be taken to remedy the cause of the incipient fault. The financial reward of such action can be substantial, for example with large generator transformers where an outage also means an attendant considerable loss in revenue.

1.14   SELECTION OF STANDARD SPECIFICATIONS

1.14.1   IEC Standards

These are issued by the International Electrotechnical Commission (IEC), 1, rue de Varembé, Geneva, Switzerland.

G1.1.   *IEC 27, Letter Symbols to be Used in Electrical Technology*, Part 1: General (1971)
G1.2.   *IEC 38, IEC Standard Voltages* (1975)
G1.3.   *IEC 50, International Electrotechnical Vocabulary* (1973 onwards) (this is issued in more than eighty individually obtainable chapters)
G1.4.   *IEC 52, Recommendations for Voltage Measurement by Means of Sphere Gaps (One Sphere Earthed)* (1960)
G1.5   *IEC 60, High-voltage Test Techniques*
G1.6.   *IEC 71, Insulation Co-ordination*
G1.7.   *IEC 76, Power Transformers* (1967)
G1.8.   *IEC 79, Electrical Apparatus for Explosive Gas Atmospheres*
G1.9.   *IEC 85, Recommendations for the Classification of Materials for the Insulation of Electrical Machinery and Apparatus in Relation to their Thermal Stability in Service* (1957)
G1.10. *IEC 99, Lightning Arresters*

G1.11. *IEC 137, Bushings for Alternating Voltages above 1000 V* (1973)

G1.12. *IEC 152, Identification by Hour Numbers of the Phase Conductors of Three-phase Electric Systems* (1963)

G1.13. *IEC 156, Method for the Determination of the Electric Strength of Insulating Oils* (1963)

G1.14. *IEC 214, On-load Tap Changers* (1976)

G1.15. *IEC 243, Recommended Methods of Test for Electric Strength of Solid Insulating Materials at Power Frequency* (1967)

G1.16. *IEC 270, Partial Discharge Measurements* (1968)

G1.17. *IEC 289, Reactors* (1968)

G1.18. *IEC 296, Specification for New Insulating Oils for Transformers and Switchgear* (1969)

G1.19. *IEC 345, Method of Test for Electrical Resistance and Resistivity of Insulating Materials at Elevated Temperatures* (1971)

G1.20. *IEC 354, Loading Guide for Oil-immersed Transformers* (1973)

G1.21. *IEC 422, Maintenance and Supervision Guide for Insulating Oils in Service* (1973)

G1.22. *IEC 475, Method of Sampling Liquid Dielectrics* (1974)

1.14.2 British Standards

These are issued by the British Standards Institution (BSI), 2 Park Street, London W1A 2BS, Great Britain.

G2.1. *BS 148, Insulating Oil for Transformers and Switchgear* (1972) (this agrees with *IEC 296*)

G2.2. *BS171, Power Transformers* (1970) (this agrees with *IEC 76* (1967))

G2.3. *BS 223, High-voltage Bushings* (1956) (see *IEC 137* (1973))

G2.4. *BS 231, Pressboard for Electrical Purposes* (1967)

G2.5. *BS 355, Mining-type Transformers*, Part 1: Dry-type transformers (1966)

G2.6. *BS 358, Method for the Measurement of Voltage with Sphere Gaps (One Sphere Earthed)* (1960) (see *IEC 52* (1960))

G2.7. *BS 601, Steel Sheet and Strip for Magnetic Circuits of Electrical Apparatus*, Part 2: Oriented steel, 0.25 mm thick and above (1973)

G2.8. *BS 638, Arc-welding Plant, Equipment and Accessories* (1966)

G2.9. *BS 923, Guide on High-voltage Testing Techniques* (1972) (this agrees with *IEC 60)*

G2.10. *BS 2757, Classification of Insulating Materials for Electrical Machinery and Apparatus on the Basis of Thermal Stability in Service* (1956) (see *IEC 85* (1957))

G2.11. *BS 2914, Surge Diverters for Alternating Current Systems* (1972) (this closely follows *IEC 99*, section 1)

G2.12. *BS 2918, Electric Strength of Solid Insulating Materials at Power Frequencies* (1957) (this substantially agrees with *IEC 243* (1967))

G2.13. *BS 3065, The Rating of Resistance Welding and Resistance Heating Machines* (1965)

G2.14. *BS 3435, Measurement of Electrical Power and Energy in Accepting Testing* (1961)

G2.15. *BS 3482, Method of Test for Desiccants used in Dynamic Dehumidification Equipment* (1962)

G2.16. *BS 3523, Silica Gel, Cobalt Chloride Impregnated* (1963)

G2.17. *BS 4571, On-load Tap changers for Power Transformers*, (1970) (see *IEC 214* (1976))

G2.18. *BS 4653, Paper-covered Copper Conductors*, Part 2: Rectangular conductors (1970)

G2.19. *BS 4683, Electrical Apparatus for Explosive Atmospheres*, Part 1: Classification of maximum surface temperatures (1971) (this agrees with *IEC 79*, section 8); Part 2: The construction and testing of flameproof enclosures of electrical apparatus (1972) (this agrees with *IEC 79*, section 1); Part 4: Type of protection 'e' (1973)

G2.20. *BS 4727, Glossary of Electrotechnical, Power, Telecommunication, Electronics, Lighting and Colour Terms* (1971 onwards) (these are issued in individually obtainable groups), Part 1: Terms common to power, telecommunications and electronics (at least eleven groups); Part 2: Terms particular to power engineering (at least eleven groups); Part 3: Terms particular to telecommunications and electronics (at least nine groups); Part 4: Terms particular to lighting and colour (at least three groups)

G2.21. *BS 4828, Guide to Partial Discharge Measurements* (1972) (this agrees with *IEC 270* (1968))

G2.22. *BS 4944, Reactors, Arc-suppression Coils and Earthing Transformers for Electric Power Systems* (1973) (this substantially agrees with *IEC 289* (1968))

G2.23. *BS CP 1009 Maintenance of Insulating Oil* (1959) (with special reference to transformers and switchgear)

G2.24. *BS CP 1010, Loading Guide for Oil-immersed Transformers* (1975) (this substantially agrees with *IEC 354* (1973))

Note. The references G2.23 and G2.24 are Codes of Practice. The annually updated sectional list of British Standards, SL26, Electrical engineering: power, electronics, telecommunications, acoustics, illumination, domestic appliances, is obtainable from British Standards Institution Sales Department, 101 Pentonville Road, London N1 9ND, Great Britain.

1.14.3 USA Standards

The Catalogue of American National Standards is issued by the American National Standards Institute (ANSI), 1430 Broadway, New York, NY 10018, USA.

The IEEE Standards are issued by the Institute of Electrical and Electronics Engineers (IEEE), 345 47th Street, New York, NY 10017, USA.

G3.1. *ANS C42.100, Dictionary of Electrical and Electronics Terms* (1972) (*IEEE Std 100* (1972))

G3.2. *ANS C57.12.00, General Requirements for Distribution, Power, and Regulating Transformers* (1973) (*IEEE Std 462* (1973), *IEC 76* (1967), *288*, section 2, *289* (1968))

G3.3. *ANS C57.12.10, Requirements for Transformers, 138 000 volts and below, 501 through 10 000/13 333/16 667 kVA, single-phase, 501 through 30 000/40 000/50 000 kVA three-phase* (1969)

G3.4. *ANS C57.12.20, Requirements for Overhead-type Distribution Transformers 67 000 volts and below, 500 kVA and smaller* (1974)

G3.5. *ANS C57.12.30, Requirements for Load-tap-changing Transformers 138 000 volts and below, 3750 through 30 000/40 000/50 000 kVA, three-phase* (1971) (*IEC 76* (1968))

G3.6. *ANS C57.12.90, Test Code for Distribution, Power, and Regulating Transformers* (1973) (*IEEE Std 262* (1973), *IEC 76* (1967), *288*, section 2, *289* (1968))

G3.7. *ANS C57.16, Requirements, Terminology, and Test Code for Current-limiting Reactors* (1958, revised 1971) (*IEC 288*, section 2, *289* (1969))

G3.8. *ANS C57.21, Requirements, Terminology, and Test Code for Shunt Reactors* (1971)

G3.9. *ANS 57.100, Test Procedure for Thermal Evaluation of Oil-immersed Distribution Transformers* (1974) (*IEEE Std 345* (1972))

G3.10 *ANS C57.92, Guide for Loading Oil-immersed Distribution and Power Transformers* (1962)

G3.11. *ANS C57.96, Guide for Loading Dry-type Distribution and Power Transformers* (1959)

G3.12. *ANS C57.97, Guide for Preparation of Specifications for Large Power Transformers, with or without Load-tap Changing* (1971)

G3.13. *ANS C57.99, Guide for Loading Dry-type and Oil-immersed Current-limiting Reactors* (1965)

G3.14. *ANS C59.2, Methods of Testing Electrical Insulating Oils* (1974)

G3.15. *ANS C59.21, Methods of Sampling Electrical Insulating Liquids* (1973)

G3.16. *ANS C59.23, Method of Test for Gas Content of Insulating Oils* (1966, revised 1971)

G3.17. *ANS C59.50, Specification for Electrical Insulating Paper and Paperboard-sulfate or Kraft Layer Type* (1974)

G3.18. *ANS C59.107, Method of Test for Approximate Acidity and Polar Contamination in Used Mineral Transformer Oil by Spot Tests* (1970, revised 1975)

G3.19. *ANS C59.129, Specification for Uninhibited Mineral Insulating Oil for Use in Transformers and Oil Circuit Breakers* (1975)

G3.20. *ANS C59.131, Guide for Acceptance and Maintenance of Insulating Oil in Equipment* (1971) (*IEEE Std 64* (1969))

G3.21. *ANS C59.144, Specification for Chlorinated Aromatic Hydrocarbons (Askarels) for Transformers* (1975)

G3.22. *ANS C62.1, Surge Arresters for Alternating-current Power Circuits* (1975) (*IEEE Std 28.* (1974), *IEC 99*, section 2)

G3.23. *ANS C68.2, Techniques for Switching Impulse Testing* (1972) (*IEEE Std 332* (1972))

G3.24. *ANS C76.1, Requirements and Test Code for Outdoor Apparatus Bushings* (1964, revised 1970) (*IEEE Std 21* (1964), *IEC 137* (1973))

G3.25. *ANS C84, Voltage Ratings for Electric Power Systems and Equipment* (60 Hz) (1970) (*IEC 38* (1975), *71*)

G3.26. *IEEE Std 507, Guide for Loading Mineral-oil-immersed Power Transformers with 55°C or 65°C Winding Rise* (1975)

1.15   UNITS OF MEASUREMENT USED IN CHAPTERS 1 TO 12

1.15.1   SI base units

Quantity	Name	Symbol
length	metre	m
mass	kilogram	kg
time	second	s
thermodynamic temperature	kelvin	K

1.15.2   SI derived units with special names

Quantity	Name	Symbol	Remark
frequency	hertz	Hz	
force	newton	N	
pressure	pascal	Pa	$1\,Pa = 1\,N\,m^{-2}$
work, energy	joule	J	$1\,J = 1\,N\,m$
power	watt	W	$1\,W = 1\,J\,s^{-1}$
voltage, potential difference	volt	V	$1\,V = 1\,W\,A^{-1}$
capacitance	farad	F	$1\,F = 1\,As\,V^{-1}$
resistance	ohm	Ω	$1\,\Omega = 1\,V\,A^{-1}$
conductance	siemens	S	$1\,S = 1\,A\,V^{-1}$
magnetic flux	weber	Wb	$1\,Wb = 1\,V\,s$
magnetic flux density	tesla	T	$1\,T = 1\,Wb\,m^{-2}$
inductance	henry	H	$1\,H = 1\,Wb\,A^{-1}$
reactance	ohm	Ω	
impedance	ohm	Ω	
temperature (Celsius)	degree Celsius	°C	$1\,°C = (273+1)\,K$

1.15.3   SI derived units expressed by means of special names

Quantity	Name	Symbol
electric field strength	volt per metre	$V\,m^{-1}$
permittivity	farad per metre	$F\,m^{-1}$
current density	ampere per square metre	$A\,m^{-2}$
resistivity	ohm metre	$\Omega\,m$
(electric) conductivity	siemens per metre	$S\,m^{-1}$
thermal conductivity	watt per metre kelvin	$W\,m^{-1}\,K^{-1}$
thermal resistance	kelvin per watt	$K\,W^{-1}$
magnetic field strength	ampere per metre	$A\,m^{-1}$
permeability	henry per metre	$H\,m^{-1}$
magnetic reluctance	1 per henry	$1\,H^{-1}$
(mass) density	kilogram per cubic metre	$kg\,m^{-3}$
apparent power	voltampere	VA
reactive power	reactive voltampere	var
specific heat capacity	joule per kilogram kelvin	$J\,kg^{-1}\,K^{-1}$
viscosity	pascal second	Pa s

1.15.4   SI supplementary unit

Quantity	Name	Symbol
phase difference	radian	rad

1.15.5   Further unit

Quantity	Name	Symbol
energy	kilowatt hour	kWh

1.15.6   SI prefixes indicating decimal multiples or sub-multiples

Multiple	Prefix	Symbol
10^9	giga	G
10^6	mega	M
10^3	kilo	k
10^{-2}	centi	c
10^{-3}	milli	m
10^{-6}	micro	μ

2

Theory of Transformer Design Principles

A. B. Crompton*

2.1 INTRODUCTION

The general problem of design may be defined as that of determining the most suitable form of equipment, if both technical considerations and cost are taken into account. A satisfactory result is not obtained simply by the solution of a series of equations. The design process is basically iterative, whether performed by hand or by computer.

It is the purpose of this chapter to review the fundamental principles and show how they apply to a specimen design. More detailed treatment of modern practice in construction and design is given in other chapters and in reference 1.

2.2 GENERAL CONSIDERATIONS: THE LOSSES

2.2.1 Specific iron loss

The specific iron loss is the loss in watts per kilogram of the material used in the form of thin laminations for assembling the core.

To a first approximation and over a limited range of sinusoidally alternating flux density of amplitude B_m, the specific iron loss of hot-rolled steel may be taken as proportional to B_m^2 for a given core material and frequency. For cold-rolled steel, however, any simple mathematical relationship assumed between the specific iron loss and B_m is very approximate.

2.2.2 Iron loss in assembled cores

The losses occurring in an assembled core cannot be taken as simply the product

* Wigan College of Technology, formerly Ferranti Limited.

of the specific iron loss measured on a specimen of material and the mass of the core. The losses in the assembled core are increased by uneven flux distribution due to joints, bolt holes and burrs on lamination edges, as described in chapter 4. In practice the core loss is calculated from curves based on tests on assembled cores of similar type and construction.

Typical curves for cold-rolled steel are given in figure 2.1, which also illustrates the reduction in loss obtained by using mitred joints.

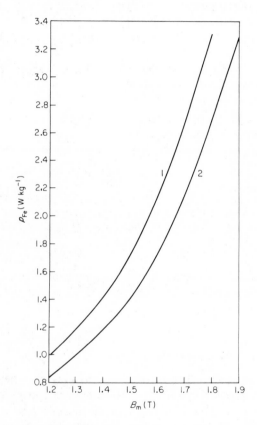

Figure 2.1   Iron loss in cold-rolled steel cores: curve 1, square-cut corners; curve 2, mitred corners

2.2.3 $I^2 R$ loss

The copper losses consist of two parts: (1) the $I^2 R$ loss, due to the load current I flowing through the conductor resistance R and (2) the eddy current loss, due to the circulating currents induced by the leakage field which links the winding conductors.

In order to find the $I^2 R$ loss, the winding resistances are calculated at an assumed mean working temperature known as the reference temperature, which, from standard specifications, is 75 °C for transformers with class A insulation. For example, the resistance of the low-voltage winding is

$$R_1 = \rho N_1 s_1 / a_1 \; \Omega \text{ per phase} \tag{2.1}$$

where the subscript 1 denotes quantities related to the low-voltage winding, irrespective of whether this is used as the primary or secondary side of the transformer, N_1 is the number of turns, s_1 is the length of mean turn and a_1 the cross-sectional area of conductor. If the dimensions are in millimetres, the value of ρ for rectangular conductors may be taken as $\rho = 21.4 \times 10^{-6} \, \Omega \text{mm}$ at 75 °C. In practice $N_1 s_1$ may be increased slightly by the length of the leads from the winding.

It can be shown[1] that minimum $I^2 R$ loss is obtained approximately when

$$J_1 = J_2 \tag{2.2}$$

where J_1 and J_2 are the current densities in the low- and high-voltage windings, respectively, and may be expressed conveniently in amperes per square millimetre.

2.2.4   Specific $I^2 R$ loss

The specific $I^2 R$ loss is the $I^2 R$ loss in watts per kilogram of conductor material and is given by[1]

$$p_R = J^2 \rho \times 10^9 / \rho_{d, \, Cu} \qquad W \, kg^{-1} \tag{2.3}$$

where $\rho_{d, \, Cu}$ is the density of copper in kilograms per cubic metre. If we take $\rho_{d, \, Cu}$ as $8890 \, kg \, m^{-3}$ and ρ as $21.4 \times 10^{-6} \, \Omega \text{mm}$ at 75 °C, this gives

$$p_R = 2.41 J^2 \tag{2.4}$$

2.2.5   Eddy current loss in conductors

For a thin copper sheet of thickness δ in a sinusoidal magnetic field of maximum flux density B_m at the frequency f and at 75 °C, equation 5.5 gives the specific eddy current loss based upon the assumption that the magnetic flux is parallel to the surface of the sheet. It can be shown that equation 5.5, when applied to rectangular conductors of the windings in a transformer, leads to the approximate expression

$$p_i = 6.2(B_{lm0} f b_C)^2 \times 10^{-8} / \rho$$
$$\approx 3(B_{lm0} f b_C)^2 \times 10^{-3} \qquad W \, kg^{-1} \tag{2.5}$$

where

$$B_{lm0} = 10^3 2^{1/2} \mu_0 I N / h \qquad T \tag{2.6}$$

denotes the peak leakage magnetic flux density (which is equal to B_m in equation 5.5) in the duct between the low-voltage and the high-voltage winding, IN is the ampere–turns of either winding, h is the winding height in millimetres and b_C (which is equal to δ in equation 5.5) represents the breadth of a conductor in millimetres; the magnetic constant is $\mu_0 = 0.4\pi \times 10^{-6}\,\text{H m}^{-1}$. The substitution of equation 2.6 into equation 2.5 gives

$$p_i \approx 9.1(fb_C IN/h)^2 \times 10^{-9} \qquad \text{W kg}^{-1} \tag{2.7}$$

In equation 2.5 ρ appears in the denominator so that, unlike $I^2 R$ losses, eddy current losses decrease with temperature rise.

For round conductors the specific eddy current loss is three-quarters of that obtained by substituting the diameter d_c for b_C in the equation.

From equations 2.7 and 2.4

$$p_i/p_R \approx 3.8(fb_C IN/hJ)^2 \times 10^{-9} = P_i/P_R \tag{2.8}$$

where P_i and P_R are the total eddy current loss and $I^2 R$ loss, respectively, in a winding.

For the design procedure it is convenient to express P_i as a percentage of P_R. Thus, from equation 2.8

$$\%P_i = 100P_i/P_R \approx 9.5(b_C IN/100hJ)^2 \quad \text{at } 75\,^\circ\text{C and } 50\,\text{Hz} \tag{2.9}$$

where the symbol $\%P_i$ represents the percentage conductor eddy current loss.

The eddy current loss is proportional to b_C^2 and may be reduced by subdividing the conductor into insulated strands, provided these are suitably transposed. Subdivision alone is insufficient because the strands are connected in parallel at the ends and currents will continue to circulate through the end connections with the result that the losses remain practically unaltered. Ideal transposition is obtained when equal lengths of each strand occupy each different position relative to the leakage field across the conductor as a whole. The strand breadth b_c can then be substituted for the conductor breadth b_C in the equations, thus dividing the eddy current losses by the square of the number of radial strands.

Transposition is effected by cross-overs as described in references 2, 3 and 4 or, for large transformers, by the use of a continuously transposed conductor, as illustrated in chapter 5. In practice, ideal transposition may not be necessary, and the eddy current losses will then have a value between that corresponding to the solid conductor and that for ideal transposition, depending upon the number and type of cross-overs.

In general, eddy current losses may be kept quite small by suitable design. In small transformers they are often no more than 5% of the $I^2 R$ loss at 75 °C, increasing to about 15% in large units.

2.2.6 Stray losses

Stray losses arise from eddy currents induced in metal parts of a transformer, such as in clamps and in tank walls. These losses are discussed in sub-sections 5.6.3 and 5.6.4.

2.2.7 Load loss and resistance voltage

From standard specifications the load loss is the power absorbed on short-circuit test related to the reference temperature. It therefore consists of the I^2R loss and eddy current loss in the copper and the stray losses; the iron loss is negligible at the very low voltage of the test. If P_{Cu} is the copper loss per phase, then for a three-phase transformer the load loss P_1 is

$$P_1 = 3P_{Cu} + \text{stray losses} \qquad (2.10)$$

where

$$P_{Cu} = (1 + \%P_{i1}/100)I_1^2 R_1 + (1 + \%P_{i2}/100)I_2^2 R_2 \qquad (2.11)$$

The subscripts 1 and 2 denote quantities related to the low- and high-voltage windings, respectively, and the values of resistance and percentage eddy current loss are at 75 °C.

The resistance voltage is the component of the impedance voltage in phase with the current. Its value is related to the reference temperature and is equal to IR, where I is rated current and R is the effective ac phase resistance at 75 °C, including an allowance for the effects of eddy current and stray losses.

Expressed as a percentage, the resistance voltage is

$$\%R = 100 \times IR/V \qquad (2.12)$$

where V is rated phase voltage referred to the same side of the transformer as I and R. Thus,

$$\%R = \frac{I^2 R}{VI} \times 100 = \frac{\text{load loss per phase in watts}}{\text{voltamperes per phase}} \times 100$$

or, for a three-phase transformer, with equation 2.10

$$\%R = 10^{-4} \times P_1/3S \qquad (2.13)$$

where P_1 is in watts and S is the rating per phase in megavoltamperes.

2.3 GENERAL CONSIDERATIONS: TRANSFORMER WINDINGS AND INSULATION

The subject of transformer windings and insulation is presented in chapter 5. Additional information, contained in subsequent chapters, is listed in sub-section 1.8.3.

Figure 2.2 illustrates the principles of the arrangement of winding insulation in one type of distribution transformer.

The ultimate limit to the life of a transformer is imposed by the life of its insulation which decreases with increase in operating temperature. To ensure an economic life, upper limits of temperature have been set for the various classes of insulation, for example, 105 °C for class A materials[G1.9]. Such limits relate to the

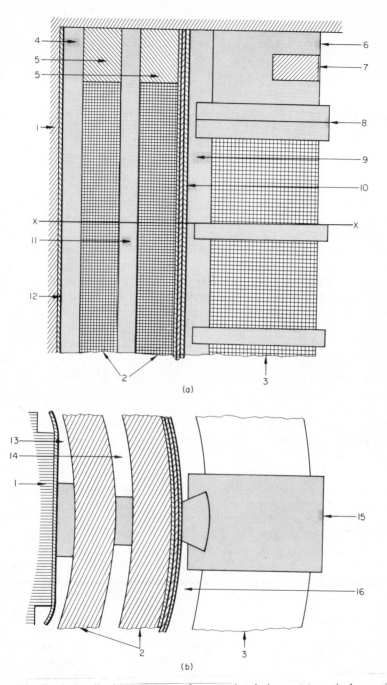

Figure 2.2 Typical distribution transformer insulation: (a) vertical section; (b) horizontal section on X–X: 1, core leg; 2, low-voltage winding; 3, high-voltage winding; 4, spacing strip; 5, edge block; 6, insulation end block; 7, end washer; 8, spacers; 9, spacing strip; 10, paper insulation; 11, spacing strip; 12, paper insulation; 13, duct; 14, cooling duct; 15, spacer; 16, duct

hottest part of the windings and are implicit in the definitions of continuous rating in standard specifications.

2.4 GENERAL CONSIDERATIONS: COOLING OF ONAN TRANSFORMERS

2.4.1 Temperature distribution in core, windings and oil

Figure 2.3 indicates the approximate temperature distribution. Figure 2.3(a) represents conditions near the top of the windings where the maximum temperatures occur. All the oil above the transformer is assumed to have the same temperature, irrespective of whether it has ascended through the ducts or past the external winding surfaces. Since the thermal resistivity of copper is negligible in comparison with that of insulation, the high-voltage winding is the most difficult to cool and usually contains the hot spot.

The core material has a greater thermal resistivity than that of copper but much less than that of paper insulation or of the insulation between laminations. A path of comparatively high thermal conductivity is thus provided to the edge of the laminations. This generally limits the temperature of the core to a value less than that of the windings when the transformer is operating at full-load condition.

Figure 2.3(b) shows the temperature distribution across the high-voltage winding to an enlarged scale. Internally, the heat passes by conduction through the layers of insulation and the curve is approximately parabolic. The coil surfaces, however, cool by convection, and this results in the characteristically steep surface temperature drops shown. The difference between the hot-spot temperature and the oil temperature is called the maximum winding temperature gradient $\Delta\theta_{wom}$. Similarly the difference between the mean coil temperature and that of the oil is the mean gradient $\Delta\theta_{wo}$. The term gradient is used merely to indicate a difference in temperature and not in the strict sense of temperature difference per unit length.

At the tank surfaces, there are also steep drops in temperature, as shown in figure 2.3(a). The external temperature drop, however, does not take place entirely at the surface. Cooling by both radiation and convection is involved, and the air temperature remains above the ambient value for some distance from the tank wall.

Figure 2.3(c) indicates approximately the temperature distribution vertically through the windings. The oil temperature may be considered to increase linearly with height up the coil stack, reaching a maximum at the top of the windings. The temperature then remains constant to the surface and is referred to as the top oil temperature. Over the full height of the windings the graph of copper temperature is parallel with that of the oil, since the temperature gradients in the body of the winding are assumed independent of vertical position, as shown. In fact, the hot spot does not occur at the very top of the windings owing to cooling

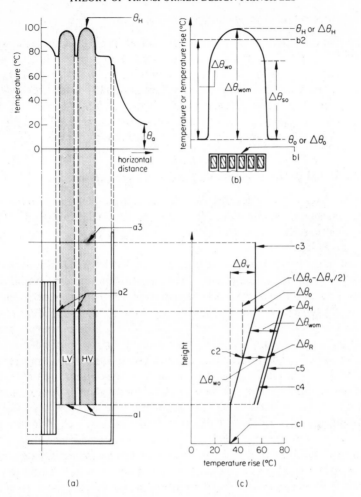

Figure 2.3 Temperature distribution in an ONAN transformer: (a) typical temperature profile on horizontal plane through winding hot spot; (b) temperature gradients in high-voltage winding to enlarged scale; (c) simplified temperature rise diagram in vertical plane: a1, sections of windings to enlarged horizontal scale; a2, ducts; a3, oil level; b1, top of high-voltage winding; c1, approximate oil temperature rise at tank bottom; c2, average duct oil temperature rise; c3, approximate oil temperature rise at tank top; c4, mean oil temperature rise; c5, maximum oil temperature rise; θ_H, hot-spot temperature; θ_a, ambient temperature; $\Delta\theta_{wom}$, maximum winding temperature gradient; $\Delta\theta_{wo}$, mean winding temperature gradient; $\Delta\theta_{so}$, surface temperature drop; $\Delta\theta_o$, temperature rise of top oil; $\Delta\theta_v$, difference in oil temperature rise between tank top and bottom; $\Delta\theta_R$, mean winding temperature rise

from the upper surface. Its vertical position is very difficult to determine, however, and the approximation implicit in the graph is accepted in calculating the winding temperature.

2.4.2 Specified limits of temperature rise

Although the hot-spot temperature is of critical importance, it can only be measured directly by a thermocouple embedded in the winding, which is obviously impracticable. Standard specifications, therefore, give limits of temperature rise above ambient temperature which, although related only indirectly to the hot-spot temperature rise, are easily measured on test. These are the mean temperature rise of the windings as measured by the increase $\Delta\theta_R$ in resistance, and the temperature rise $\Delta\theta_0$ of the top oil measured by thermometer.

2.4.3 Tank configuration and winding temperature gradients

The specified limits of temperature rise considered in conjunction with the design of the tank imply maximum values for the winding temperature gradient $\Delta\theta_{wo}$, and the windings should be formed in such a way that these values are not normally exceeded.

In figure 2.3(c), the difference in temperature rise between the oil at the top and bottom of the tank is denoted by $\Delta\theta_v$. This depends upon the tank design, and a low value is preferable as the slope of the graphs is then increased. Thus the maximum copper temperature rise at the top of the windings is reduced for the same specified mean temperature rise, which, for the linear graphs assumed, occurs at half the winding height. A lower value of $\Delta\theta_v$ corresponds to a more vigorous rate of oil circulation, resulting in an increased rate of heat transfer at the coil and tank surfaces and thus reducing the surface temperature drops.

In ON-type transformers the rate of circulation of oil depends upon the difference in densities between the hot oil, which rises through and above the transformer, and the cool oil which descends at the tank surfaces. Ideal conditions for ON-type cooling are thus obtained by placing the transformer in a relatively tall plain tank. The bulk of the ascending oil then has the maximum temperature rise $\Delta\theta_0$, corresponding to the lowest possible density and thus giving the most vigorous rate of oil circulation. Furthermore, a plain surface is the most efficient for heat dissipation.

Except for transformers rated at less than about 50 kVA, however, such an arrangement is uneconomical, as the tank would have to be very large relative to the size of transformer to provide sufficient cooling surface area. Radiators or tubes are, therefore, provided to increase the cooling area without a corresponding increase in oil quantity but at the expense of some loss in cooling efficiency.

The relationship between $\Delta\theta_v$ and $\Delta\theta_0$ may be expressed in the form

$$\Delta\theta_v = k_v\Delta\theta_0 \qquad °C \tag{2.14}$$

where the factor k_v has values ranging from about 0.3 for plain tanks or for tanks equipped with radiators to 0.5 for three-row tubular tanks. From figure 2.3(c), the mean duct oil temperature rise is $\Delta\theta_0 - \Delta\theta_v/2$; thus the mean winding temperature rise is

$$\Delta\theta_R = (\Delta\theta_0 - \tfrac{1}{2}\Delta\theta_v) + \Delta\theta_{wo} \qquad °C \tag{2.15}$$

Substituting for $\Delta\theta_v$ and rearranging, we obtain

$$\Delta\theta_{wo} = \Delta\theta_R - \Delta\theta_o(1 - \tfrac{1}{2}k_v) \qquad °C \qquad (2.16)$$

Thus values of $\Delta\theta_{wo}$ compatible with the specified limits of temperature rise may be calculated.

It may be assumed[1] that

$$\Delta\theta_{wom} = 1.1\Delta\theta_{wo} \qquad °C \qquad (2.17)$$

thus the hot-spot temperature rise is given by

$$\Delta\theta_H = \Delta\theta_o + \Delta\theta_{wom} = \Delta\theta_o + 1.1\Delta\theta_{wom} \qquad °C \qquad (2.18)$$

and the hot-spot temperature is

$$\theta_H = \Delta\theta_H + \theta_a \qquad °C \qquad (2.19)$$

where θ_a is the ambient temperature in degrees Celsius.

2.5 PRACTICAL CONSTRAINTS ON THE DESIGN

2.5.1 Specific electric and magnetic loadings

Equation 2.4 indicates that the specific I^2R loss is directly proportional to J^2 and figure 2.1 relates the specific iron loss at 50 Hz to B_m for a given core material and type of construction. For these reasons, the values of current density J and flux density B_m are often termed the specific electric and magnetic loadings, respectively. The choice of specific loadings is commonly regarded as the starting point for preliminary design work on a transformer. Provided the losses and reactance are not specified, the highest acceptable values may be chosen for both J and B_m, thus reducing the cost of materials to a minimum.

The cooling method determines the maximum value of current density, which in class A insulated transformers varies from about $3.2\,A\,mm^{-2}$ for distribution transformers to $5.5\,A\,mm^{-2}$ for large transformers with forced cooling.

The limit on the value of B_m is imposed by distortion of the magnetising current and generation of noise, as described in sub-section 1.2.4 and section 1.9, respectively. For generator transformers, these factors are relatively unimportant, and flux densities as high as 1.8 T have been used. In other cases, however, it is considered good practice to keep B_m below 1.6 T.

2.5.2 Relationship between current density and flux density

If we assume that the percentage eddy current losses in the low-voltage and high-voltage windings on normal tapping, $\%P_{i1}$ and $\%P_{i2}$, may be taken as each approximately equal to the average percentage eddy current loss $\%P_i$,

$$1 + \%P_{i1}/100 \approx 1 + \%P_{i2}/100 \approx 1 + \%P_i/100 = k_i$$

and equation 2.11 for the copper loss per phase may be written

$$P_{Cu} = (I_1^2 R_1 + I_2^2 R_2)k_i \times 10^{-3} \qquad kW \qquad (2.20)$$

Substituting for R_1 and R_2 from equation 2.1, we obtain

$$P_{Cu} = (I_1^2 \rho N_1 s_1/a_1 + I_2^2 \rho N_2 s_2/a_2)k_i \times 10^{-3} \qquad kW$$

Therefore,

$$P_{Cu} = IN(I_1 s_1/a_1 + I_2 s_2/a_2)\rho k_i \times 10^{-3} \qquad kW$$

where IN is the ampere–turns in either winding.

The assumption of equal current densities in the windings in accordance with the condition for minimum $I^2 R$ loss expressed in equation 2.2 gives

$$P_{Cu} = 2INJs\rho k_i \times 10^{-3} \qquad kW \qquad (2.21)$$

where s is the mean value of s_1 and s_2. Thus

$$J = 10^3 P_{Cu}/2INs\rho k_i \qquad A\,mm^{-2} \qquad (2.22)$$

Multiplying equation 1.4 by I and rearranging, we get

$$IN = VI/4.44 f B_m A_{Fe}$$

where A_{Fe} is the nett core cross-sectional area.

If S is the rating per phase in megavoltamperes, $VI = 10^6 S$ and

$$IN = 10^6 S/4.44 f B_m A_{Fe}$$

Substituting this value in equation 2.22 and taking ρ as $21.4 \times 10^{-6}\,\Omega\,mm$ at $75\,°C$, we find

$$J = 104 f B_m A_{Fe} P_{Cu}/k_i s S \qquad A\,mm^{-2} \qquad (2.23)$$

Expressed as a percentage of the transformer rating, the copper loss is

$$\%P_{Cu} = 0.1 P_{Cu}/S \qquad (2.24)$$

where P_{Cu} is in kilowatts. By substituting in equation 2.23

$$J = 10.4 f \frac{B_m}{k_i} \frac{A_{Fe}}{s} \%P_{Cu} \qquad A\,mm^{-2} \qquad (2.25)$$

Thus, if J and B_m are chosen independently, the transformer will have a natural value of copper loss depending on the ratio A_{Fe}/s. Conversely, if the losses are guaranteed, the choice of J must correspond to that of B_m and possible values of A_{Fe}/s.

2.5.3   The significance of reactance

Approximate formulae developed for a two-winding transformer, on the assumption that the leakage flux density is parallel to the winding surfaces for their full height[2, 5], may be arranged in the form

$$\%X = 59.4IN \times 10^{-9}(3b_0 s_0 + b_1 s_1 + b_2 s_2)/\phi_m h \tag{2.26}$$

where s_0 is π multiplied by the mean diameter or 'the length of mean turn' of the duct between the high- and low-voltage windings in millimetres, b_0 is the radial width of the gap between the high- and low-voltage windings in millimetres, b_1 and b_2 are the radial widths of the low- and high-voltage windings, respectively, in millimetres, and h is the axial length, assumed equal for both low- and high-voltage windings in millimetres.

To a first approximation s_0 is equal to s, the mean value of s_1 and s_2, and

$$b_1 s_1 + b_2 s_2 = (b_1 + b_2)s$$

Therefore, rearranging equation 2.26, we obtain

$$\%X = 178IN \times 10^{-9} s\{b_0 + \tfrac{1}{3}(b_1 + b_2)\}/\phi_m h \tag{2.27}$$

Let the expression in brackets be b_X. Then b_X can be regarded as an effective width through which the total leakage flux passes. The cross-sectional area of the leakage flux path is represented by sb_X and its approximate length by h. Therefore, the larger the quantity sb_X/h the smaller is the reluctance of the leakage flux path and the larger is the percentage reactance.

From equation 1.4, $\phi_m = V/4.44fN$. The phase current I, expressed in terms of megavoltamperes, is equal to $10^6 S/V$. Substitution in equation 2.27 gives

$$\%X = 0.79SfN^2 sb_X/V^2 h \tag{2.28}$$

or, expressing in terms of the voltage per turn and the ratio h/s, we find

$$\%X = \frac{0.79Sf b_X}{(V/N)^2 h/s} \tag{2.29}$$

From these equations, it can be seen that an increase in b_X, caused by an increase in b_0 which depends mainly on the voltage, increases the reactance, and for this reason high-voltage transformers have an inherently higher reactance than low-voltage transformers.

If we assume that all the other quantities could remain constant while the turns were altered, reactance is also seen to be inversely proportional to the square of the voltage per turn or directly proportional to the square of the turns for a given voltage. Similarly, if we assume that s and the voltage per turn could be kept constant, the reactance is inversely proportional to the coil height h. In practice, the quantities in equations 2.28 and 2.29 are all interrelated, and in an economic design alteration of one implies alteration in the others.

An increase in voltage per turn corresponds to an increase in the ratio A_{Fe}/A_{Cu} where A_{Cu} is the total copper cross-section in one phase. It is shown in reference 1 that, in general, a reduction in reactance is associated with increased mass of iron and increased iron loss and reduced mass of copper and copper loss. Also the ratio h/s may be increased. Conversely, an increase in reactance is associated with reduced iron loss and increased copper loss, and the ratio h/s may be decreased.

Reactance is thus of fundamental importance in determining the initial dimensions of a design and should be considered at this stage. This is possible, provided an approximate value can be found for b_X.

Substituting $V/N = 4.44 f \phi_m$ in equation 2.29 and rearranging, we find

$$\phi_m^2 h/s = 0.04 S b_X/f \,\%X \tag{2.30}$$

The quantity $\phi_m^2 h/s$ is of major importance in transformer design. Its evaluation is beyond the scope of this chapter, but in chapter 3 an expression is given relating $\phi_m^2 h/s$ to the quantities known from the transformer specification. It is also shown in chapter 3 that, provided the reactance and copper loss are both specified and therefore constant, $\phi_m^2 h/s$ must remain practically constant irrespective of any changes in dimensions made when forming the design. Thus it follows from equation 2.30 that b_X also remains constant under these conditions, and its value can be found if $\phi_m^2 h/s$ is known.

In conjunction with standard insulation clearances the value of b_X determines the total width of the windings and of the window in the frame required to accommodate them. For this reason, transformers which nearly all have similar values of percentage reactance and percentage copper loss, such as distribution transformers, can usually be built on frames with standard distances between leg centres. These standard centres are related to the core cross-sectional areas and are known from past design experience. Thus, if the required core area is known, it is often possible to estimate a reasonably accurate value of b_X without previous knowledge of $\phi_m^2 h/s$.

Equation 2.30 may be rearranged by substituting $B_m A_{Fe}$ for ϕ_m. Thus

$$A_{Fe}^2 h/s = 0.04 S b_X / B_m^2 f \,\%X \tag{2.31}$$

where the nett core cross-sectional area A_{Fe} is in square metres, b_X in millimetres and B_m in teslas. Therefore, when reactance and copper loss are specified, $A_{Fe}^2 h/s$ is a constant for a particular value of B_m.

2.5.4   Standardisation

In the manufacture of transformers it is possible to standardise for a few years on some designs, especially with distribution transformers which are produced in large quantities. In general, however, standard designs suitable for use over a long period are not possible because of alterations in customers' requirements and new developments in materials and in manufacturing methods. Standardisation on cores, insulation, tanks and fittings, however, must always be practised in order to make manufacture at all economic and this may be regarded as semi-standardisation.

Figure 2.4 shows the main dimensions of a three-phase core-type frame. The number of steps, method of construction and dimensions of the core, d, b_{Fe} and

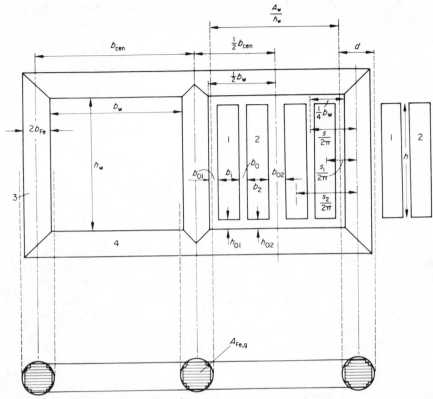

Note: This diagram is not to scale

Figure 2.4   Three-phase core-type frame: b_{cen}, distance between leg centres; h_w, window height; b_w, window width; b_{Fe}, half-width of widest core-plates; $A_{Fe,g}$, gross core cross-sectional area; d, diameter of circumscribing core circle; A_w, window area taken between core circles

$A_{Fe,g}$ are fixed, but the window height h_w is variable. The nett core cross-sectional area is

$$A_{Fe} = k_{Fe} A_{Fe,g} \qquad (2.32)$$

where k_{Fe} is the space factor of the core laminations used.

Because the percentage values specified for losses and reactance in distribution transformers are similar for most units of a particular size, it is generally possible to use a standard window width b_w and hence standard leg centres b_{cen}. This means that, for the majority of distribution transformers, standard frames are used, variable only in height. Provision is, however, made for non-standard centres when required in transformers with unusual characteristics.

For larger transformers the values of reactances and losses specified vary widely, and it is not usually possible to standardise on the leg centres, although the use of standard cores is normal.

The effect of standardisation, especially for cores, is to impose limitations on the design. Once the dimensions of the standard cores have been fixed, it is nearly always cheaper to design the transformer on the most appropriate standard core, even though this may not be exactly the optimum size. In practice, the existence of standard cores and frames often simplifies the process of obtaining a design by restricting the field of choice.

Standard conductors are also employed. Their use means that it is not always possible, when forming the windings, to obtain optimum values for the length and radial width or for the current density.

2.6   FRAME AND WINDING PROPORTIONS

2.6.1   Size and shape of frame

The selection of frame dimensions is the first and probably the most critical stage in the design of a transformer, because the best design will only be obtained on a frame of the correct size and proportions. The dimensions of this frame are determined by the full technical specification for the transformer in conjunction with considerations of cost.

An indication of the size of frame required is obtained from the output equation developed in reference 1, which may be written

$$S = 2.22 f B_m J A_{Fe} A_{Cu} \qquad \text{MVA} \qquad (2.33)$$

Alternatively, the output may be expressed in terms of the window area A_w instead of the copper cross-section A_{Cu}. For a three-phase transformer,

$$S = 1.11 f B_m J A_{Fe} k_w A_w \qquad \text{MVA} \qquad (2.34)$$

where the window space factor k_w is given by

$$k_w = 2 A_{Cu}/A_w \qquad (2.35)$$

with $2A_{Cu}$ representing the total cross-section area of copper in the window.

The output equation thus relates the transformer rating per phase to the size of frame required, in so far as this is indicated by the product of areas $A_{Fe} \times A_{Cu}$ or $A_{Fe} \times A_w$. It does not, however, enable the areas to be separated without making certain assumptions of a very approximate nature, nor does it give any indication of the shape of window area required.

The electrical characteristics of the transformer depends upon its shape. For example, reactance depends upon the ratios A_{Fe}/A_{Cu} and h/s. These ratios, in conjunction with appropriate space factors and insulation clearances, also establish the relative volumes of copper and iron used in the design. Consequently, if fixed values of specific loadings and therefore specific losses can be assumed, the ratio of total copper loss $P_{Cu,t}$ to iron loss P_{Fe} is determined.

Considerations of material costs also influence the shape of the frame. The price of copper per unit mass is approximately three times that of iron. Thus the cost of active iron and copper employed in the construction of the transformer depends upon the relative volumes of these materials as well as upon their total volume. For a given size of transformer, the shape of the frame affects both the total and relative volumes of materials used.

2.6.2   Lengths of active iron and copper

If the product $B_m \times J$ in the output (equation 2.33) is assumed to remain constant as the dimensions of a design are varied, A_{Cu} is proportional to $1/A_{Fe}$. From this relationship the effect of changes in frame geometry may be investigated by calculating varying proportions for typical designs, the corresponding changes in electrical characteristics being ignored. For this purpose the lengths of iron and copper are required which, when multiplied by the cross-sectional areas A_{Fe} and A_{Cu}, give the total volumes of active materials in the different frames.

   If we use the notation of figure 2.4 the total length l_{Fe} of the cores and yokes for a three-phase core-type transformer is given by

$$l_{Fe} = 3h_w + 4b_w + 12b_{Fe} \qquad \text{mm}$$

but

$$h_w = h + h_0 \qquad \text{mm}$$

where h_0 is the total axial clearance between ends of the windings and the yoke on the assumption that the low- and high-voltage windings are of equal length. Therefore

$$l_{Fe} = \{3(h + h_0) + 4(b_{cen} + b_{Fe})\} \times 10^{-3} \qquad \text{mm} \qquad (2.36)$$

where $b_{cen} = b_w + 2b_{Fe}$. If we express this in terms of the winding dimensions and radial clearances, the width of the window is

$$b_w = 2(b_{01} + b_1 + b_0 + b_2 + \tfrac{1}{2}b_{02}) \qquad \text{mm} \qquad (2.37)$$

where b_{01} is the radial width of the gap between the core and the low-voltage winding and b_{02} the radial width of the gap between high-voltage windings.

   The total length s_t of copper cross-section in the transformer is the mean circumference of the windings on each leg multiplied by the number of legs. Thus for the three-phase core-type transformer.

$$s_t = 3s \times 10^{-3} \qquad \text{mm} \qquad (2.38)$$

and from figure 2.4, if we assume that $b_1 = b_2$,

$$s = 2\pi(b_{Fe} + b_{01} + b_1 + \tfrac{1}{2}b_0) \qquad \text{mm} \qquad (2.39)$$

If it is also assumed that $b_{01} = b_{02}/2$,

$$s = 2\pi(b_{Fe} + \tfrac{1}{4}b_w) \qquad \text{mm} \qquad (2.40)$$

The diameter of the core circumscribing circle is

$$d = 2(A_{Fe} \times 10^6/\pi k_c)^{1/2} \qquad \text{mm} \qquad (2.41)$$

where the core-circle space factor k_c is given by

$$k_c = 4k_{Fe} A_{Fe,g}/\pi d^2$$

Also $b_{Fe} = k_s d/2$, where k_s is a factor depending upon the number of steps in the core. Thus

$$b_{Fe} = k_s(A_{Fe} \times 10^6/\pi k_c)^{1/2} \qquad (2.42)$$

For a given type of core construction, therefore, b_{Fe} is proportional to $A_{Fe}^{1/2}$.

Using this relationship between b_{Fe} and A_{Fe} and assuming also that A_{Cu} is proportional to $1/A_{Fe}$, we may employ equations 2.36 to 2.40 to evaluate l_{Fe} and s_t as the dimensions of an existing design are varied.

In the calculation, a common and constant value of winding space factor k may be assumed for both low- and high-voltage windings, thus enabling the overall winding dimensions to be related directly to A_{Cu}. The individual space factors are

$$k_1 = a_1 N_1/b_1 h_1 \qquad (2.43)$$

and

$$k_2 = a_2 N_2/b_2 h_2 \qquad (2.44)$$

where $a_1 N_1$ and $a_2 N_2$ are the cross-sectional copper areas of the low- and high-voltage windings. If both windings have the same current density, $a_2 N_2 = a_1 N_1$; if we also assume a common winding height h and that $b_2 = b_1$, then

$$k_2 = a_1 N_1/b_1 h = k_1 = k$$

and

$$k = 2a_1 N_1/2b_1 h = A_{Cu} \times 10^6/2b_1 h \qquad (2.45)$$

2.6.3   Practical limits of A_{Fe}/A_{Cu} and h/s

Calculations based on typical designs and equations 2.36 to 2.45 indicate that the minimum lengths and most economic utilisation of active materials occur within a limited range of h/s from about 0.3 to 1.0. Within this range, wide variations in A_{Fe}/A_{Cu} do not cause much increase in length above the minimum. In practice, the limits of A_{Fe}/A_{Cu} for nearly all classes of power transformers lie between 1 and about 3.5, and distribution transformers tend to have lower values than high-voltage transformers.

It may appear that distribution transformers should have the higher values of A_{Fe}/A_{Cu} since their specified reactance is low and low reactance is associated with high values of A_{Fe}/A_{Cu}. The effect of the difference in voltage, however, must be considered. Transformers designed without explicit consideration of particular parameters may be said to have natural values of these quantities. The natural values of reactance for high-voltage transformers are much larger than those for distribution transformers, owing to the increased value of b_0 required for insulation purposes. The specified reactances are also higher but tend to be less than the natural values. In these circumstances A_{Fe}/A_{Cu} must be increased to reduce the reactance to the specified value.

Distribution transformers are designed for low iron losses. The amount of iron used in the design and hence the value of A_{Fe}/A_{Cu} is thus restricted. Low values of A_{Fe}/A_{Cu} may be used because the inherent reactance is very small due to the low voltage.

For small transformers the limits of h/s of 0.3 to 1, stated above, are very approximately correct. As the size of transformer increases, however, the range of h/s in practical use narrows. Figure 2.5, based on an analysis of modern designs, indicates the form of the practical range of h/s plotted to a base of megavoltamperes.

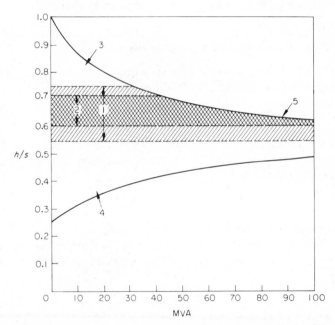

Figure 2.5   Practical limits of h/s for three-phase core-type transformers: 1, most designs with normal losses and reactance; 2, designs based only on minimum cost of materials for $c_{Cu}/c_{Fe} \approx 3/1$; 3, low reactance, increased iron loss, reduced copper loss for the same value of A_{Fe}/A_{Cu}; 4, high reactance, reduced iron loss, increased copper loss for the same value of A_{Fe}/A_{Cu}; 5, height restriction limits h/s on large transformers

The upper limit is associated with transformers having low values of specified reactance, high iron loss and low copper loss, as explained in section 2.5.3. Conversely, transformers having reactances higher than their natural value combined with very low iron loss and high copper loss are close to the bottom limit.

The majority of transformers with normal values of losses and reactance lie between these extremes, the values of h/s used ranging very approximately from about 0.55 to 0.75. For these transformers the exact value of h/s depends on the circumstances of a particular design. In some cases depending on the value of A_{Fe}/A_{Cu}, the conditions at the boundaries of figure 2.5 may be reversed: transformers with lower reactance have the higher value of h/s and vice versa.

For the largest transformers, height restrictions due to transport difficulties limit the value of h/s to less than about 0.6 in the majority of units.

The limiting geometrical proportions within which a design should be produced thus follow from the factors outlined in this section. Within these limits the exact dimensions of a particular design must be fixed by more detailed considerations of reactance and of losses and costs.

2.7　CONSIDERATIONS OF EFFICIENCY AND COST

2.7.1　Loss and mass ratios for maximum efficiency

It can be shown[2, 5] that maximum efficiency occurs at a fractional load k_S such that the copper loss is equal to the iron loss. Thus

$$k_S^2 P_{Cu,t} = P_{Fe} \tag{2.46}$$

If the mean working load of the transformer is known from the system load factor, this gives a value for k_S, and the transformer can be designed with a ratio of full-load copper to iron loss such that

$$P_{Cu,t}/P_{Fe} = 1/k_S^2 \tag{2.47}$$

It will then operate at maximum efficiency at its mean working load.

Let m_{Fe} be the total mass of active iron and m_{Cu} be the total mass of copper in the transformer. The specific iron and copper losses are p_{Fe} and p_{Cu}, respectively. Thus

$$P_{Fe} = m_{Fe} p_{Fe}$$

and, if we assume that the presence of extra turns for positive tappings may be ignored for the purpose of establishing approximate equations,

$$P_{Cu,t} = m_{Cu} p_{Cu}$$

Substituting in equation 2.47, we obtain

$$m_{Fe}/m_{Cu} = k_S^2 p_{Cu}/p_{Fe} \tag{2.48}$$

A study of typical modern transformer designs[1] indicates that for distribution and rural transformers there is reasonable correlation between the actual loss and mass ratios and those derived from the above equations.

2.7.2 Mass ratios for minimum cost of materials

Variations in material costs between different designs are determined mainly by the proportions of active iron and copper in the transformer. Depending on the relative costs of these materials, an optimum mass ratio for minimum cost can be obtained, giving an alternative basis for design to that of maximum efficiency.

Let c_{Fe} and c_{Cu} be the specific costs per unit mass of iron and copper, respectively, and C_t be the total cost of active materials in the transformer. Let $k_C C_t$ be the proportion of the total cost due to the copper and $(1 - k_c)C_t$ be that due to the iron. Then

$$k_C C_t = m_{Cu} c_{Cu}$$

and

$$(1 - k_C)C_t = m_{Fe} c_{Fe}$$

Thus

$$k_C(1 - k_C)C_t^2 = c_{Fe} c_{Cu} m_{Fe} m_{Cu}$$

or

$$C_t = \{c_{Fe} c_{Cu} m_{Fe} m_{Cu}/k_C(1 - k_C)\}^{1/2}$$

It is shown in reference 1 that the numerator of this equation remains approximately constant as the dimensions of a design are varied. It is therefore possible to write

$$C_t = K_C\{k_C(1 - k_C)\}^{-1/2} \tag{2.49}$$

where, as a first approximation, the coefficient k_C may be assumed constant. C_t is thus a minimum when $k_C(1 - k_C)$ is a maximum, that is when

$$d(k_C - k_C^2)/dk_C = 1$$

or

$$k_C = 0.5$$

Therefore, for minimum cost of active materials, the cost of the copper is equal to the cost of the iron, or the cost ratio

$$\text{cost of copper/cost of iron} = 1 \tag{2.50}$$

In terms of specific costs, the cost of copper is $m_{Cu} c_{Cu}$ and that of iron $m_{Fe} c_{Fe}$. Thus, for minimum cost

$$m_{Fe}/m_{Cu} = c_{Cu}/c_{Fe} \tag{2.51}$$

Market fluctuations, of course, affect the ratio of c_{Cu} to c_{Fe}. At the time of writing, it is impossible to quote a stable value, but a figure of about 3 to 1 could be assumed for transformers with cold-rolled steel cores.

The achievement of a design based solely on minimum cost of materials is often impracticable, because such a design would have natural values of losses and reactance, corresponding to any specified values only by coincidence. It may be shown, however, that the conditions established in equation 2.51 are not critical; considerable variations in m_{Fe}/m_{Cu} are possible without much alteration in total cost.

It is found[1] that modern generator and primary transmission transformer designs show good correlation between the actual and optimum values of cost ratio. In secondary transmission, distribution and rural units the actual cost ratios are high and the mass ratios are lower than the optimum, indicating that the amount of copper relative to iron in the design is greater than that required for minimum cost of materials.

2.7.3   Capitalisation

The cost of active materials is only the most easily evaluated part of the total cost of the transformer, and the above theory ignores the cost of labour and overheads. Many modern designs, however, are based on capitalisation formulae which take account of the total cost of building the transformer and the cost of the losses and allow for interest and depreciation on the money involved. The formulae give a total annual cost, and the best design is that having the minimum annual cost.

2.8   TRANSFORMER WINDING SPACE FACTORS

2.8.1   Application

Values of winding space factor indicate the amount of window area required to accommodate the copper of the windings. In the determination of preliminary dimensions for a transformer, therefore, data are required from which the space factors may be found. These can be obtained from similar designs with the same type of windings and cooling requirements, and such information represents one way in which past design experience may be stored.

Space factors employed in programmes of computer-aided design need not be very accurate since the computer forms the windings and the values initially assumed are modified during the design process. If the frame dimensions are to be selected without forming the windings, however, a more accurate estimation of space factors is required, especially in low-voltage transformers where the windings occupy a larger proportion of the total window area than in high-voltage units.

It is convenient to express the space factors in the form

$$k_f = 1/k_1 + 1/k_2 \tag{2.52}$$

Thus, from equations 2.43 and 2.44,

$$k_f = b_1 h_1/a_1 N_1 + b_2 h_2/a_2 N_2$$

If we assume a common current density in the low- and high-voltage windings

$$a_1 N_1 = a_2 N_2 = \tfrac{1}{2} A_{Cu} \times 10^6$$

Therefore,

$$k_f = 2(b_1 h_1 + b_2 h_2)/A_{Cu} \times 10^6$$

or, for the area occupied by windings,

$$(b_1 h_1 + b_2 h_2) \times 10^{-6} = \tfrac{1}{2} A_{Cu} k_f \qquad (2.53)$$

Thus a predicted value of k_f enables the window area required to be found directly from A_{Cu}.

In reference 1, formulae are developed relating k_f to $(J/S)^{1/2}$. These formulae are not of general application because the coefficients of $(J/S)^{1/2}$ may be treated as constants only over a limited range of similar designs with the same winding structure. They do, however, indicate a convenient method of storing data from existing designs.

2.8.2 Data for distribution transformers with helical windings

Figure 2.6(a) shows values of k_f plotted to a base of $(J/S)^{1/2}$ for each arrangement of vertical ducts in a series of distribution transformers with ratings ranging from 0.15 to 1 MVA. Both the low- and high-voltage windings of all these units consist of helical coils; all have similar values of flux density within the range 1.55 to 1.6 T and of percentage reactance between 4.1 and 4.7%. The percentage resistance, however, varies from 1.83 to 0.99%, although no transformer has an impedance outside the range 4.5 to 5%.

All the transformers are constructed with standard tappings of $\pm$ $2\tfrac{1}{2}$ and 5% and the high-voltage winding cross-sectional areas thus all include approximately the same proportion of extra space to accommodate the turns required for the positive tappings. This enables the values of k_2 to be calculated so that they relate the copper cross-sections on normal tapping position, $a_2 \times$ (high-voltage turns for normal tapping), to the overall winding areas obtained from the full dimensions, $b_2 h_2$. It is convenient to express k_2 in this form since the output (equation 2.33) is derived by neglecting tappings, and the value of A_{Cu} obtained from it is thus that for normal tapping position. The corresponding values of k_f, therefore, include an allowance for tappings and can be substituted directly in equation 2.53.

The graphs may be used to predict with reasonable accuracy the space factors for designs similar to those for which they are drawn. The flux density should be approximately the same; otherwise the number of turns required for a given core area will be altered considerably, and some error may result. In practice, this problem seldom arises since there are normally only slight variations in flux density between different transformers within a particular class.

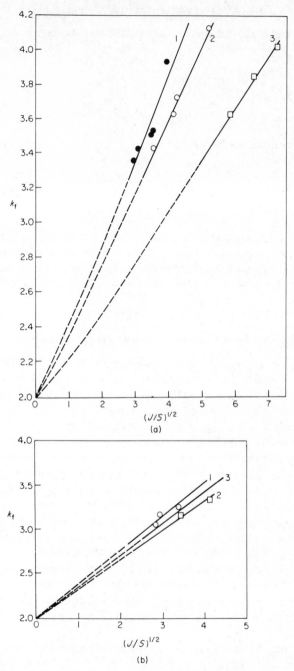

Figure 2.6   Values of k_f as a function of $(J/S)^{1/2}$: (a) distribution transformers with helical windings: curve 1, range 0.6 to 1 MVA; curve 2, range 0.33 to 0.5 MVA; curve 3, range 0.15 to 0.25 MVA; (b) distribution transformers with disk windings: curve 1, n_d/n_h within range 1.1 to 1.4 mm; curve 2, h_d/n_h within range 0.6 to 0.7 mm; curve 3, mean values from 1 and 2

2.8.3 Effect of variation in reactance

The values of impedance specified for distribution transformers lie normally within the range 4 to 5%; therefore the reactance is approximately constant. In exceptional cases impedances outside this range are required, and the reactances specified for larger transformers may differ considerably, even between units of the same size.

Figure 2.6(a) cannot be used directly to predict accurate values of k_f for transformers of abnormal reactance. The magnitude of the error involved is about $+5$ to $+7\%$ of the true value of k_f for distribution transformers with similar winding structures, copper losses and flux densities to those forming the graphs but with reactances of up to approximately twice the normal figure. The percentage by which the predicted value of k_f exceeds the actual value is greatest in designs with the highest reactance.

2.8.4 Transformers with disk windings

Some distribution transformers have disk-type high-voltage windings. This type of construction is not now very common, but figure 2.6(b) summarises the data obtained from five designs ranging in size from 0.5 to 1 MVA. All the transformers have a vertical duct in the centre of the low-voltage winding and between the low- and high-voltage windings. They have values of percentage reactance between 3.90 and 4.95%, and all the flux densities lie within the range 1.53 to 1.62 T. The percentage resistance varies from 1.34 to 0.95%.

For such designs the problem of predicting the value of k_f is complicated by variations in the number of horizontal cooling ducts in the disk windings and by variations in the vertical dimension h_d of these ducts, and there is considerable scatter in the values of k_f plotted to a base of $(J/S)^{1/2}$. If the transformers can be classified according to the ratio duct height per strand h_d/n_h where n_h is the number of conductor strands between horizontal cooling ducts, quite accurate straight lines of best fit can be drawn as shown in figure 2.6(b). At the preliminary design stage, however, the value of h_d/n_h will not be known within close limits, and the mean line of best fit must be used to estimate k_f. Thus the accuracy, although adequate, is less than that obtainable for helical coils.

2.9 ESTIMATION OF FRAME DIMENSIONS

2.9.1 Output coefficients

Very approximate frame dimensions may be obtained by means of output coefficients. These are based on previous design experience but are imprecise in nature.

TABLE 2.1 Output and winding coefficients for typical core-type transformers

Class of transformer	Percentage reactance		Output coefficient $K_{AS} = A_{Fe}/S^{1/2}$		Average winding coefficient $K_{VS} = (V/N)/S^{1/2}$ ($K_{VS} = 2.22 B_m K_{AS} \times 10^2$)
	Approximate range	Average value	Approximate range	Average value	
generator	12.0–18.0	15.0	0.056–0.075	0.071	26.8
primary transmission (including auto-transformers)	12.0–22.5	17.5	0.045–0.060	0.055	19.0
secondary transmission	10.0–26.0	14.2	0.051–0.069	0.058	20.0
distribution	4.75	4.75	0.054–0.060	0.057	19.6

TABLE 2.2 *Comparison of theoretical and actual output coefficients*

Class of transformer	Current density $J (A\,mm^{-2})$		A_{Fe}/A_{Cu} for transformers in table 2.1		Average flux density $B_m(T)$ for transformers in table 2.1	$K_{AS} = \left(\dfrac{1}{2.22fB_mJ} \times \dfrac{A_{Fe}}{A_{Cu}}\right)$ Calculated from average values for each class of transformer	Average actual value of K_{AS}
	Approximate range for class of transformer	Average value for transformers in table 2.1	Approximate range	Average value			
generator	2.6–4.0	2.9	2.1–3.8	2.9	1.70	0.073	0.071
transmission (including auto-transformers)	3.7–5.0	4.3	1.4–2.6	2.2	1.55	0.055	0.055
distribution	2.7–3.2	2.8	1.2–2.8	1.8	1.55	0.061	0.058
final distribution	2.4–3.1	2.6	1.4–1.7	1.6	1.55	0.060	0.057
rural (single-phase, wound cores)	2.0–2.5	2.3	0.65–0.85	0.8	1.55	0.045	0.042

The output (equation 2.33) may be rearranged in the form

$$A_{Fe} = \{(1/2.22fB_mJ)(A_{Fe}/A_{Cu})\}^{1/2}S^{1/2} \qquad m^2 \qquad (2.54)$$

Let

$$K_{AS} = \{(1/2.22fB_mJ)(A_{Fe}/A_{Cu})\}^{1/2} \qquad (2.55)$$

It is shown in reference 1 that K_{AS} can be regarded as very approximately constant for transformers of a particular class with approximately standard values of losses and reactance.

Thus

$$A_{Fe} = K_{AS}S^{1/2} \qquad m^2 \qquad (2.56)$$

where K_{AS} is an output coefficient relating the core area to the rating.

Average values of K_{AS} for a number of typical transformers designs are given in table 2.1, together with the range of variation for the transformers in each class. These are current values applying to the transformers of one manufacturer and may differ appreciably from the corresponding values for other firms. The approximate nature of the output coefficients is indicated by the range of variation in their values.

Changes in reactance, associated with changes in A_{Fe}/A_{Cu}, have considerable effect on the output coefficients. Thus, where the range of reactance for a particular class of transformer is large, as for the generator and transmission transformers, the variation in K_{AS} is also large. Where the specified reactance is the same for all the units in a particular class, as for the distribution and rural units considered, the range of variation in K_{AS} is very much smaller.

In table 2.2 the theoretical values of output coefficient calculated from the average figures for B_m, J and A_{Fe}/A_{Cu} for the designs considered in each class of transformer are compared with the average values of K_{AS} calculated from A_{Fe} and $S^{1/2}$. The correlation is quite good and would appear to indicate that the use of different specific loadings, or values of A_{Fe}/A_{Cu}, to those listed can be allowed for by modifying the output coefficients accordingly.

Once A_{Fe} has been determined from K_{AS}, A_{Cu} can be calculated from the output (equation 2.33) by using appropriate values of B_m and J. The nett window area A_w may then be found from equation 2.35 and the window space factor k_w.

Typical values of k_w for modern designs are listed in table 2.3. The values are obviously very approximate, since they cover the range of megavoltamperes corresponding to each class of transformer. More precise tabulation is not practicable, because, in evaluating k_w, no distinction is made between conductor and major insulation. Therefore no account can be taken of variations in the relative proportions of conductor to major insulation in designs with the same rating but different values of A_{Fe}/A_{Cu}.

The value found for A_w is thus very approximate but, in conjunction with A_{Fe}, gives an indication of the size of frame required. The window height h_w and width b_w are unknown; therefore additional information is required to find the shape of the frame. If, however, a standard range of frame sizes is available, the procedure

TABLE 2.3 Approximate values of window space factor k_w

Class of transformer	Equivalent megavoltampere range	Approximate range of k_w				
		400 kV	275 kV	132 kV	33 kV	11 kV
generator	340—144 144—72	0.14	0.21—0.17	0.25—0.16		
primary transmission (auto-transformer)	161—156 125—93.6		0.15—0.14			
primary transmission (two-winding)	120—100 90—45		0.11	0.20—0.14		
secondary transmission	20—10				0.40—0.30	
distribution	1.0—0.1					0.40—0.30

is to calculate A_{Fe} and then to select the frame with a nett core area nearest to that calculated. This will have standard centres, and so b_w is determined. The value of A_{Fe} for the standard frame is used in calculating A_w, which is the nett window area between core circles. Thus, for a three-phase transformer, from figure 2.4,

$$h_w = A_w \times 10^6 / \{b_w - (d - 2b_{Fe})\} \qquad \text{mm} \qquad (2.57)$$

It is sometimes required to relate the voltage per turn to the transformer rating

$$V/N = 4.44 f B_m A_{Fe}$$

Substituting for A_{Fe} from equation 2.54, we find

$$V/N = \{(8.88 f B_m / J)(A_{Fe} / A_{Cu})\}^{1/2} S^{1/2}$$

and thus

$$V/N = K_{VS} S^{1/2} \qquad (2.58)$$

where the winding output coefficient

$$K_{VS} = \{(8.88 f B_m / J)(A_{Fe} / A_{Cu})\}^{1/2} \qquad (2.59)$$

can be regarded as constant to the same degree of accuracy as K_{AS}. Average values of K_{VS} are given in table 2.1.

For three-phase transformers it may be convenient to express equations 2.56 and 2.58 in terms of the total megavoltamperes instead of the megavoltamperes per phase, S. In this case, the values of k_{AS} and K_{AS} listed in the tables are divided by $3^{1/2}$.

The method described above for finding the frame proportions takes no explicit account of reactance or losses or of economic consideration. It requires the use of the window space factor k_w, which can be very inaccurate, and depends upon a knowledge of standard frames to obtain the window height h_w and thus the winding length $h = h_w - h_o$.

In practice, reactance and losses are nearly always specified, and the values obtained by this method are unlikely to meet the guaranteed figures with sufficient accuracy. It may, therefore, be necessary to undertake a long process of modification by trial and error, involving the formation of the windings on alternative standard frames.

2.9.2 Inclusion of copper loss and reactance

Where the specified values of reactance and losses are fairly normal for the type of transformer considered, it is possible by the use of equations 2.25 and 2.31 to confine most of the trial and error process to the preliminary design stage, instead of during the calculation of the windings. The method depends initially upon output coefficients to select a standard frame, but the use of a window space factor k_w is avoided.

As discussed in sub-section 2.5.3, when the percentage reactance and copper loss are both specified, the window width b_w is fixed. Many transformers for which a particular standard core area is suitable have specifications such that they all require approximately the same window width and can, therefore, be built on a frame having standard leg centres b_{cen}. This is generally true of distribution transformers and means that, for most designs of a similar rating, the reactive width b_X of the windings is approximately constant and can be determined from the dimensions of the appropriate standard frame.

To apply the method, an approximate value of A_{Fe} is found from the output coefficient, and the nearest standard core is selected. This gives the actual value of A_{Fe} to be used in the equations, the number of steps in the core, the core circle diameter, the plate widths, the spacing of the standard centres and the corresponding width of window opening. The average length s of mean turn may be calculated for the standard value of b_w from equation 2.40

$$s = 2\pi(b_{Fe} + \tfrac{1}{4}b_w) \qquad \text{mm}$$

An approximate value of B_m is known from table 2.2. When A_{Fe} is fixed, however, it is generally not possible to use the value first selected. Possible values of voltage per turn must always correspond to whole numbers of turns and are, therefore, determined from the winding with least turns. Thus the values of B_m and A_{Fe} must be compatible in the voltage equation 1.4 with integral numbers of low-voltage turns. In large distribution transformers the voltage per turn is 4 to 5% of the total voltage of the low-voltage winding. Therefore, for a standard value of A_{Fe}, the actual value of B_m may differ considerably from that first assumed.

The procedure is first to substitute A_{Fe} and the initial value of B_m in the voltage equation 1.4 to give an approximate figure for the low-voltage turns, N_1. For three-phase transformers the nearest whole number of turns may then be selected. The value of B_m for subsequent use can thus be calculated from the voltage equation 1.4. This procedure is not applied in the method described in sub-section 2.9.1, because the approximate nature of k_w means that small errors in B_m are of little consequence in the calculation of A_w.

Equation 2.25 states

$$J = 10.4f \, \frac{B_m}{k_i} \, \frac{A_{Fe}}{s} \%P_{Cu} \qquad \text{A mm}^{-2}$$

The value of B_m is known, and the eddy current loss factor k_i can be regarded as approximately constant for a particular class of transformer. Thus the value of J which is compatible with the specified copper loss may be calculated.

The value obtained should lie approximately within the limits of table 2.2. If it is too low, a larger core area may be required, and, if too high, a smaller frame is indicated. Adjustments of this nature are required because the larger core has the higher value of J for the same copper loss. If B_m is constant, the voltage per turn is directly proportional to the core area. An increase in A_{Fe}, therefore, means an increase in voltage per turn and a reduction in turns. Although the length of mean turn is increased slightly, the amount of copper required in the winding is reduced, and a higher value of J is used to meet the guaranteed copper loss.

When the value obtained for J is satisfactory, the next step is to find the width of the windings from the frame dimensions. A knowledge of standard clearances at the required voltage is necessary. The total winding width $b_1 + b_2$ may then be calculated from equation 2.37

$$b_w = 2\{b_{01} + b_0 + \tfrac{1}{2}b_{02} + (b_1 + b_2)\}$$

and the reactive width found from

$$b_X = b_0 + \tfrac{1}{3}(b_1 + b_2) \tag{2.60}$$

Equation 2.31 states

$$A_{Fe}^2 h/s = 0.04\, S\, b_X / B_m^2 f\, \%\, X$$

Thus $A_{Fe}^2 h/s$ is obtained from b_X, B_m and the specified reactance. The initial values of h/s and h are therefore determined, since A_{Fe} is known. A_{Cu} may also be calculated from the output equation, or by evaluating I_1, by dividing by J and by multiplying by $2N_1$.

For accurate results, the theory should take account of the extra turns required for positive tappings. In the derivation of equations 2.25, 2.31 and the output equation, equality of ampere–turns is assumed; thus these extra turns are not explicitly allowed for. All the values substituted in the equations should, therefore, be those for normal tapping position.

The guaranteed values of copper loss and reactance relate to normal tapping and will thus give the correct result. In many disk coils, the tapping sections are situated entirely in the body of the winding and occupy its full depth. The winding width is therefore independent of tapping position, and the full value of b_X obtained from the dimensions of the standard frame may be used in the calculations. In layer-type windings the tapping sections occupy the external layers of the high-voltage coils. To find b_X, the width available for the windings should thus be reduced by an amount depending upon the extra turns required for the positive tappings. Since the width of the coils is small compared with their diameter, the effect upon s of small changes in $b_1 + b_2$ is negligible.

From the considerations in sub-section 2.6.3, the value of h/s will give some idea of the viability of the initial design. It is, however, essential to check whether the winding will fit within the space available. The practical value of k_f can be estimated from data such as that given in sub-section 2.8.2, for similar transformers with the same type of windings and cooling requirements. From equation 2.53 the area available for the windings is $\tfrac{1}{2}A_{Cu}k_f$, and the condition to be satisfied is

$$(b_1 + b_2)h \times 10^{-6} \approx \tfrac{1}{2}A_{Cu}k_f \tag{2.61}$$

where $(b_1 + b_2)h \times 10^{-6}$ is the space allowed for the windings within the preliminary dimensions of the transformer window.

The value of A_{Cu} calculated from the output equation is that on normal tapping position, and an allowance for the positive tappings is required. For distribution transformers with standard tappings, this may be included in the value of k_f, as described in sub-section 2.8.2. Any non-standard tapping arrangement must be allowed for by rewriting equation 2.61:

$$(b_1 + b_2)h \times 10^{-6} \approx \tfrac{1}{2}A_{Cu}k_f(100 + 0.5 \times \text{max. \% positive tapping})/102.5 \tag{2.62}$$

The values used for the standard frame centres are chosen by experience to accommodate those designs requiring the larger values of b_w; other designs have a slight increase on the minimum clearance required between windings. Thus, if $(b_1 + b_2)h \times 10^{-6}$ is slightly greater than $\tfrac{1}{2}A_{Cu}k_f$, the preliminary dimensions may often still form a satisfactory basis for further work. More space is available on the standard frame than is actually required, which means that the insulation clearances are increased by a small amount. Some iron is being wasted, but against this can be set the advantages of standardisation.

If $(b_1 + b_2)h \times 10^{-6}$ is slightly less than $\tfrac{1}{2}A_{Cu}k_f$ all the previous calculations, which are based on the exact load loss and reactance specified, are not necessarily invalidated. Standard specifications allow tolerances of $+1/7$ of each component loss, provided that a tolerance of $+1/10$ of the total losses is not exceeded, and $\pm 1/10$ of the impedance voltage. Consequently it may be possible to obtain a satisfactory compromise by making, within these limits, appropriate alterations in the calculation.

In practice, the majority of designs should fit the standard frames, but provision is usually made for alteration of the leg centres to accommodate designs with very unusual specifications. For example, if $(b_1 + b_2)h \times 10^{-6}$ is considerably less than $\tfrac{1}{2}A_{Cu}k_f$, the use of non-standard centres or a larger standard frame may be indicated. If it is possible to meet the specification, it is often preferable to employ a larger standard frame, particularly when only a few transformers of a given design are to be manufactured.

2.9.3 Iron loss

The method described in sub-section 2.9.2 does not take explicit account of iron loss, but a standard value is initially assumed for B_m. Thus, if the calculated frame dimensions are within normal limits, the design will probably meet the guaranteed iron loss, provided this is also within normal limits. The iron loss must, however, be calculated before proceeding further with the design.

For a three-phase core-type transformer, the total length of iron in the frame may be obtained conveniently by substituting for b_{cen} in equation 2.36:

$$l_{Fe} = \{3(h + h_0) + 4b_w + 12b_{Fe}\} \times 10^{-3} \qquad \text{mm} \tag{2.63}$$

If we assume that the yoke has the same cross-sectional area as the legs, the mass of iron is

$$m_{Fe} = A_{Fe} l_{Fe} \rho_{d, Fe} \qquad \text{kg} \tag{2.64}$$

where $\rho_{d, Fe}$ is the density of the core material in kilograms per cubic metre. From figure 2.1 the specific iron loss p_{Fe} is found for the type of core construction used, and the total iron loss is, therefore, obtained.

If the result exceeds the guaranteed no-load loss by more than the standard tolerance of $+1/7$ of component losses, the preliminary design work must be repeated with the object of reducing the total mass of iron in the core, or with a reduced flux density. In practice, owing to such manufacturing uncertainties as the effect of joints and burrs on lamination edges, it is usually not advisable at the final design stage to accept a calculated loss which is more than slightly in excess of the guaranteed value.

2.10   FRAME DIMENSIONS FOR A 0.75 MVA THREE-PHASE DISTRIBUTION TRANSFORMER

2.10.1   Specification

To illustrate the application of the methods described in sub-sections 2.9.2 and 2.9.3, preliminary frame dimensions will be obtained for a three-phase ONAN class A insulated distribution transformer design. The specification for this transformer is as follows.

Rating	0.75 MVA, three-phase, 50 Hz
Nominal voltage	11000 to 433–250 V
Connections	high-voltage delta, low-voltage star
Tappings	$\pm 2\frac{1}{2}\%$ and $\pm 5\%$ on high-voltage to switch
Impedance	4.75 %
Load loss	9500 W
No-load loss	1420 W

The transformer is not constructed entirely in accordance with the best modern practice as the core has square-cut interleaved joints and not mitred joints, and a disk winding is used for the high voltage, whereas a layer-type winding would now be more common. The limits of temperature rise are 60 °C and 50 °C for $\Delta\theta_R$ and $\Delta\theta_0$, respectively, but those currently specified for class A insulated transformers are 65 °C and 55 °C, respectively. These factors do not affect the principles involved in the determination of the preliminary dimensions in this section. Their effect is fully discussed in chapter 9, where a complete design is obtained, including the formation of the windings and calculation of temperature gradients and tank dimensions.

2.10.2   Approximate method

Preliminary dimensions may be obtained very approximately by the use of window space factors, as described in sub-section 2.9.1.

Applying this method, we find

$$S = 0.75/3 = 0.25 \, \text{MVA}$$

From table 2.1 the average output coefficient K_{AS} for distribution transformers is 0.057. Thus from equation 2.56 the nett core section

$$A_{Fe} = K_{AS} S^{1/2} = 0.057 (0.25)^{1/2} = 0.0285 \, \text{m}^2$$

In a range of standard frames with square-cut corners, the nearest core sections have nett areas of $0.0273 \, \text{m}^2$ and $0.0307 \, \text{m}^2$. The designs in table 2.1 from which the output coefficients are obtained have frames with mitred joints. Larger cross-sections are permissible with mitred frames because the iron loss is reduced so that the specified value can be obtained when a comparatively large frame is used. This means that the core area derived from table 2.1 will tend to be too large, and the smaller standard core is probably the correct one.

In the subsequent calculations, however, both standard frames will be considered. With distribution transformers the ratio of the cost of copper to the cost of iron is normally greater than the optimum value of one. The use of a larger core cross-section has, therefore, the advantage that the number of turns and the amount of copper required is reduced. Also the natural value of reactance is reduced, since fewer turns are used. For the same copper loss a higher value of current density is required, which may lead to difficulty in cooling the windings.

The calculations for the two standard frames may be summarised as follows.

		Frame I	Frame II
Cross-sectional area	$A_{Fe} \, (\text{m}^2)$	0.0273	0.0307
From table 2.2, average value	$B_m \, (\text{T})$	1.55	1.55
From table 2.2, average value	$J \, (\text{A mm}^{-2})$	2.6	2.6
From equation 2.33	$A_{Cu} \, (\text{m}^2)$	0.0205	0.0182
From table 2.3, average value	k_w	0.35	0.35
From equation 2.35	$A_w \, (\text{m}^2)$	0.1171	0.1040
For the standard frame	$b_w \, (\text{mm})$	152	165
For the standard frame	$d \, (\text{mm})$	197	211
For the standard frame	$2b_{Fe} \, (\text{mm})$	2×95	2×102
From equation 2.57	$h_w \, (\text{mm})$	808	658

Two sets of approximate frame dimensions are thus obtained, between which it is impossible to distinguish without considering more detailed aspects of the design.

2.10.3 More accurate method for standard frames

Determination of current densities

Employing the method described in sub-section 2.9.2, we find the following.

		Frame I	Frame II
From equation 2.40	s (mm)	837	900
From specification	P_1 (W)	9500	9500
From equation 2.13	$\%R\,(\%)$	1.27	1.27
Allowance for stray losses	(W)	500	500
From equation 2.24	$\%P_{Cu}\,(\%)$	1.20	1.20
From equation 1.4 with $B_m = 1.55\,\mathrm{T}$	V/N (V)	9.4	10.57
Corresponding low-voltage turns per phase	N_1 (turns)	26.6	23.62
Nearest whole number of low-voltage turns	N_1 (turns)	27	24
Revised voltage per turn	V/N (V)	9.26	10.42
Revised flux density	B_m (T)	1.53	1.53
From equation 2.35 with $k_i = 1.05$	$J\,(\mathrm{A\,mm^{-2}})$	2.97	3.10

The value of J for frame I is near the upper limit of table 2.2 and that for frame II is just at the upper limit. These values are considerably higher than the average value quoted in the table, and this illustrates a major source of inaccuracy in the previous method. For each class of transformer the range of J is wide; thus the use of an average value must lead to considerable error in the initial dimensions of those designs where a value near the ends of the range is appropriate.

Estimation of winding heights

The winding arrangement used in the construction of this transformer consists of a simple two-layer helical low-voltage winding with a disk-type high-voltage winding. Vertical ducts are provided between the low- and high-voltage windings. No special provision is made for a cooling duct between the low-voltage winding and the core, although the mechanical clearances between the winding and the flats of the core section do allow some circulation of oil. The end-insulation consists of blocks and spacers, no bent washers being employed.

Suitable clearances for this arrangement are

$$b_{01} = 6.4\,\mathrm{mm} \qquad b_{02} = 12\,\mathrm{mm}$$
$$b_0 = 9.1\,\mathrm{mm} \text{ excluding conductor insulation}$$

The equivalent total axial end clearance, based on the assumption of equal high- and low-voltage winding lengths, may be taken as

$$h_0 = 60\,\mathrm{mm}$$

		Frame I	Frame II
From equation 2.37	$b_1 + b_2$ (mm)	54.5	61
From equation 2.60	b_X (mm)	27.3	29.4

		Frame I	Frame II
From specification			
and equation 1.16	$\%X\,(\%)$	4.58	4.58
From equation 2.31	$A_{Fe}^2 h/s$	5.1×10^{-4}	5.484×10^{-4}
Hence	h/s	0.684	0.582
Therefore	$h\,(\text{mm})$	573	524
From equation 2.33	$A_{Cu}\,(\text{m}^2)$	0.0182	0.0155

Compatibility

Both initial values of h/s are within the practical range of 0.3 to 1 discussed in sub-section 2.6.3. Therefore, the next step is to check each set of dimensions for compatibility. The transformer has standard positive tappings of $2\frac{1}{2}$ and 5 %; thus equation 2.61 may be used in conjunction with data from figure 2.6(b).

		Frame I	Frame II
For values of J			
from figure 2.6(a)	$(J/S)^{1/2}$	3.45	3.52
From curve 3 in			
figure 2.6(b)	k_f	3.22	3.25
For equation	$(b_1 + b_2)h \times 10^{-6}\,(\text{m}^2)$	0.0313	0.0320
2.61	$A_{Cu}k_f/2\,(\text{m}^2)$	0.0293	0.0252

For frame I, approximately 7 % more space than required appears to be available for the windings. Such a result is normally satisfactory for the initial dimensions. It also indicates that the value of b_X corresponding to the standard window width is appropriate to the design under consideration, although slightly larger than that actually required. The value of h/s obtained from equation 2.31 is thus high with the result that h is overestimated by a small amount. A more precise value will be obtained by repeating the calculation for a slightly reduced winding depth although, in practice, this would hardly be justified owing to the limits of accuracy of k_f and the limitations imposed on the winding formation by the use of standard conductor sections.

For frame II, the area available is 27 % greater than that required to accommodate the windings, and the window dimensions of the standard frame are incompatible with the transformer specification. Thus, if frame II is used, it will require non-standard centre spacings to give the same width of window as that of the smaller frame.

Revised dimensions for frame I

Reducing $b_1 + b_2$ in the ratio $(100:107)^{1/2}$, we obtain

$$b_1 + b_2 = 52.7\,\text{mm}$$

Thus

$$b_X = 26.7\,\text{mm}$$

From equation 2.31,

$$A_{Fe}^2 h/s = 4.99 \times 10^{-4}$$

and

$$h/s = 0.67$$

The slight reduction in $b_1 + b_2$ has a negligible effect upon s. Thus

$$h = 561 \, mm$$

Applying the test of compatibility, from equation 2.61

$$(b_1 + b_2)h \times 10^{-6} = 0.0296 \, m^2$$

$$A_{Cu} k_f/2 = 0.0293 \, m^2$$

The space available is now only 1% greater than that required, and the final value of h for frame I can thus be taken as approximately

$$h = 560 \, mm$$

Revised dimensions for frame II

From equation 2.40 the value of s for the non-standard centres is

$$s = 2\pi(102 + 152/4) = 880 \, mm$$

Since s appears in the denominator of equation 2.25, the reduction in s implies an increase in J, the new value of which is

$$J = 3.10 \times 900/880 = 3.17 \, A \, mm^{-2}$$

This is about 2.3% above the upper limit of table 2.2 and may result in high winding temperature gradients. By itself, an excess of this magnitude in the preliminary value of J is not, however, sufficient to preclude further consideration of the larger core.

The value of B_m is the same as that for frame I; thus both designs must have practically the same value of $A_{Fe}^2 h/s$. For the larger core, therefore, if we use the revised figures for $A_{Fe}^2 h/s$ and $b_1 + b_2$,

$$h/s = 4.99 \times 10^{-4}/0.0307^2 = 0.53$$

Thus

$$h = 0.53 \times 880 = 466 \, mm$$

For the increased value of J,

$$A_{Cu} = 0.0155 \times 3.10/3.17 = 0.0152 \, m^2$$

$$(J/S)^{1/2} = 3.17/0.5 = 3.56$$

From figure 2.6(b) the corresponding value of k_f is 3.26. The components of equation 2.61 are

$$(b_1 + b_2)h \times 10^{-6} = 52.7 \times 466 \times 10^{-6} = 0.0246 \, \text{m}^2$$

$$A_{Cu} k_f/2 = 0.0152 \times 3.26/2 = 0.0248 \, \text{m}^2$$

In this case the space available is less than that required, but only by 0.8 %. The dimensions are therefore satisfactory, and h may be taken as approximately 470 mm.

2.10.4 Calculation of iron loss

For the two possible sets of preliminary dimensions obtained, the iron loss may be found from equation 2.63, equation 2.64 (if we take $\rho_{d,Fe}$ as 7700 kg m^{-3}) and figure 2.1.

		Frame I	Frame II (non-standard centres)
From equation 2.63	l_{Fe} (m)	3.608	3.422
From equation 2.64	m_{Fe} (kg)	758	809
At $B_m = 1.53\,\text{T}$, from figure 2.1	p_{Fe} (W kg^{-1})	1.84	1.84
Therefore	P_{Fe} (W)	1395	1489

For frame I the result is only 1.8 % below the guaranteed iron loss of 1420 W and is quite satisfactory. For frame II the calculated loss is 4.9 % above the guaranteed value. This is well within the allowable tolerance of + 1.7 which must not be exceeded on test, but, for the reasons explained in sub-section 2.9.3., it may be considered inadvisable to accept a preliminary result which is nearly 5 % in excess of the specified loss.

It is of interest to note that a frame with the same dimensions as the non-standard frame II but constructed with mitred joints, has a specific loss p_{Fe} of 1.5 W kg^{-1}, giving a total iron loss of only 1214 W.

2.10.5 Assessment of results

The larger frame has a high current density, a high iron loss and non-standard construction. For these reasons, frame I appears preferable, although, as explained in sub-section 2.10.2, it is more expensive in materials.

The existing design for the transformer specified in sub-section 2.10.1 is, in fact, based on frame I. The window height of this design is

$$h_w = 624 \, \text{mm}$$

and the mean specific loadings

$$B_m = 1.53\,\text{T} \qquad J = 2.89 \, \text{A mm}^{-2}$$

The calculations of sub-section 2.9.3 indicate that for this frame the window height should be

$$h_w \approx 560 + 60 = 620 \, \text{mm}$$

and the corresponding specific loadings are

$$B_m = 1.53 \, \text{T} \qquad J = 2.97 \, \text{A} \, \text{mm}^{-2}$$

In the example, therefore, very accurate results are achieved, owing principally to the recalculation of the dimensions after the initial list of compatibility.

To attempt to attain such accuracy in the preliminary calculations is normally neither necessary nor justified by the nature of the data upon which they are based. Between the initial and final results for frame I the change in h is only 2.4 % and in $b_1 + b_2$ only 3.3 %. Variations of similar magnitude to these may often be imposed on the winding shape by the use of standard conductor sections.

The limitations of the approximate method are indicated clearly by a comparison between the results obtained from sub-sections 2.10.2 and 2.10.3. For frame I, this reveals a discrepancy of 12.5 % in the values of J and 30 % in those of h_w. These errors arise from the selection of an average figure for J from table 2.2 and from the very approximate nature of k_w. The method gives a tentative indication of the correct standard frame, but to find an accurate value for the core height it is necessary to form the windings by trial and error; there is no provision for including losses and reactance in the preliminary calculations.

Although the method of sub-section 2.10.3 overcomes these limitations, it does not form a complete solution from first principles because it relies upon knowledge of the dimensions of standard frames. The shape of these frames has been evolved by experience to accommodate designs of normal specification; thus the method lacks flexibility. Furthermore, there is no provision to enable explicit consideration to be given to economic factors in the initial calculations. These are, however, implicit in the evolution of the standard frames, and knowledge of the principles of section 2.7 will enable the best choice to be made between alternative designs.

In practice, existing design sheets or computer programmes form the basis of nearly all design work undertaken by manufacturers, and it is unlikely that details of standard frames will be available without access to the corresponding design files. The example given is, therefore, chiefly of interest as an illustration of the application of theoretical principles.

2.11   DESIGN OF WINDINGS AND TANK

2.11.1   Design of windings

The problem of winding design is to form, within the required dimensions, windings which are adequately insulated and have the correct values of reactance, I^2R losses and eddy current losses.

To design the insulation from first principles would be extremely difficult and would require a quantitative knowledge of many of the factors outlined in chapter 5. Fortunately, much insulation is semi-standardised, and insulation lists based on previous experience are normally available which give general arrangements of windings and insulation for the various sizes and types of transformer. The axial clearances b_{01}, b_0 and b_{02} and the end clearances h_{01} and h_{02} depend upon the arrangement of insulation, and minimum values are usually given in conjunction with the standard insulation lists. These are the clearance values employed in determining the preliminary frame dimensions.

Empirical formulae[2] have been developed which give a guide to safe values of clearances, but these formulae are necessarily approximate since, if they are to be of reasonably general application, they must allow for different arrangements of ducts and solid insulation. The clearances associated with standard insulation lists, having been proved on test, should be used whenever possible. Furthermore, the standard lists allow for such factors as the mechanical requirements of assembly, which in some transformers may decide the clearance b_{01} between the low-voltage winding and the core.

The thickness of insulation on paper-covered conductors may also be determined more by the mechanical properties of the paper than by voltage. Normal practice is to define the thickness of the covering as the amount by which the bare conductor dimension is increased, which is twice the wall thickness of the insulation. Typical values for the thickness of covering on double-paper-covered rectangular conductors range from about 0.36 to 0.5 mm.

Paper insulation between winding layers must withstand the effect of winding the upper layer over it, so that again mechanical considerations may partly determine its thickness, which is seldom less than about 0.25 mm in distribution or larger transformers. Greater thicknesses may be required when the voltage between layers is high, and, as an approximate guide to the thickness required, the normal working rms voltage gradient between layers at the point of greatest stress should not exceed $2 \, \text{kV mm}^{-1}$. The insulation thickness is measured between bare conductors and thus includes the conductor insulation.

The formation of windings suitable for the 0.75 MVA transformer with the preliminary dimensions found in the previous section is described in detail in sub-sections 9.9.3 to 9.9.8.

When windings with satisfactory dimensions and values of reactance and losses have been obtained, the winding temperature gradients must be calculated as described in sub-section 9.11.1. If these gradients are too high, the winding formation must be re-examined with the object of reducing the thermal power to be dissipated per unit of cooling surface area; it is only when satisfactory cooling has been obtained that the winding design can be finalised.

2.11.2 Design of tank

Some of the considerations underlying the design of tank for an ONAN transformer are discussed in sub-section 2.4.3. The rate of circulation of the oil depends upon the vertical position of the heat sources of the transformer relative to the tank cooling surfaces and to the type of tank, whether plain or equipped with radiators. As in other aspects of design, empirical constants are used, and these factors are implicit in their values.

The tank surface cools by both radiation and convection, but it is shown in reference 1 that, within the range of temperature differences involved, the heat transfer from such a surface may be calculated approximately from

$$M_{et} = K_{et}\Delta\theta^{1.25} \quad W\,m^{-2} \tag{2.65}$$

where M_{et} is the total rate of heat transfer from unit tank surface, K_{et} an empirical coefficient and $\Delta\theta$ the temperature difference between the surface and cooling medium. It is generally more convenient, however, to express the relationship in the form

$$\Delta\theta = K_t M_{et}^{0.8} \quad °C \tag{2.66}$$

where K_t is $(1/K_{et})^{0.8}$ and has empirical values dependent upon the type of tank.

The effective tank cooling surface area A_T must be sufficient to dissipate the losses of the transformer without temperature rises exceeding those given in standard specifications. A relationship is therefore required between A_T and the top oil temperature rise $\Delta\theta_o$. The problem is complicated by the fact that the temperature of the tank surface and, therefore, its rate M_{et} of dissipation of heat per unit area changes according to height from the bottom of the tank. A simple approximate solution may, however, be obtained by assuming that the average temperature rise of the tank cooling surface is equal to the mean duct oil temperature rise $\Delta\theta_{om}$ where

$$\Delta\theta_{om} = \Delta\theta_o - \tfrac{1}{2}\Delta\theta_v \quad °C \tag{2.67}$$

It can be seen from figure 2.3(a) and figure 2.3(c) that this assumption will be approximately true in practice. When the transformer occupies the vertical position of figure 2.3(c) with respect to the tank cooling surfaces, the average oil temperature rise for the tank as a whole is slightly above $\Delta\theta_{om}$, but this will tend to be compensated for by the internal temperature drop at the oil to tank surface shown in figure 2.3(a). Thus the accuracy of the solution depends upon the tank configuration. The method gives satisfactory results when applied, in conjunction with appropriate values of empirical constants, to tanks of reasonably normal design.

Let the average value of M_{et} over the effective tank cooling surface be

$$M_{eT} = (P_{Fe} + P_1)/A_T \quad W\,m^{-2} \tag{2.68}$$

Substituting the average values into equation 2.66, we obtain

$$\Delta\theta_{om} = \Delta\theta_0 - \tfrac{1}{2}\Delta\theta_v = K_t M_{eT}^{0.8} \quad °C \tag{2.69}$$

From equation 2.14, $\Delta\theta_v$ may be replaced by $k_v\Delta\theta_o$. Thus

$$\Delta\theta_o(1 - k_v/2) = K_t M_{eT}^{0.8}$$

or

$$\Delta\theta_o = K_T M_{eT}^{0.8} \quad °C \tag{2.70}$$

where K_T is equal to $K_t/(1 - k_v/2)$ and is an overall coefficient for the tank which applies only to a limited range of transformers of similar construction. Typical values of K_T used in the design of ONAN distribution transformers range from 0.27 for plain tanks to 0.7 for tanks fitted with radiators. The substitution of M_{eT} from equation 2.68 gives

$$\theta_o = K_T\{(P_{Fe} + P_1)/A_T\}^{0.8} \quad °C \tag{2.71}$$

Semi-standardisation is usually employed for distribution transformer tanks; thus the general arrangement of a suitable tank will be known, and appropriate values can be selected for the empirical constants k_v and K_T. The mean low- and high-voltage winding temperature gradients can be compared with the value of $\Delta\theta_{wo}$ found from equation 2.16 to be compatible with the specified limits of temperature rise. If neither winding has a gradient exceeding that obtained from equation 2.16, the specified limit for $\Delta\theta_o$, usually 55 °C, may be substituted in equation 2.71, from which a minimum value can then be calculated for A_T.

Equation 2.16 may, however, give a value of $\Delta\theta_{wo}$ which is so low that it is unduly restrictive on the winding design. For this case, rearranging the equation, we get

$$\Delta\theta_o = (\Delta\theta_R - \Delta\theta_{wo})/(1 - k_v/2) \quad °C \tag{2.72}$$

from which a value may be found for $\Delta\theta_o$ less than the specified limit but compatible with the mean temperature gradient of the hottest winding, which, as discussed in chapter 9, should normally be limited to about 21 °C. The corresponding value of M_{eT} can then be obtained from equation 2.70, or A_T can be found directly from equation 2.71.

When a value has been obtained for A_T, the tank dimensions can be calculated. With this approximate method normal procedure is to regard only those external surfaces in direct contact with the oil as contributing to the effective cooling area. The overall dimensions of the transformer core and windings determine the internal width and length of the tank. The clearances allowed are usually standardised and are considerably greater than those required purely for insulation purposes to ensure adequate space for free circulation of the oil, as well as for the leads. The total height of the transformer core, including any tap-change gear mounted on top of it, determines a minimum oil level, which must be such that the equipment is covered adequately by the oil when cold. Details of the design procedure for a tank fitted with radiators are given in sub-section 9.12.3.

Experience has shown that the thermal data obtained by the use of appropriate empirical constants in the above equations are reasonably accurate. The calculated temperature rises tend to be rather higher than those measured on test, but this provides a safety margin to ensure that the transformer will meet the specification.

Reference 4 gives more exact formulae for the thermal calculations, in which radiation and convection are treated separately. In reference 3, a method of tank design is described which takes explicit account of the dimensions and position of the heating elements of the core and windings with respect to the tank cooling surfaces; these surfaces are divided into a number of separate zones, each with a temperature rise related empirically to that of the hottest oil.

ACKNOWLEDGEMENTS

The author gratefully acknowledges the supply of data by Bruce Peebles Limited, Distribution Transformers Limited and Ferranti Limited which made it possible to complete this chapter.

REFERENCES

(Reference numbers preceded by the letter G are listed in section 1.14.)
1. Crompton, A. B., and Rowe, K. S., Principles of Transformer Design, to be published
2. Waterhouse, T., *Design of Transformers*, Draughtsman Publishing, London (1945–6)
3. Bean, R.L. *et al.*, *Transformers for the Electric Power Industry*, McGraw-Hill, New York, Toronto, London (1959)
4. Blume, L. F. *et al.*, *Transformer Engineering*, Wiley, New York, London (1951)
5. Say, M. G., *Alternating Current Machines*, Pitman, London, 4th edn, (1976)

3

The Use of the Automatic Electronic Digital Computer as an Aid to the Power Transformer Designer

K. Rowe*

3.1 INTRODUCTION

The automatic electronic digital computer is a device that can be programmed to perform automatically a sequence of arithmetic operations on numbers presented in digital form at very high speeds. The basic elements which constitute such a device are as follows.

(a) Input equipment, such as paper tape, card reader or magnetic tape equipment. This allows the operator to communicate with the computer.
(b) Ouput equipment, such as paper tape, card punch or printer. This equipment allows the computer to communicate with the outside world.
(c) A store for holding information which can be in the form of instructions or data. The store, for example, could be a magnetic drum, magnetic tape or disk.
(d) An arithmetic unit for operating on the numbers.
(e) A control unit for directing the operations of the computer, that is controlling the storage and extraction of numbers.

The whole process of preparing information and instructions for the computer is called programming. This is a skilled art since one must formulate the problem accurately, explicitly and unambiguously. The programming task can be broken down into three distinct parts as follows.

(a) Systems analysis. This is the formulation of the problem, which in the case of transformer design is usually carried out by someone who is familiar with the technical problems involved.

* Ferranti Engineering Limited.

(b) The next stage is to interpret the systems analysis into a form which is suitable for the computer being used.

(c) The final stage is to transcribe the above information into computer language. This part is usually called coding and will vary depending upon the size and speed of the computer employed.

3.2 CUSTOMER'S SPECIFICATION

The transformer designer has to have certain information before a design can be created; the major items are as follows.

(a) Rating of the transformer.
(b) Primary and secondary line voltages.
(c) Primary and secondary winding connections (star, delta, etc.).
(d) Number of phases.
(e) Frequency.
(f) Type of cooling (ON, OB, OFB, etc.).
(g) Tapping type, range and steps.
(h) Temperature rise limits.
(i) Test conditions.
(j) Mass and dimensional restrictions.

With this information a design can be produced which would give the designer a choice to decide upon the reactance, load loss and iron (core) loss. To minimise the cost of the transformer he would invariably work to the maximum permissible flux and current densities in keeping with good design practice for the size and type of transformer being considered. The resultant design would, therefore, have natural values for the reactance and losses in keeping with the particular manufacturer's design techniques. The problems involved in arriving at such a design are considerably reduced where the reactance and losses are specified by the customer.

In practice it is found that the majority of customers specify the reactance since this has to comply with the electrical system in which the transformer is to be installed. Sometimes the losses are specified, or alternatively the capitalised cost of the losses. In the latter case the designer has the problem of minimising the total cost, that is the prime cost of the transformer plus the cost of the losses.

In recent years the maximum noise level of power transformers has also been specified by the customer, and this has created a further restriction, since the noise level is primarily a function of the core mass and flux density. Therefore, there are four more parameters which could be specified at the quotation stage.

(k) Reactance or impedance.
(l) Load loss or capitalised cost of the loss per kilowatt.
(m) Iron loss or capitalised cost of the loss per kilowatt.
(n) Noise level.

3.3 DESIGNER'S SPECIFICATION

These specifications are governed by the particular manufacturer's code of practice or standards for the design being considered. A list of some of these is given below.

(a) General arrangement of the windings with respect to the core leg.
(b) Type of winding design (helical, layer, disk, etc.).
(c) Type of winding conductor (copper, aluminium, transposed cable or normal strand).
(d) Type of core (one-, two-, three-, four- or five-limbed mitred or non-mitred joints).
(e) Type and grade of core steel.
(f) Type of cooling arrangement (radiators on the tank or separate bank, pumps, fans, etc.).
(g) Maximum thermal gradient in the windings.
(h) Maximum current densities in the windings.
(i) Maximum winding eddy current losses expressed as a percentage of the winding (I^2R) loss.
(j) Maximum flux density in the core leg.
(k) Major insulation clearances (interwinding and windings to core insulation).
(l) Minor insulation thickness (paper covering on the conductor, radial duct widths in the windings, etc.).
(m) Type of tap-change gear and position of the tapping winding.
(n) Type of material for the tank (steel or aluminium).

When an order is placed the impedance and losses become guaranteed values, and the designer has to ensure that all the specifications are adhered to. The restrictions imposed by the customer and his own firm's code of practice limit the resultant size and appearance of the transformer.

3.4 FORMULATION OF THE PROBLEM

The designer usually selects a core size (core leg area) from a standard table as his starting point. This table can be derived by using a computer programme to calculate the most economical number of steps to employ for a given core area and also to decide on the change in area required, from one core to the next. A structure of windings and insulation is then built up by using standard wire tables and standard insulation clearances depending upon the test voltage conditions. Once this general arrangement has been established, then the various inherent characteristics such as load loss, iron loss, impedance and temperature rise can be calculated and compared with the desired results. A process of modification by trial and error now takes place which could change the winding proportions or even the core size. This trial and error technique is influenced by the designer's experience and intuition to ignore those things which are obviously wrong. The design so obtained to meet the specification may not prove to be the most economical one, but it can be used as the basis for restarting the design cycle.

A computer does not have intuition, and therefore this part of the design is the most difficult to overcome. The computer can calculate the difference in magnitude of two numbers and can then give the resultant magnitude and sign. This simple test gives the computer the possibility of a form of intuition since it can answer yes or no to a particular limiting condition set inside the programme. Experience and flexibility can be built into the computer programme by careful thought as to the input data and the various limiting conditions. Where possible the designer should use his experience in deciding the general configuration of the design and should employ the computer as a fast slide rule. This philosophy may not apply in the future when the cost of using high-speed digital computers becomes a more economical proposition.

The computer is exceptionally fast and hence economical when solving mathematical equations, where large iterative processes are reduced to a minimum. Therefore, in pursuing a method of approach this fact should be borne in mind. There are numerous methods of approach which could be adopted to arrive at the desired characteristics, but they fall into three main categories.

(a)   A systematic mechanisation of the normal hand calculations carried out by the designer.
(b)   A fundamental mathematical method of approach expressing all the design parameters in the form of equations.
(c)   A hybrid of (a) and (b).

The first method is a very difficult and laborious approach from the computer programmers' point of view and tends to give rather large specialised programmes. The lack of flexibility means that more programmes have to be written to cover another type or range of design. The fact that they are of a specialised nature means that the restricting limits inside the programme can be set fairly close, and this tends to reduce computer time.

At first sight the second method seems to be the most logical approach since a high degree of flexibility can be achieved. It also rationalises the whole design technique and gives a clearer picture of where experience and intuition are required. Unfortunately, it is virtually impossible to express the whole of the design technique in the form of equations which can be solved simultaneously.

The third method of approach is one which is likely to give the best overall solution to the problem, since it should give a high degree of flexibility, together with a tendency to reduce the programming time. It is this method of approach which is described in the following text.

There are other factors which can modify one's ideas, apart from those of economics. For example, one should take into account the type of person who is going to be responsible for the running of the programme and vetting the results obtained. If the programme is written in a form where minor modifications in logic can be introduced easily, then the flexibility of the programme is increased so as to deal with non-standard type designs. The size and speed of the computer to be used can also modify the approach considerably.

Before the computer programmer can start to code the instructions, the complete method of approach has first to be expressed in detail, stating the logical sequence of operations to be carried out. This will certainly involve many logical decisions, together with the necessary boundary limitations to ensure convergence to the desired results. A flow diagram is drawn which shows in pictorial form the method of approach. This comprises two parts: (1) the master programme and (2) the routines or sub-routines. The master programme is the control programme which determines how the various routines have to be utilised. The routines (sub-routines) are self-contained programmes which calculate a specific value, such as iron loss, load loss, reactance, etc.

The whole design process can, therefore, be written down as a master programme with several sub-routines. The master programme deals with the method of convergence, that is it takes the results from the various sub-routines and decides which path to take when these results are tested against the limits set in the master programme.

The task of arriving at the overall programme can be shared by several persons. A designer or system analyst deals with the drawing up of all the flow diagrams, and then several programmers can be employed to transcribe these diagrams into computer language.

3.5   METHOD OF CALCULATION

3.5.1   The basic equation

For simplicity in describing the method which has been adopted for using the computer, a basic equation is employed which pertains to a core-type three-phase transformer with two loaded windings per leg, one the low-voltage and the other the high-voltage winding. Any tappings are assumed to be in the body of the windings.

The basic equation interrelates the major factors which govern the shape and size of a transformer. With the notation in figure 3.1 and if we refer to the list of quantity symbols and associated dimensions at the beginning of this book, the basic equation is represented by the expression*

$$\frac{\phi_m^2 h_1}{s} = \frac{0.02007\,S b_0 k_{sX}}{\%X f k_{hX}}\left(1+\frac{36\,000\,\%X S k_F k_{hX} k_{i1} k_{Js}}{P_{Cu} f b_0^2 k_{sX}}\right)^{1/2} \qquad (3.1)$$

* Details of the derivation and further development of the method for more complex designs, such as split concentric windings and three loaded windings per leg, will be described in a book on transformer design (see chapter 2, reference 1).

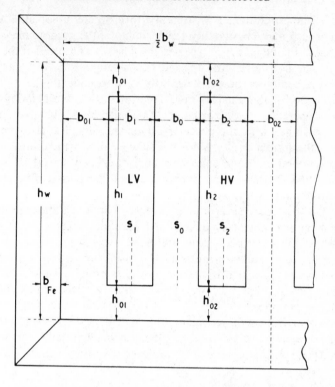

Figure 3.1   Winding arrangement for one transformer leg (for explanation of symbols see list of quantity symbols at the beginning of this book)

for which equation 1.4 gives ϕ_m and $NI = S \times 10^6/4.44f\phi_m$ with $S = VI \times 10^{-6}$. Equation 3.1 is derived from the reactance equation

$$\%X = \frac{5.94NI(3b_0 s_0 + b_1 s_1 + b_2 s_2)}{\phi_m \times 10^8\{h_x + 0.32(b_0 + b_1 + b_2)\}} \tag{3.2}$$

The factor 5.94 applies theoretically to no-core concentric windings only, but it may assume a slightly higher value in practice for iron-core windings depending upon the way in which the windings are arranged with respect to the core.

The various factors in equation 3.1 represent the following expressions.

$$k_{sX} = (3b_0 s_0 + b_1 s_1 + b_2 s_2)/s(3b_0 + b_1 + b_2) \tag{3.3}$$

$$k_{hX} = \{h_x + 0.32(b_0 + b_1 + b_2)\}/h_1 \tag{3.4}$$

(this factor corrects the axial winding height to a reactive height)

$$k_{Js} = (s_1 + s_2 k_J k_e)/s \tag{3.5}$$

$$k_{i1} = 1 + 0.01\%P_{i1} \tag{3.6}$$

and

$$k_F = k_{t1}/k_1 + k_{t2}/k_2 k_h k_J \qquad (3.7)$$

where

$$k_1 = N_1 s_1 k_{t1}/b_1 h_1 = N_1 I_1 k_{t1}/J_1 h_1 b_1 \qquad (3.8)$$

$$k_2 = N_2 s_2 k_{t2}/b_2 h_2 = N_2 I_2 k_{t2}/J_2 h_2 b_2 \qquad (3.9)$$

The expression for the winding load loss P_{Cu} in equation 3.1 is

$$P_{Cu} = P_{Cu1} + P_{Cu2} \qquad (3.10)$$

where, considering that the resistance per millimetre length of copper winding conductor at $75\,^{\circ}C$ is $2.14 \times 10^{-5}/a_1\,\Omega$, we obtain

$$P_{Cu1} = 2.14 I_1 N_1 J_1 s_1 k_{i1} \times 10^{-8} \qquad (3.11)$$

and

$$P_{Cu2} = 2.14 I_2 N_2 J_1 s_2 k_{i1} k_J k_e \times 10^{-8} \qquad (3.12)$$

with k_{i1} given in equation 3.6 and

$$k_e = k_{i2}/k_{i1} = (1 + 0.01\,\%P_{i2})/(1 + 0.01\,\%P_{i1}) \qquad (3.13)$$

From equation 3.10 with equations 3.11, 3.12 and 3.5 and with $I_1 N_1 = I_2 N_2 = IN$

$$P_{Cu} = 2.14 SJ_1 s k_{i1} k_{Js} \times 10^{-2}/4.44 f\phi_m \qquad (3.14)$$

All the terms on the right-hand side of equation 3.1 can be specified by the designer from the customer's specification or from design experience. The values $S, f, \%X, P_{Cu}$ and b_0 are readily available, and the other terms can be obtained from tables based on existing design experience.

The factor k_{sX} (see equation 3.3) usually lies between 0.95 and 1.05 and is initially set at 1.0. The factor k_{hX} (see equation 3.4) usually lies between 1.0 and 1.1 and is initially set at 1.0.

For minimum copper mass the value of $k_J = J_2/J_1$ should be almost unity; therefore, it is set at 1.0, unless otherwise stated by the designer. Equation 3.13 shows that $k_e = 1.0$ if $\%P_{i2} = \%P_{i1}$. Therefore, k_e is initially set at 1.0. Thus, from equation 3.5 with $s = \frac{1}{2}(s_1 + s_2)$ and $k_J = k_e = 1.0$, $k_{Js} = 2.0$ is obtained.

The factor k_{i1} (see equation 3.6) usually lies between 1.05 and 1.20 and is initially set at 1.1.

For the factor k_F (see equation 3.7) an initial value can be obtained fairly easily as follows.

(1) The ratio $k_h = h_2/h_1$ usually lies between 0.90 and 1.0 and is initially set at 1.0.

(2) The values of k_{t1} and k_{t2} in equations 3.8 and 3.9 depend upon the tapping range and the position of the tapping windings. If no tappings are specified, then $k_{t1} = k_{t2} = 1.0$. If tappings are specified, say on the high-

voltage winding for high-voltage variation, then the value of k_{12} depends upon where the tapping turns are situated. If the tappings are in the body of the high-voltage winding, then

$$k_{t2} = 1 + 0.01 \times (\text{maximum positive per cent tapping})$$

and, if the tappings are in a separate layer, then

$$k_{t2} = 1 - 0.01 \times (\text{minimum negative per cent tapping})$$

Therefore it follows that k_{t1} and k_{t2} can be specified initially. If the tappings are in a separate layer, then the value of $\%X$ should be modified to the value on the minimum tapping position and equation 3.12 requires modification to allow for the fact that the tappings will be wound on a different diameter from that of the main winding.

(3) The initial values for the winding space factors k_1 and k_2 (see equations 3.8 and 3.9) can be obtained from tables formed from historical design records. Therefore, the initial value for k_F can be calculated.

Since all the terms on the right-hand side of equation 3.1 can be specified, then

$$\phi_m^2 h_1 / s = (B_m A_{Fe})^2 h_1 / s = K \tag{3.15}$$

where $K = \text{const}$. It follows that the radial widths of the windings

$$b_1 = 2.14 S^2 k_{t1} k_{i1} k_{Js} \times 10^4 / 4.44^2 f^2 P_{Cu} k_1 K \tag{3.16}$$

and

$$b_2 = b_1 k_1 k_{t2} / k_J k_h k_2 k_{t1} \tag{3.17}$$

have fixed values independent of core area, for specified values of reactance and copper loss, provided k_1, k_2, k_{i1} and k_{i2} remain constant with change in core area. In practice it is found that they do remain almost constant with changes in core area A_{Fe}; therefore the window width b_w is a constant since all the radial clearances b_{01}, b_0 and b_{02} are independent of core area.

3.5.2 Calculation of b_1, b_2, s, h_w and b_w

Equations 3.16 and 3.17 with equation 3.15 give the expressions for b_1 and b_2.

The maximum value of B_m either is specified by the customer or is fixed by the designer. From equation 3.15 the quantity $A_{Fe}^2 h_1 / s$ is therefore also a constant. A_{Fe} can be specified by the designer or can be selected by the computer from standard core area tables. Once it has been selected, then the value of s can be calculated with the expression

$$s = \pi(2b_{Fe} + 2b_{01} + b_0 + 1.5b_1 + 0.5b_2) \tag{3.18}$$

where $2b_{Fe}$ is the maximum core-plate width of the core area selected. Since s, A_{Fe} and B_m are now known, h_1 can be calculated from equation 3.15.

The core window height is

$$h_w = h_1 + 2h_{01} \tag{3.19}$$

For a three-phase three-limbed core-type transformer the core window width is

$$b_w = 2b_1 + 2b_2 + 2b_0 + 2b_{01} + b_{02} \tag{3.20}$$

Thus the winding heights, radial widths and core dimensions are known.

The maximum and minimum practical values for h_1/s for varying rated power is stored inside the computer from previous design experience. Because of restriction in the number of core sizes, the computer has to search through to select the most economical one.

3.5.3 Calculation of J, N and a

The expressions for J_1 and J_2 are

$$J_1 = 4.44 f \phi_m P_{Cu} \times 10^2 / 2.14 S s k_{i1} k_{Js} \tag{3.21}$$

$$J_2 = k_J J_1 \tag{3.22}$$

The turns N_1 and N_2 can be calculated from the leg voltages and the voltage per turn V/N (see equation 1.4). The leg voltages are calculated from the specified line voltages and the connections of the windings in the usual way. The leg currents are obtained from the line currents. Therefore, the winding cross-sectional areas can be calculated from the values of J_1 and J_2 since $a_1 = I_1/J_1$ and $a_2 = I_2/J_2$.

3.5.4 Winding formation

At this stage the winding height, radial width, turns, currents and turn cross-sectional area are all known. The problem is to fit the turns into the space allocated to meet the voltage, eddy current loss and thermal characteristics required. If the original space factors were exactly right, then the windings should fit into the space allocated, thus giving the required values of $\%X$ and P_{Cu}. This part of the computer programme is most complex, but by using the values k_1 and k_2 from previous design experience the computer has already homed onto a feasible design. This means that the limits or boundary conditions inside the winding formation sub-routines can be fixed fairly closely. To streamline the winding formation, each type of winding (helical, disk or layer) has an individual sub-routine which is selected by the designer. Each of the sub-routines uses add-routines such as the looking up and selection of standard wires from a wire table or the calculation of eddy current losses, which are common to all the winding formation sub-routines.

The method of approach in forming the winding is to assume the height to be fixed and then to see what the residual difference in radial width is compared with the desired value when the winding is completed.

Each manufacturer will use different techniques to form the windings; therefore, the computer programmes will be different. The overall convergence technique, however, could be used by anyone.

3.6 EXAMPLE

Suppose the following specification is given for a transformer.

Rated power	60 MVA
Number of phases	3
Frequency	50 Hz
Voltage ratio	132 kV to 33 kV
Connections	star – delta
Percentage reactance	15.5 %
Load loss	405 kW
Iron loss	46 kW

The designer selects the following criteria.

Winding arrangement	low-voltage winding nearest to the core
A_{Fe}	0.2875 m^2
B_m	1.54 T
b_0	64 mm
b_{01}	30 mm
b_{02}	68 mm
h_{01}	77.5 mm
h_{02}	77.5 mm
k_J	1
$\%P_{i1}$	13 %
$\%P_{i2}$	13 %
k_1	0.36
k_2	0.36
Stray loss	15 kW

The computer would normally set

$$k_{sX} = k_{hX} = k_h = k_e = k_{t1} = k_{t2} = 1 \qquad k_{Js} = 2$$

Therefore, $k_{i1} = k_{i2} = 1.13$

$$P_{Cu} = (405 - 15)/3 = 130 \, kW$$

$$S = 60/3 = 20 \, MVA$$

$$k_F = 1/k_1 + 1/k_2 = 5.55$$

Substitution in equation 3.1 gives

$$h_1 \phi_m^2 / s = 0.116$$

Substitution in equation 3.16 and 3.17 gives

$$b_1 = b_2 = 72 \, mm$$

Because

$$\phi_m = A_{Fe}B_m = 0.44275 \, \text{Wb}$$

therefore

$$h_1/s = 0.592$$

The maximum plate width for the core selected is 660 mm. From equation 3.18

$$s = \pi(660 + 60 + 64 + 108 + 36) = 2920 \, \text{mm}$$

Hence

$$h_1 = 0.592 \times 2920 = 1730 \, \text{mm}$$

$$h_w = 1730 + 155 = 1885 \, \text{mm}$$

$$b_w = 144 + 144 + 128 + 60 + 68 = 544 \, \text{mm}$$

Thus

$$h_w/b_w = 3.46$$

From equation 3.21

$$J_1 = 4.54 \, \text{A mm}^{-2}$$

Because

$$I_1 = 20\,000\,000/33\,000 = 606 \, \text{A}$$

hence

$$a_1 = 606/4.54 = 133.8 \, \text{mm}^{-2}$$

$$J_2 = 4.54 \, \text{A mm}^{-2}$$

$$I_2 = 20\,000\,000 \times 3^{1/2}/132\,000 = 262.5 \, \text{A}$$

$$s_2 = 262.5/4.54 = 58 \, \text{mm}^2$$

From equation 1.4

$$V/N = 4.44 \times 50 \times 0.44275 = 98.3 \, \text{V turn}^{-1}$$

Hence

$$N_1 = 33\,000/98.3 = 336 \, \text{turns}$$

$$N_2 = 132\,000/98.3 \times 3^{1/2} = 775 \, \text{turns}$$

A check on the calculations can be made by evaluating k_1 from equation 3.8.

$$k_1 = 336 \times 606/4.54 \times 1730 \times 72 = 0.36$$

At this stage all the parameters are known to enable the winding formation to start. The problem is to design the windings in the height and width calculated to meet the eddy current loss, voltage and thermal characteristics. If the windings can be formed in the space allocated, this means that the choice of winding space factors was correct, and inherently the reactance and load loss will also be correct. The iron loss can be calculated and checked against the specified value. Changes in A_{Fe} and B_m can now be made if the calculated value exceeds the specified value.

3.7   GENERAL FLOW DIAGRAM

3.7.1   General

The flow diagram (figure 3.2) shows the convergence technique used to arrive at the internal design of the transformer using a medium-range computer. It is

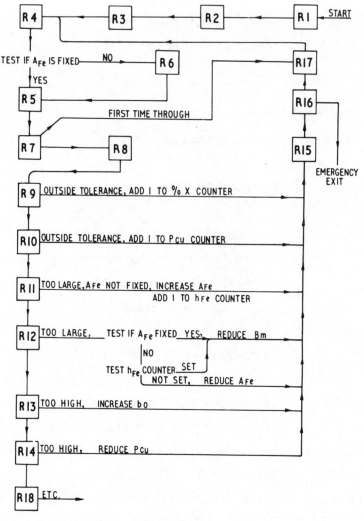

Figure 3.2   General flow diagram (for explanation of sub-routines R, see text)

assumed that only one programme is computed at a time. Therefore a certain amount of designer interference to modify the input data is allowed, depending upon the results that are calculated.

The following information is stated on the input data sheet for the computer.

(1) Arrangement of the windings with respect to the core leg, including tapping windings.
(2) Continuous maximum rating of the transformer.
(3) Number of phases.
(4) Number of core legs with windings.
(5) Frequency.
(6) Low- and high-voltage line voltages.
(7) Low- and high-voltage winding connections.
(8) Unloaded tertiary winding rating, if any.
(9) Maximum percentage low-voltage tapping, if any.
(10) Minimum percentage low-voltage tapping, if any.
(11) Number of low-voltage tapping steps, if any.
(12) Maximum percentage high-voltage tapping, if any.
(13) Minimum percentage high-voltage tapping, if any.
(14) Number of high-voltage tapping steps, if any.
(15) Type of low-voltage winding.
(16) Type of high-voltage winding.
(17) Type of low-voltage winding conductor.
(18) Type of high-voltage winding conductor.
(19) Maximum percentage low-voltage winding eddy current loss.
(20) Maximum percentage high-voltage winding eddy current loss.
(21) Maximum low-voltage current density.
(22) Maximum high-voltage current density.
(23) Ratio of current densities (high-voltage/low-voltage).
(24) Approximate low-voltage winding temperature rise by resistance measurement above the mean oil temperature.
(25) Approximate high-voltage winding temperature rise by resistance measurement above the mean oil temperature.
(26) Maximum temperature rise by resistance measurement.
(27) Maximum temperature rise of the oil.
(28) Maximum core height.
(29) Maximum core leg flux density.
(30) Type of core and material.
(31) Core leg area, if selected by the designer; otherwise set at zero.
(32) Core loss, if known; otherwise set at zero.
(33) Percentage reactance at continuous maximum rating.
(34) Load loss at continuous maximum rating.
(35) Type of cooling.
(36) Radial distance between the low- and high-voltage windings.
(37) Breakdown insulation level of low-voltage winding.

(38)   Breakdown insulation level of high-voltage winding.
(39)   Cost of load loss per kilowatt, if known; otherwise zero.
(40)   Cost of iron loss per kilowatt, if known; otherwise zero.

3.7.2   Sub-routine 1 (R1)

The first step is to read, scale and store the input data, clearing all the relevant working and output locations allocated to the programme. These data are also checked for validity by testing the magnitudes of the numbers with respect to one another. This eliminates some errors which could occur on the input data. When all the data have been processed and stored, the various sub-routines to be used are now selected and stored in the appropriate locations by transferring them from a magnetic tape library of design sub-routines to the internal computer store.

3.7.3   Sub-routine 2 (R2)

Sub-routine 2 (R2) is used to calculate

$$I_1, I_2, V_1 \text{ and } V_2$$

From the rated power, the number of legs wound and the line voltages and winding connections, the leg currents and voltages can now be calculated. This routine is not included in the major convergence loop since the values calculated are independent of such quantities as core size, flux density and losses.

3.7.4   Sub-routine 3 (R3)

Sub-routine 3 (R3) is used to calculate

$$k_{t1}, k_{t2}, k_{i1}, k_{i2}, k_e, k_J$$

Set

$$k_{sX} = k_{hX} = k_h = 1 \text{ and } k_{Js} = 2$$

All the constants pertaining to the type of design under investigation are set at this stage. They are obtained from tables stored inside the computer or are set as pure numbers. Various counters are also set as controls to the master programme. For example, counters are set which total the number of iterations in each convergence loop. If difficulty arises in converging, owing to 'hunting' for a solution, a warning is printed out so that the designer can modify the input data or can set a handswitch on the computer. The setting of handswitches produces a print-out of the major parameters giving the designer the necessary information to decide upon which course of action to take. This only occurs when the designer has asked the computer to formulate a peculiar type of design where the various limiting factors are being encountered in each convergence loop. For example, if low iron loss and load loss are required with high impedance, then eddy current loss difficulties will arise, and the design itself might be impossible to achieve.

3.7.5 Sub-routine 4 (R4)

Sub-routine 4 (R4) is used to calculate

$$k_F, \Phi_m^2 h_1/s, b_1, b_2 \text{ and } b_w$$

This is the first routine to be used inside the major convergence loop. All the values are initially calculated from the constants set in sub-routine 3 (R3), but subsequently they are derived from the values calculated in forming the design (R17).

3.7.6 Sub-routines 5 and 6 (R5 and R6)

Sub-routine 5 (R5) is used to calculate

$$B_m, N_1, N_2, h_1, h_w \text{ and } s$$

and sub-routine 6 (R6) is used to select

$$A_{Fe}$$

(a) If the core leg area has been fixed either by the designer or by the computer in the process of forming the design, then the computer does not have to select an add-routine which calculates the core area for minimum cost of active materials.

(b) If A_{Fe} is fixed, then B_m is rounded off so that the turns ratio on all tapping positions is within tolerance, and the values N_1, N_2, h_1, h_w and s are calculated.

(c) If A_{Fe} is not fixed, then the computer automatically goes through the optimisation routine. This is done by selecting the smallest practical value of A_{Fe} from tables of the maximum and minimum values of h_1/s against S stored in the computer from historical design records. It is important to note that the computer must differentiate between A_{Fe} fixed and A_{Fe} selected. This is done by counters set in the programme.

The sequence of calculation in the optimisation routine is the following.

(a) Select A_{Fe}.
(b) Round off B_m.
(c) Calculate s, h_1 and h_w.
(d) Calculate the core mass, since A_{Fe}, h_w and b_w are known.
(e) Calculate the core cost of active materials.
(f) Calculate winding current densities.
(g) Calculate watts per kilogram from current densities.
(h) Calculate winding material mass from winding losses and watts per kilogram.
(i) Calculate winding material cost.

The process is repeated by selecting each larger standard core area in turn until the minimum cost condition is achieved.

If the capitalised cost of the losses is stipulated, then the cost of the losses can also be calculated and added to the active material cost since the load loss is known and since the iron loss can be quickly obtained if we know the iron mass and the watts per kilogram at the current value of core flux density.

The optimisation routine can also be used for obtaining the minimum mass or minimum volume design, which is sometimes of more importance with very large power transformers.

If the core labour cost is stored in the computer and if a function of how the winding labour cost varies with turns is stored, then the major labour costs can also be added to the above material values to give a better choice of core size.

It must be pointed out that b_w is only a constant independent of core area provided that the winding radial depths remain constant as A_{Fe} varies. Over a small change the winding space factors can be assumed to remain constant since the winding current densities do not vary greatly. If, however, the core size is changed over a wide range, then, as the core size increases, the current density increases for a fixed load loss, and therefore the winding space factors tend to become smaller.

From equation 3.1, as the space factors decrease, k_F increases, and $\Phi_m^2 h_1 / s$ also increases slightly.

From equation 3.16, the radial widths of the winding vary inversely as $k_1 \Phi_m^2 h_1 / s$; therefore, as k_1 decreases, b_1 increases but at a slower rate than the change in k_1.

However, with high-voltage transformers where b_0, b_{01} and b_{02} are quite large, b_w can be assumed to be constant as A_{Fe} changes for a fixed load loss and reactance.

The computer does not set the core area as being fixed in this routine since minor modifications to the various parameters may change the core selected during the next iteration.

3.7.7 Sub-routine 7 (R7)

Sub-routine 7 (R7) is used to calculate

$$J_1, J_2, a_1, a_2 \text{ and } h_2$$

The values calculated in this routine are derived from the equations for current density. As the leg currents have been calculated in R2, the winding conductor cross-sectional area can be derived. A counter is tested to see whether R7 has been used for the first time. If so, the master programme directs the computer to R17. The reason for this is to modify the constants which were set in R3 so that they will be more exact before starting to form the windings.

3.7.8 Sub-routine 8 (R8)

Sub-routine 8(R8) is used to calculate

$$b_1, b_2, \Delta\theta_{R1}, \Delta\theta_{R2}, k_1, k_2, \%P_{i1}, \%P_{i2}, J_1 \text{ and } J_2$$

This routine can be in two forms: (1) either a simplified routine which does not form the windings in detail or (2) a complex and detailed routine for forming the windings completely. The choice of routine to be selected is decided by the designer on the input data sheet. The simplified formation is used when the designer wants to investigate the change of various parameters on a design and when the results required need not necessarily be very accurate. The complex routine takes up most of the computer time since it is in this routine where most of the design logic is required.

At this stage all the necessary information is known to enable a winding formation to commence. The winding heights are known, and the initial assumption is that no radial cooling ducts will be required unless otherwise specified owing to voltage reasons. The whole of the winding height is now filled with covered winding conductor, ensuring that the strand sizes selected meet the maximum eddy current loss conditions set. The thermal gradient is now calculated. If this is too high, then radial ducts are introduced, leaving less winding height in which to fit the conductor. This process is repeated until all the ducts for thermal and voltage reasons have been introduced.

When selecting strands from the standard wire table, the computer always tries to select the largest possible conductor size since this inherently gives the highest wire space factor and hence the highest winding space factor. It is only forced into selecting smaller wire sizes and more strands by the fact that the eddy current losses are too high. All the normal design limiting conditions are built into the programme, for example the maximum and minimum ratios of the axial to radial depths of the wire strands and the maximum number of strands in parallel for a particular former size. Because there are several mechanical as well as electrical boundary conditions, warning print-outs are given when a winding formation cannot be achieved inside the limitations set. The winding heights have been fixed in this routine; therefore, if the windings cannot be formed in the space allocated, all the error will appear in the radial depths of the windings.

3.7.9   Sub-routine 9 (R9)

Sub-routine 9 (R9) is used to calculate

$$\%X, \Phi_m, s_1, s_2 \text{ and } s_0$$

The mean turns are formed, and the percentage reactance is calculated. If the $\%X$ is within tolerance, then the computer proceeds to R10; otherwise it is diverted to R15. Failure at this stage is caused by variation in winding space factors and hence winding radial depths formed as against those allocated in R4.

3.7.10   Sub-routine 10 (R10)

Sub-routine 10 (R10) is used to calculate

$$P_{Cu_1}, P_{Cu_2} \text{ and } P_{Cu}$$

If P_{Cu} is less than the positive tolerance, the computer proceeds to R11; otherwise it is diverted to R15. Failure is caused for the same reasons as given in R9, together with the fact that the values of a_1 and a_2 formed in R8 will not be exactly the same as those calculated in R7 because the wire table has discrete steps.

3.7.11 Sub-routine 11 (R11)

Sub-routine 11 (R11) is used to calculate

$$m_{Fe} \text{ and } h_{Fe}$$

The core mass m_{Fe} for the type of core construction required, together with the core dimensions, are calculated. The overall core height h_{Fe} is tested against the maximum permissible for transport purposes. If h_{Fe} is acceptable, the computer proceeds to R12; otherwise it tests to see whether A_{Fe} is fixed. If so, it ignores the height restriction. If A_{Fe} is not fixed, the next larger core is selected, and a counter is set. The reason for setting a counter is to ensure that the computer realises that A_{Fe} has been increased owing to a height restriction failure when R5 is used again. This counter is also checked if failure occurs in R12.

3.7.12 Sub-routine 12 (R12)

Sub-routine 12 (R12) is used to calculate

$$P_{Fe}$$

From curves relating flux density to the watts per kilogram for electrical core steel, stored inside the computer, the specific watts per kilogram can be calculated for the grade of core iron selected. There are two methods of calculation. The first is to store an equation in terms of flux density B_m so that p_{Fe} can be calculated, once B_m is fixed. The second method is to store the p_{Fe} at fixed intervals of B_m and then to use linear interpolation to select the correct value for the given B_m.

The core building factor is obtained from tables relating the building factor to core size and flux density. The iron loss P_{Fe} can now be calculated by multiplying the iron mass by the watts per kilogram and the building factor. If the specified iron loss value is set to zero in the input data or if the iron loss is less than the maximum tolerance, then the computer proceeds to R13. If the iron loss is too high, then a check is made to see whether A_{Fe} has been fixed. If so, B_m is reduced, and the computer proceeds to R15. If A_{Fe} is not fixed, then the height restriction counter is tested. If this counter is set, then the computer proceeds as if A_{Fe} had been set; otherwise the next smaller core size is selected.

3.7.13 Sub-routine 13 (R13)

Sub-routine 13 (R13) is used to test

$$\%P_{i1} \text{ and } \%P_{i2}$$

The winding eddy current loss percentages are now tested against the maximum values set by the designer. If they are smaller, then the computer proceeds to R14. Otherwise the only alternative is to increase the interwinding gap b_0. When b_0 is increased, the winding height is increased, and the radial depths of the windings are reduced to maintain the same reactance. This automatically reduces the eddy current losses since they vary inversely as the square of the winding height. A failure at this stage can only occur under the following conditions.

(1) The maximum values are set far too low. For large power transformers a reasonable setting would be 20% at $75\,^{\circ}\mathrm{C}$ depending upon the type of winding conductor chosen by the designer, that is transposed cable or normal strand.

(2) The load loss is far too low or a combination of low load loss and high impedance.

3.7.14   Sub-routine 14 (R14)

Sub-routine 14 (R14) is used to calculate

$$J_1 \text{ and } J_2$$

The current densities are now tested against the maximum values set by the designer (if these are inputted as zero, the computer proceeds to R18). If they are smaller, then proceed to R18; otherwise the load loss is reduced and the design process restarted. This approach is used when the load loss is not specified by the customer but when the designer decides that a certain current density would be permissible in the windings. The designer sets the value of the load loss on the high side from experience, and then the computer will automatically reduce the load loss to a value compatible with the current densities specified.

3.7.15   Sub-routine 15 and 16 (R15 and R16)

Counters are added to every time this route is selected. Once these counters have reached their maximum values, an emergency exit from the programme is used. A print-out is given with each iteration to show which sub-routine has caused the failure. It also shows the way in which the convergence is proceeding. The emergency exit only occurs if the input data is incorrect or the design requirements are impossible to achieve.

3.7.16   Sub-routine 17 (R17)

All the constants which were set in R3 are now calculated from the design which has been formed and are used in calculating the values in R4.

3.7.17   Summary

The internal convergence loop includes all the sub-routines R4 to R17, with several emergency exits. At first sight the order of calculating R9 to R14 could be rearranged without changing the end product.

The whole convergence technique relies on the fact that the windings will fit exactly into the space allocated by using the winding space factors calculated during the previous iteration, since all the other values such as k_h, k_e, k_J, etc., remain almost constant after the first iteration. Therefore, any discrepancy in the values of k_1 and k_2 from one iteration to the next can only cause errors in $\% X$ or P_{Cu}. Since $\% X$ and P_{Cu} are specified values, it is imperative that these be correct before proceeding further with the design. $\% X$ was chosen in preference to P_{Cu} because there is usually a more stringent tolerance on reactance compared with load loss. The order of calculating R11, R12, R13 and R14 is less relevant, except that R11 should precede R12. The reason for deciding upon the described order was to calculate the routines in order of failure rate so as to reduce the computing time. Once the convergence loop R4 to R17 has been completed, then the computer proceeds to R18 to calculate tank masses, oil quantities, etc.

3.8   COST OPTIMISATION

This process can be carried out in several ways, but it is necessary to ensure that each sub-routine that is used is oriented in such a way so as to produce sub-optimisation at each step during computation. If the reactance, iron loss and load loss are fixed, which they will be at the order stage, then the problem is to select the most economical core size, once the general configuration of the design has been settled. This can be done as described in sub-section 3.7.6. Alternatively, a complete cost programme can be written which takes into account the labour as well as the cost of the materials. The latter method is more exact since it covers all the aspects of the transformer cost. To ensure that the most economical design has been achieved, designs on a range of core sizes must be completed and costed.

Variation in cost with change in core area is flat near the minimum, provided the ratio of the copper to core iron cost is less than 2 to 1. When the ratio is higher than this value, the selection of the core area becomes critical, and a well-defined minimum appears. At this stage the use of aluminium as an alternative to copper must be considered.

If, at the quotation stage, the losses are not specified but values are given for the capitalised cost of the losses, then the problem of achieving the most economical design overall becomes more difficult. This can be solved in a similar way to that described above by varying the load loss until a curve can be drawn to indicate the minimum overall cost. In this case the designer would use the simplified winding formation sub-routine to narrow the field of search for the best design. From transformer theory it is possible to select the most economical core area and current density to use for the minimum cost of active materials plus the cost of the losses, but this does not include labour costs and therefore does not necessarily give the overall minimum cost. It would, however, reduce the search for the optimum to a narrower bandwidth.

With the aid of a large computer it would be possible to set a range of losses and to let the computer only print out the overall minimum cost design. If the computer is also allowed to print out the other computed designs, this would avoid losing them and would provide additional results that could be useful for other quotations.

3.9   ADVANTAGES AND LIMITATION OF USING COMPUTERS

The advantages of using a digital computer for design work are fairly obvious. Several quotation designs can be produced for one enquiry in order to arrive at the most economical solution. Tables can be made up which show the change in cost for variation in loss, reactance and flux density for a given rated power and voltage. These tables can be modified quickly if a large change in the cost of the active materials or labour occurs. The variation in cost tends to follow a smoother curve than it would if several designers were employed to achieve the same results, since the programme follows the same technique for each design. When several quotation designs are required in a short time, then the computer is the only method of providing the necessary information accurately. It gives the designer more time to read the specification and to ensure that all the requirements have been taken into account. Most of the advantages in using a computer are based on the speed and accuracy of calculation which ensures that a greater volume of work can be achieved in a given time.

However, there are limitations in using a computer. The computer cannot read the specification and is therefore dependent upon the designer to input the correct information for the design. The designer decides upon the general configuration of the transformer, and the computer obeys the necessary instructions inside the boundary conditions set. If the initial concept is incorrect in any way, then the final computer design will also have these faults. To overcome this, careful inspection of the input data and the results is necessary.

The way in which the computer is used and its geographical location, with respect to the design office, can cause disadvantages. For instance, if the computer is heavily loaded and the designer has to wait in a queue before he can have access to it, then he feels that he would be better employed carrying out a manual operation. The situation could be overcome to some extent if the design office scheduled their requirements and used the computer as an integral part of their environment.

If the computer is not sufficiently large to store all the programmes required to enable a complete design to be achieved, then again the designer feels he may as well calculate by hand the part done by the computer, since he will have to finalise the results for cost purposes by hand anyway. Therefore, the computer should be capable of providing all the information required to finalise the design for quotation purposes.

To write a programme to finalise the design at the order stage is most complex and can only be achieved provided a high degree of rationalisation and standardisation is built into the programme. It must be remembered that all the designer's time is not spent with slide rule calculations to finalise a quotation or design but that the majority of his time is spent in dealing with queries from drawing office, planning, works, test and the customer, as well as reading specifications. Therefore that part of his time in which the computer can help may only be 15 to 25 %.

3.10   ALTERNATIVE METHODS OF APPROACH

The number of ways in which the design problem can be solved by a computer are virtually unlimited. The method evolved is dependent upon which parameters of the design are considered to be the most important. From the viewpoint described in this chapter, the reactance has been assumed to be the prime factor which determines the final physical dimensions of the transformer. The iron (core) loss has been ignored in the basic equation and is treated as a parasitic loss, that is a factor which becomes a natural value once the reactance and load loss have been fixed. If the core loss is stipulated as a guaranteed value, then the answer given will be less or equal to the guaranteed value, if we use the maximum value of B_m compatible with the solution.

If a method were employed in which the iron loss was the prime factor, that is a fixed value, then it is quite conceivable that uneconomical designs could be produced. For instance, if we ignore the iron loss guarantee, the optimum cost design to meet the reactance and load loss could give an iron loss 10 % below the guaranteed value at maximum B_m. The design to meet the same reactance and load loss, together with the guaranteed iron loss, would therefore have to be on a larger core size. This would increase the core mass and would reduce the winding mass, and hence this would cause deviation from the optimum cost solution.

One of the simplest approaches is to store in a sorted order all the design records of units which have been completed. The computer can use this information as a starting point to meet the requirements for the design in question. This follows very closely the method which the designer employs, since he has access to this information in the design office. The library of information will increase and be kept up-to-date with respect to new design modifications as time passes. The computer programme will therefore consist of two parts. The first is to search the store to retrieve the necessary starting information. The second part of the programme will modify the various physical characteristics, in small steps, until the required design is achieved. This method of approach can certainly be employed in large manufacturing organisations, where the library would be quite extensive.

If the approach is to be more fundamental, that is if each quotation is to be treated as a separate entity without relying on previous design experience, then empirical relationships between the major design parameters must be found. If the computer is given a free hand to arrive at the required characteristics, then the empirical relationships must hold good even under exceptional circumstances to ensure that the computer can achieve convergence. Several papers must have been written on this subject, and invariably the method of approach has been different. The reason for this is that each manufacturer has different views as to what are the major factors which govern his type of transformer construction.

3.11   OTHER ASPECTS OF TRANSFORMER DESIGN

The computer can be used in nearly all the other aspects of transformer design where the fundamental principle is understood and where mathematical equations have to be solved. In many cases empirical equations have been used, for simplicity, in design offices for a number of years without the designers consciously realising their significance. The computer can, therefore, be used as a research and development tool to solve these problems from first principles.

There are numerous ways in which the computer can be employed. The question as to whether a particular routine should be programmed is an economical one. Before the cost of running the programme is compared with the cost of the hand method, the usage should be evaluated against the cost of writing and testing the programme.

ACKNOWLEDGEMENT

The author wishes to express his thanks to Ferranti Engineering Limited for permission to publish the information.

4

Transformer Cores

S. Palmer*

4.1 INTRODUCTION

The cores of today's power transformers still retain the essential features developed over 80 years ago: they are built up from thin flat laminations of soft iron.

Nevertheless, over the years the magnetic properties of the iron have been improved very considerably, as have the manufacturing processes used to produce both the basic material and the individual laminations. Many changes in core design have also been introduced to allow full advantage to be taken of improvements in iron quality.

This chapter is devoted to an account of power transformer cores as they are today, with an occasional backward glance to see how particular practices developed. Most attention is given to three-phase three-limb cores of cruciform cross-section, since the majority of transformers now use cores of this kind.

4.2 MATERIALS

4.2.1 Magnetic materials

Only iron–silicon alloys of high purity are used in power transformers at the present time. Before the Second World War, these alloys were produced in sheet form 0.35 mm thick from slabs by successive rollings at a red heat, followed by annealing at about 900 °C. The sheets were folded over between successive rollings to keep the size manageable, so that the final rolling would usually be done on a pack of eight sheets. This process required the silicon content to be

---

* Canadian Westinghouse Company Limited.

limited to $4\frac{1}{2}\%$, in order to avoid an unworkable degree of brittleness, and it produced sheets in which the constituent crystals were almost randomly oriented.

The magnetic properties of single crystals of silicon – iron vary with the direction of magnetisation. In 1935 Goss[1] described a process of cold rolling and high-temperature annealing which induces a substantial number of the crystals to align themselves relative to the direction of rolling. This alignment is shown in figure 4.1(a), and the resulting variation of hysteresis loss with the direction of magnetisation is illustrated in figure 4.1(b). The performance along the direction of rolling represents a considerable improvement on the randomly oriented material.

The cold-reduction process limits the maximum silicon content to about 3%, which raises the saturation intensity slightly and decreases the resistivity. Nevertheless, the overall effect is that core operating flux densities can be increased and yet still give a substantial reduction in core loss over the non-oriented hot-rolled material. The cold rolling also gives a smoother surface finish and thereby improves the space factor.

The improved material became available in commercial quantities in the USA in the early 1940s and in the UK about 10 years later.

Almost all power transformer cores are now constructed from cold-rolled oriented silicon – iron, although it is substantially more expensive than the older material owing to the more elaborate method of manufacture. It is normally supplied with a thickness of 0.33 mm, but there is a tendency in the better grades to reduce this to 0.3 or 0.28 mm in order to offset the lower resistivity and to reduce the total loss further.

Figure 4.2 shows the iron losses of the best grades of core material which are currently available from UK and US sources, and table 4.1 summarises the improvements in total loss since 1890.

4.2.2 Insulating materials

With the reduction in the hysteresis loss of magnetic materials over the years, it has become more important to ensure that eddy current losses are kept to a minimum by the provision of adequate insulation between the laminations of the core. As power transformers have increased in size, the duty imposed on this interlaminar insulation has become more onerous; higher voltages are induced as a result of larger core sections, and the areas of insulation stressed by these voltages are greater. Furthermore, the limiting effect of transport restrictions on mass and dimensions has forced designers to try and improve core space factors by reducing the insulation thickness. The only mitigating factor has been the improvement in surface finish that results from the use of cold-rolled core material, which makes it practicable to use thinner insulating coatings.

With cold-rolled grain-oriented silicon – iron, some degree of surface insulation is provided by the application of magnesia during the manufacturing process. This is necessary to avoid adjacent turns of the coils fusing together during the high-temperature anneal. It is usual to form a supplementary coat of

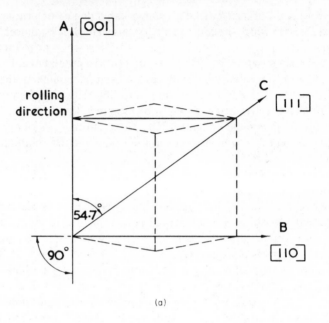

(a)

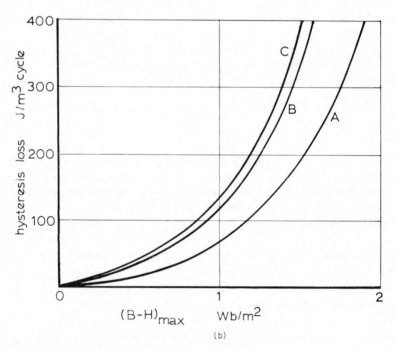

(b)

Figure 4.1   Hysteresis loss in grain-oriented silicon-iron: (a) preferred orientation of crystals relative to the rolling direction of sheet; (b) hysteresis loss measured in the three directions relative to rolling direction

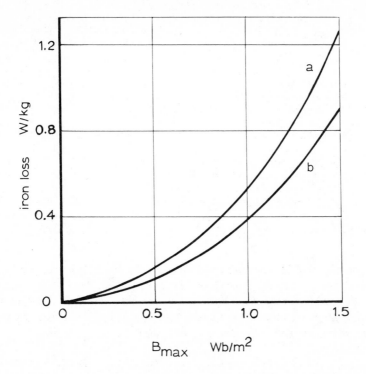

Figure 4.2   Iron loss of core materials: curve a, iron loss of Silectron W.58 measured at 60 Hz on annealed samples 0.3 mm thick parallel to the direction of rolling; curve b, iron loss of Unisil 41 measured at 50 Hz on annealed samples 0.28 mm thick parallel to the direction of rolling (curve a, by courtesy of Allegheny – Ludlum Steel Corporation; curve b, by courtesy of Steel Company of Wales)

insulation by treating the iron with acid phosphates prior to the final anneal. Provided that care is taken to minimise the production of burrs during the cutting of the laminations, this phosphate insulation alone can be satisfactory for cores up to a rated value of about 30 MVA. The coating also serves to prevent rusting of the core material during storage.

On core-plate for larger transformers, some additional insulation will normally be required. Traditional insulations are paper, mixtures of china clay and flour, and varnishes. The first two of these suffer from the disadvantage that a worthwhile increase in insulation resistance is only achieved by a relatively thick coating in excess of 0.0125 mm. Varnishes, though thinner, are apt to soften slightly at core operating temperatures, so that under pressure the insulation resistance between laminations is liable to fall. This behaviour is illustrated in figure 4.3, which shows a typical distribution of resistance values obtained by testing single core-plates coated with varnish at 6V and a force of 2.3 N on a contact area of $6.45\,cm^2$, that is a pressure of 3.58 kPa, both hot and cold. Furthermore, all of the substances mentioned are not particularly heat resistant,

TABLE 4.1   *Improvement in transformer core materials since 1890*

Year	Type of material	Thickness (mm)	Total loss (W kg^{-1})
1890	unalloyed steel	0.35	5.9
1900	Swedish charcoal iron	0.35	3.5
1910	$3\frac{1}{4}\%$ silicon–iron	0.35	1.75
1925	$4\frac{b}{0}\%$ silicon – iron	0.35	1.4
1940	4.2 % silicon – iron	0.35	1.2
1945	4.3 % silicon – iron	0.35	1.08
	3.2 % grain-oriented silicon – iron (USA)	0.33	0.57
1954	2.9 % grain-oriented silicon – iron (UK)	0.33	0.46
1970	3 % grain-oriented silicon – iron (UK)	0.28	0.375
	3 % grain-oriented silicon – iron (USA)	0.3	0.4

Total losses stated for $f = 50$ Hz and $B = 1.0$ T.

which means either they will slowly deteriorate at normal core hot-spot temperatures of 115 to 120 °C or the core has to be either underrun or overventilated to compensate. The trend of current practice is, therefore, towards the application of some kind of additional phosphate treatment to the core-plates, which can give a thinner and more heat-resistant coating.

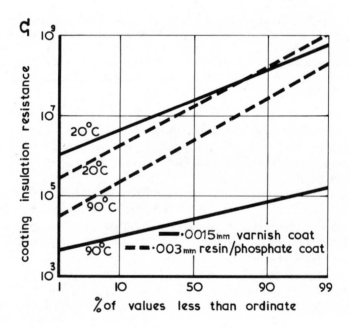

Figure 4.3   Insulation resistance of interlaminar insulation measured on a contact area of 6.45 cm^2 under a pressure of 3.58 kPa at 6 V

Various kinds of phosphate treatment are in use. Figure 4.3 also shows the results of some single-plate tests on a coating formed by treating the iron surface with a combination of acid phosphates and polyvinyl butyral resins. Thicknesses of 0.003 mm are easily achieved with this material. It will be seen that with a thinner coating a superior performance to that of the varnish is obtained.

The additional loss in a laminated core caused by leakage currents in the interlaminar insulation can be shown[2] to be proportional to the square of the plate width and inversely proportional to the mean value of insulation resistance per unit area throughout the core. Figure 4.4 shows how this mean value must increase as the core size increases if this additional loss is to be limited to 1 % of the iron loss. Although these values of resistance are low in relation to the measurement displayed in figure 4.3, it must be remembered that the test values relate to an electrode area of 6.45 cm^2. Testing with larger areas gives lower values, and the contact area between core-plates in a large transformer is very large.

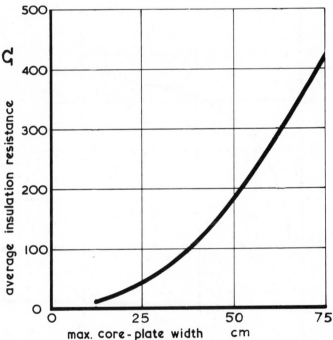

Figure 4.4   Minimum insulation requirements for core laminations

4.3   FORM

4.3.1   Cross-section

With some exceptions in smaller ratings, the windings for core-form transformers are cylindrical in shape, so that the cross-section of the core limb inside the

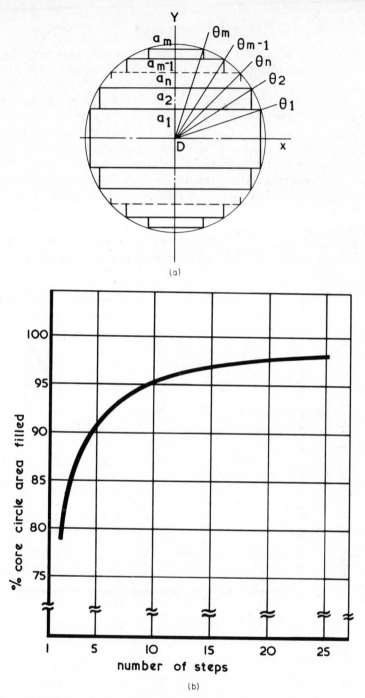

Figure 4.5   Stepped-core relationships: (a) generalised cruciform core section

$$\text{core area} = D^2 \sum_{n=1}^{m} (\tfrac{1}{2}\sin 2\theta_n - \sin\theta_{n-1}\cos\theta_n)$$

(b) optimum fill factors for cruciform-section cores

smallest winding has to be made as nearly circular as possible if space is not to be wasted. By making a stepped pattern of different widths of strip, it is possible to achieve a close approximation to a circular contour, limited only by the number of different strip widths which the manufacturer is prepared to cut and build.

For the generalised stepped-core section shown in figure 4.5(a) it is possible to solve a number of partial differential equations equal to the number of steps and to obtain the proportions of plate width and step thickness which result in the maximum area being filled[3]. Figure 4.5(b) shows how the percentage of the unit circle area filled with core material varies with numbers of steps between 2 and 25. It is clear from this that with 9 steps a fill factor of practically 95 % is achieved and that it becomes difficult to justify a greater number of steps than this.

Table 4.2 lists the plate widths for these core sections of maximum area, with odd numbers of steps varying from 3 to 11. Most transformer designers work

TABLE 4.2   *Plate widths for maximum areas*

No. of steps	Percentage of circle area filled	Plate widths (see figure 4.5(a))										
		a_1	a_2	a_3	a_4	a_5	a_6	a_7	a_8	a_9	a_{10}	a_{11}
3	85.1	0.906	0.707	0.424	—	—	—	—	—	—	—	—
5	90.8	0.949	0.846	0.707	0.534	0.314	—	—	—	—	—	—
7	93.4	0.967	0.901	0.814	0.707	0.581	0.434	0.254	—	—	—	—
9	94.8	0.976	0.929	0.868	0.796	0.707	0.605	0.497	0.37	0.216	—	—
11	95.8	0.982	0.943	0.893	0.832	0.762	0.707	0.648	0.555	0.45	0.333	0.19

The widths are stated as fractions of a core circle diameter of unity.

with a series of core sections of fixed circumscribing circle diameters, all with the same number of packets. The steps of diameter are chosen to give suitable increments of cross-sectional area. In these circumstances it is difficult to use the maximum area proportions of table 4.2 without a large number of different plate widths, few of them recurring throughout the range. It is therefore usual to depart slightly from the optimum proportions and to work with a series of standard plate widths, so that each width can be used in a number of different core circle sizes. This is particularly convenient for the manufacturer when, as is common practice, the first operation on the rolls of core material is to slit them to width. As an example, table 4.3 gives a short extract from a series of core sections based on 14 steps, 10 mm increments of core circle diameter and plate widths that increase by 5 mm increments up to 200 mm and by 10 mm increments beyond that. It can be seen that, although the four cores tabulated require a total of 56 different plate widths, because of the standardised widths only 29 are actually needed. The gross core areas obtained are hardly distinguishable from the theoretical maxima.

This sort of achievement is only possible with particular forms of mechanical construction, as will be examined further in sub-section 4.4.5. By reducing the number of steps, by adjusting the actual core circle diameters and by increasing the increment of plate width, it is possible to achieve an even greater simplification of manufacture. The core cross-sectional area for any given core circle diameter will then be correspondingly less.

TABLE 4.3   *Practical core sections*

Core circle diameter (mm)		375	385	395	405
Percentage of circle area filled		96.37	96.33	96.33	96.33
Plate widths (see figure 4.5(a)) (mm)	a_1	370	380	390	400
	a_2	360	370	380	390
	a_3	350	360	370	380
	a_4	340	350	360	370
	a_5	330	340	350	360
	a_6	320	330	330	340
	a_7	300	310	310	320
	a_8	280	290	290	300
	a_9	260	270	270	280
	a_{10}	240	250	250	260
	a_{11}	210	220	220	230
	a_{12}	180	185	190	195
	a_{13}	145	150	150	155
	a_{14}	95	95	100	100

4.3.2   Pattern

Figure 4.6 summarises in diagrammatic form the most usual core patterns currently in use: the patterns shown in figure 4.6(a), (b) and (c) apply to single-phase transformers, and those in figure 4.6 (d) and (e) to three-phase transformers. The commonest is the three-phase three-limb design of figure 4.6(d). Other three-phase patterns are only used in special circumstances, since, generally speaking, the three-limb pattern gives the minimum mass and the lowest iron loss.

When hot-rolled non-oriented core material was in use, it was common practice to make the cross-sectional area of the yoke greater than that of the limbs by increasing its plate widths, and therefore its vertical height. It can be shown[4] that the reduction in effective flux density is not proportional to the increase in cross-section because of the non-uniformity of flux distribution. The changes of cross-section at the joints between limbs and yokes were apt also to cause flux to transfer from step to step across the plane of the laminations, thus increasing the eddy current losses.

With the advent of cold-rolled grain-oriented core material the rather doubtful gains to be made by increasing the yoke cross-section were entirely outweighed by the greater expense of the extra material required. Furthermore, the increased conductivity of the core material increased the magnitude of any eddy current losses due to cross-fluxing. For these reasons, it is now normal practice in three-phase three-limb cores to make the yoke cross-section and plate widths exactly equal to those of the limbs.

For very large three-phase power transformers the five-limb pattern of figure 4.6(e) is commonly adopted, since this reduces the yoke depth and thus the overall height of the transformer. This is particularly useful if rail transport is specified, since heights are usually restricted considerably by tunnels and bridges. To offset

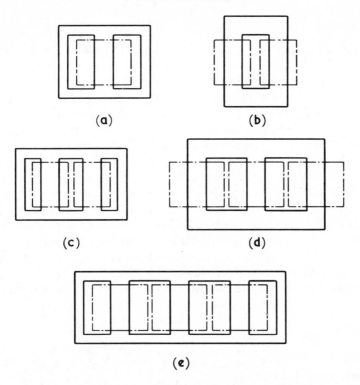

Figure 4.6 Forms of core built from flat laminations: (a) one-phase, one limb wound with side yokes; (b) one-phase, two limbs wound; (c) one-phase, two limbs wound with side yokes; (d) three-phase three-limb; (e) three-phase three-limb

this, the overall length is increased markedly by the presence of the unwound limbs. Another advantage is that the cooling of the yokes is greatly improved without providing special ducts. Where the core circle diameter is over 750 mm, it also allows the limb plate widths to be split in two with a central duct without undue complication of the joints with the yoke. This again provides supplementary limb cooling and avoids any limitation of core circle diameter by the core material roll widths.

The flux densities in the various parts of a five-limb three-phase core depend on the relative reluctances of the various paths in the iron, so it is difficult to determine their proper relative cross-sectional areas. The unwound limbs offer an easy path for third-harmonic fluxes, so there is a greater degree of harmonic distortion in the fluxes in this pattern of core than in the three-limb three-phase pattern. This again makes it difficult to assess the proper proportions. Experience has shown, however, that, if the main portion of the yoke is made about 58 % of the total wound limb cross-section and if the unwound limb comprises between 40 and 50 % of it, then the flux densities in the various parts will be substantially equal, and a reasonably low iron loss will be achieved.

4.3.3 Joints

The diagrams in figure 4.6 do not indicate any joints between limb and yoke; it is implied that the laminations are represented in plan view. This need not be so for the single-phase cores, where the laminations can be at right angles to the plane of the paper, and the cores develop into single loops lapped around rectangular mandrels as in figure 4.7(a) and (b). This arrangement gives the lowest possible iron loss and is widely used in small distribution transformers. It has also been applied to larger power transformers up to 3.3 MVA in the USA[5], but, since the entire core requires annealing after forming, expensive manufacturing plant is required. In very small transformers the principle has been extended to three-phase three-limb cores by using three interwound loops, as shown in figure 4.7(c). This pattern is not feasible on large cores, since, although there are no corner or joint losses, flux cannot readily transfer from one loop to another and since the individual loop fluxes develop large third-harmonic contents in order to produce a nett sinusoidal flux within each winding. This results in an increase in the iron loss of the entire core of approximately 33 %.

The majority of power transformer cores are built up from laminations laid

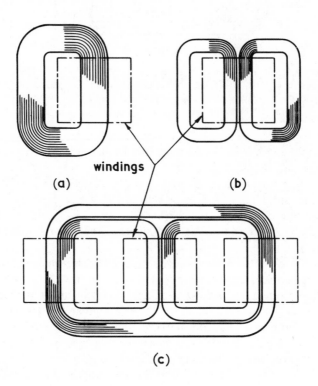

windings

(a)   (b)

(c)

Figure 4.7   Forms of core built from wound loops: (a) one-phase one-loop; (b) one-phase two-loop; (c) three-phase three-loop

flat, as indicated in figure 4.6, and some kind of joint must be made between the plates of the limbs and those of the yoke which lie at right-angles to one another. If proper care is taken to minimise the resulting air gaps and to prevent vibrations, there is little reason why butt joints should not be employed, and some manufacturers have in fact used them. It is usually considered, however, that the manufacturing complications of grinding the joint faces cannot be justified on large cores. This is despite the attractive possibility that to assemble the windings on the core only requires the butt-jointed yoke to be lifted off, the windings to be lowered over the limbs and the yoke to be lowered back in place. The majority of cores are therefore assembled with some kind of interleaved joint between limbs and yokes, formed by an alternating disposition of different lengths of lamination. The three commonest forms in current use are shown in figure 4.8. That of figure 4.8(a) requires only plain rectangular laminations and was universally used for hot-rolled non-oriented core material. With grain-oriented material an extra loss is incurred where the flux traverses the plates at the joint in directions other than parallel to the rolling direction.

To obtain the maximum benefit from cold-rolled grain-oriented material, it is therefore necessary to design some kind of mitred joint, as in figure 4.8(b) and (c). Other more complicated forms of mitred joint have been devised, but the slight

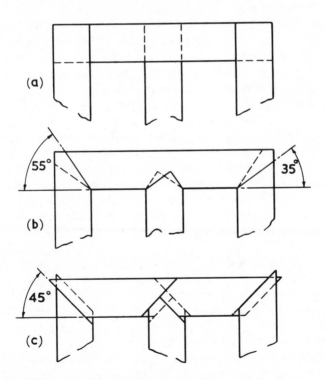

Figure 4.8 Forms of interleaved joint: (a) square; (b) 35°/55° mitre; (c) 45° mitre

gain in performance over that of the forms shown does not usually justify the extra cost of manufacture. There is little difference in the performance of the mitred joints illustrated, and the choice between them is usually made according to the designer's preference and the nature of the manufacturer's core-cutting machinery.

With any kind of interleaved joint it is important to minimise the gaps between abutting plates; otherwise excessive sideways transfer of flux into adjacent plates will cause extra eddy current losses in the iron. This factor also limits the number of identical plates that can be stacked together before changing to the overlapping plate; this will be examined further in sub-section 4.5.1.

4.4 CONSTRUCTION AND MANUFACTURE

4.4.1 Cutting

Cold-rolled grain-oriented core material is normally supplied from the steel mill in rolls several tonnes in mass and 750 to 850 mm wide, according to how much edge trimming has been done at the mill. The transformer manufacturer either must first slit it to width over its full length, or must crop it to length over its full width, before laminations can be made. Most manufacturers and the steel makers themselves, if the transformer manufacturer has asked them to supply finished laminations, prefer to slit to width first.

If the production programme for the cores is sufficiently firm, as, for example, when making transformers for stock, it is possible to analyse the cores to be made and the roll widths available and to calculate exactly how the rolls should be slit to minimise scrap. The slitting operation is generally carried out on a rotary slitter, the strip being pulled through the machine and the cutters sometimes driven as well. Perhaps the most difficult part of the operation is to ensure freedom from edge bow, since, if the strip is not slit straight, it will be difficult to ensure accuracy in the subsequent cutting processes. Burrs on the slit edges must also be controlled carefully to avoid interlaminar contacts in the finished core; commonly, when the maximum burr exceeds 0.035 mm, the tools are reground. Experience has shown that it is better to do this than to allow excess burrs to form and then to grind them off.

The slit rolls can be cut to length in several ways, depending upon the complexity of the machinery which the manufacturer is prepared to install. Generally speaking, the faster the laminations are produced and the tighter the dimensional tolerances, the more expensive the plant will be. The simplest method is to install one or more guillotines controlled by hand. These can be used for either straight-cut or angle-cut laminations. The most elaborate machines take the roll of slit strip in at one end and eject completely finished laminations from the other at the rate of several per minute, all cutting, feeding and notching operations being carried out automatically under electrohydraulic control.

4.4.2 Annealing

Inevitably the various cutting operations described above strain parts of the material above the elastic limit, and at the edges grains may be sheared through. These strains have an adverse effect on the magnetic properties, so that to obtain the best performance from the laminations they must be subjected to a stress-relieving operation before being built into cores. This is carried out by slowly raising the laminations to a temperature of about 800 °C, by maintaining them at this temperature for a short time and then by slowly cooling them again.

It is necessary to keep the variation in temperature across each lamination small; otherwise the sheets will develop edge waves instead of emerging flat. If attempts are made to build cores with wavy laminations, then either the proper space factor will not be achieved or the compression applied to the laminations to obtain it will induce strains and will spoil the magnetic properties. Should the annealing process be prolonged, it will be necessary to exclude oxygen from the vicinity of the plates; otherwise oxidation of the iron will occur.

The two basic annealing methods employed are the batch anneal and the continuous anneal. In the first of these, a load of laminations is built up on a special hearth, and a cover containing radiant electric heating elements is lowered over it. This cover also serves to contain a protective atmosphere of nitrogen with a small proportion of hydrogen. Because of the length of time taken to stack and unload the charge, several hearths are needed to achieve a reasonable throughput.

The second method, the continuous annealing process, allows uniform heating and cooling to be more easily achieved and is now the commonest method. Either by an endless belt or by a series of rollers, laminations are carried through a succession of controlled temperature zones usually comprising pre-heating and high-temperature soaking zones in the actual furnace, followed by a series of cooling zones. The last of these are usually artificially cooled by water or air blast.

With the belt furnaces, where packets of laminations stacked several plates high are passed through in succession, the speed of traverse requires that a protective gas be fed into the furnace, again consisting of nitrogen and hydrogen. Careful design of the pre-heating and cooling sections is necessary to avoid the development of edge waves in wide laminations, and the overall length of the plant may be as much as 30 m. Because of the difficulty of obtaining long belt life and because of the limit to the speed of travel set by having to heat and to cool the belt as well as the charge, the trend is towards using roller-hearth furnaces. Single laminations are passed through these in succession at a relatively high speed, so that with proper design the protective gas may be dispensed with. This has a beneficial effect on the phosphate coating, which can be adversely affected by a reducing atmosphere.

4.4.3 Insulating

The wide variety of coatings commonly employed to provide supplementary insulation on the finished core-plate is matched by an equal variety of methods of application.

Spray or roller applicators are commonly used, together with some kind of oven to dry or cure the coating. Chemical degreasing before application of the coating is not normally necessary if the process immediately follows an annealing furnace. Care must be taken that the laminations are not bent or subjected to undue mechanical strain in the plant, for example by pinch rollers, since most coatings will prevent any further stress relief by heat treatment being carried out.

Whatever the nature of the coating, or its method of application, it is necessary to keep a close check on its quality. To check the thickness of the thin coatings a magnetic gauge or similar instrument is commonly employed. The insulating value of the coating is usually measured by applying a low voltage to one or more plates held under pressure between electrodes and by converting the resultant current to ohms per unit area of coating. The lower limit for this insulation resistance is fixed by taking account of the factors described in sub-section 4.2.2. An upper limit of thickness is necessary to prevent the finished core from being oversize, or short of section, according to the method of building. A minimum thickness need not be specified, since the resistance test will detect coatings of inadequate thickness. It is usually sufficient to make these checks at intervals on selected samples rather than on every plate.

4.4.4 Building

The assembly of the cut, annealed and insulated laminations into the specified size of core is the stage at which the accuracy of all the preceding operations will be revealed and the performance of the finished core made or marred.

The supply of the core material from the steel mill to the transformer maker is usually governed by a national standard of the country concerned. These documents prescribe standard thicknesses and tolerances on these thicknesses, for example[G2.7] a tolerance of 0.05 mm on the standard nominal thickness of 0.33 mm, although in fact it is unusual for the material to vary to this extent. The thickness of the insulating coating will vary, and the flatness of the finished laminations may also vary, depending upon the work put into the material and the efficiency of the stress-relief anneal. Consequently, the assumption which the designer made as to the space factor or stacking factor that would be achieved in the finished core may be wrong. The building process must be chosen to take this into account.

Basically, the laminations are laid down flat on a building jig which accurately locates them to form the desired pattern (see figure 4.6). If the core has bolt holes, this makes the design of the jig simpler. There are three ways of judging when to stop laying down laminations of a particular width: (1) when a sufficient mass has been added, (2) when a sufficient number have been added or (3) when a given stack dimension has been reached. Method (1) guarantees that the specified cross-sectional area is achieved but may result in an unreasonably large-diameter winding being necessary to allow for build-up. It is also the most difficult of the three. Method (2) is very easy but may result in either inadequate cross-section or wrong dimensions. Method (3) is quite common but can easily result in inadequate cross-sectional area. Each transformer manufacturer has to decide which procedure, or combination of procedures, suits his factory best.

In my experience it is quite feasible to build by numbers of plates so long as a running check of dimensions is used to adjust continuously the design value of the stacking factor and so long as adequate control is maintained over the annealing and insulating processes. An occasional accurate check of finished core mass is also helpful.

The core material is potentially capable of giving a stacking factor of 0.97 or better on small samples, but practical power transformer cores with additional insulation applied to the core-plates are more likely to give stacking factors in the range 0.95 to 0.97.

The precision with which the plates have been cut, particularly the mitred corners, will be revealed by the size of the air gaps between butting laminations in the interleaved joints. The building jig is preferably arranged to locate the core-plates firmly at the joints and to allow any deviation from pattern dimensions to occur at the centres of legs and yokes. This reduces the extra iron losses due to the gaps to a minimum, but the resulting small distortions of the limb cross-section have to be taken into account when deciding upon the minimum diameter of winding which can be accommodated on a given core circle.

Associated with the joint iron loss is the number of identical laminations which can be laid down on top of one another before changing to the complementary laminations which form the interleaving. The smaller the number, the lower is the eddy current loss due to cross-fluxing at the joints, as can be seen from some experimental results plotted in figure 4.9, but the longer it takes to build the core. Commonly, a compromise is made, and cores are built by interleaving two or even three laminations at a time.

The final duty of a building jig is to allow the core to be raised from the horizontal position to its final vertical position without undue strain on the laminations and without allowing the joint gaps to enlarge themselves. Figure 4.10 illustrates this; it shows the core of a 180 MVA auto-transformer being raised in this manner with the aid of an overhead crane.

4.4.5 Clamping

When power transformer cores were built from hot-rolled non-oriented material, it was universal practice to clamp the limbs and yokes of the larger sizes together by means of bolts passing through holes punched in the laminations. This arrangement, although giving excellent clamping, suffers from several disadvantages.

(a) The bolts are potential sources of core faults.
(b) Either the bolt insulation has to be expensive temperature-resistant

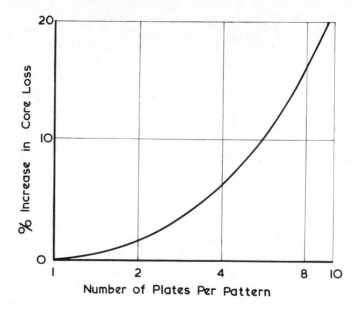

Figure 4.9   Effect on core loss of building with more than one lamination at a time

Figure 4.10   Use of building jig to raise a core to vertical position

material or transverse cooling slots have to be provided.

(c)   The bolt holes and cooling slots (if any) cause local increases in flux density and iron loss.

(d)   The punching of holes and slots complicates the cutting of the core.

(e)   The unequal reduction of section in the plate widths making up the core section causes cross-fluxing between packets at the location of each bolt and results in extra eddy current losses.

(f)   The room taken up by the bolt head, insulation and stiffening plate (if fitted) makes it difficult to achieve the optimum cross-sectional areas considered in sub-section 4.3.1.

With the introduction of cold-rolled grain-oriented core material these disadvantages become more serious. Because of the higher working flux density, the extra losses of (c) and (e) become greater in relation to the total iron loss. The lower resistivity of the material also increases the loss in (e). Furthermore, a new source of extra loss appears because the flux diverging round the bolt holes is no longer parallel to the low-loss direction of the laminations. Because of the greater cost of the core material, the consequences of (f) above also become more serious. It has been suggested[6] that with 14 steps on a bolted core a circle fill factor of only 83 % is practicable, in contrast with the 96.3 % displayed in table 4.3 for a bolt-less core design.

The tendency has been not only to reduce the number and size of the clamping bolts, for example by the use of high-tensile steel, and to eliminate cooling slots by the use of newer forms of high-temperature insulating tube but to make cores without any bolts whatever passing through the active iron section. This has been accomplished by the development of high-strength resin-glass bands[7] applied to the periphery of the core section. A further improvement has been to eliminate separate metallic limb clamping plates by bonding together the outer packets of laminations with epoxy resin. This not only allows the introduction of more active core material but eliminates eddy current losses in the clamping plates.

The remainder of the core clamping structure usually consists of beams clamped along the sides of the top and bottom yokes and interconnected by vertical tension members. These are also connected, directly or indirectly, to cross-members under the bottom yoke, so that the mass of the iron is supported cradle-wise and so that undue strain on the laminations is avoided. The vertical tension members may be high-tensile or stainless steel bolts, positioned either outside or inside the windings. The inside position wastes space but can result in fewer electric problems than external bolts. The latter are, however, better suited to the application of winding clamping pressures by thrust screws or blocks fixed to the top yoke clamps and for withstanding axial short-circuit forces. With the increasing size of power transformers there has been a corresponding tendency to use non-magnetic materials for the top and bottom clamps, in order to reduce eddy current losses due to the leakage field of the windings[7]. Materials as diverse as aluminium and laminated wood have been used for this purpose.

Figure 4.11 shows a completed core for a 180 MVA auto-transformer, featuring limbs and yokes without through bolts.

Figure 4.11   Completed bolt-less core for a 180 MVA 275 to 132 kV auto-transformer

4.5 PERFORMANCE

4.5.1 Losses

Ideally, to obtain the iron loss of a transformer core the designer would only have to multiply the specific loss of the core material (figure 4.2) at the operating flux density and frequency by the total mass of the core. However, because of the imperfections introduced by the design and manufacture of the core, this ideal remains practically unattainable. The ratio of the actual measured core loss to this ideally calculated loss is sometimes called the building factor. The objective of the transformer manufacturer is to make this as near unity as he can possibly achieve without unduly complicating design or manufacture of the core.

The first uncertainty to be contended with is that of knowing the mean specific loss of the material in any core. The various national standards referred to in subsection 4.4.4 have set several grades of quality for grain-oriented core material. At the steel mill, the grade of each coil or part coil is established by measuring the iron loss of 25 cm double-lap Epstein square specimens taken from each end. The samples are cut with their length parallel to the direction of rolling and are annealed after cutting. Grading is usually on the basis of measured watts per kilogram at 1.5 T and 50 or 60 Hz, and the coils are sold by grade number. There is, however, approximately 10% difference between the best and worst of any grade, so that there can be a corresponding uncertainty as to the specific loss for any core built from supplies of any one grade of material. However, the steel mill is usually obliged to supply test certificates for each consignment, so that the transformer manufacturer can, if he so desires, keep a running check on the actual specific loss of material being cut for any given core. Some manufacturers go further and take their own Epstein samples as a check, not only of the quality of the incoming material but of the functioning of the annealing plant through which the samples are passed.

The largest single component of the building factor is usually that attributable to the core joints. Here the flux crosses the low-loss direction of the laminations, and supplementary eddy current losses will arise owing to the passage of flux from plate to plate at the butt joints within the interleaving. The most satisfactory method of deducing the relative contributions of limbs or yokes and of the interleaved corners to the total core loss is to make and test a representative series of cores with differing proportions of corner mass. If required, this can be done for each grade of core material, but it is usually sufficiently accurate to make the tests on the worst grade which the manufacturer normally uses and then to apply experience factors to the results to allow for other grades. Figure 4.12 shows results deduced from a typical series of tests, for 56 grade (British Standard 601) material and a three-limb three-phase pattern without bolts. It will be observed that the building factor for limbs and yokes is approximately 1.07, whereas for a complete core with 20% of its mass in the corners it would be 1.17 with mitred joints and 1.29 with square-cut interleaving.

Generally speaking, building factors tend to be a little larger for three-phase cores than for single-phase cores. Provided that care is taken with the design of the joints and the proportioning of the yoke sections, the same basic design curve can also be used for five-limb three-phase cores. Built differently, with different methods of cutting, annealing and insulating, the same grade of material might yield smaller or larger building factors.

If bolts are used, building factors increase. Morris[4] has compared various methods of calculating the effect of bolt holes in rectangular-section cores made from non-oriented material, but this kind of calculation is of little direct help to the designer. Again, the most satisfactory method of judging the effect of bolt holes is to compare the results of tests made on cores with and without the holes.

With the trend towards the omission of bolts, this last-mentioned problem diminishes in importance. Once basic data have been established for the prediction of core iron losses, a running check of building factors as cores are made and tested serves as the ultimate control on the whole process of core manufacture.

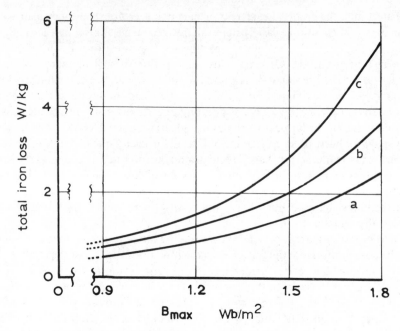

Figure 4.12   Core iron losses at 50 Hz: curve a, limbs and yokes; curve b, corners with mitred joints; curve c, corners with square joints

4.5.2   Magnetising currents

To some extent, the calculation of magnetising current from the characteristics of the material measured with the Epstein square is analogous to that of iron loss. However, with grain-oriented core material the change in permeability with angle of deviation from the rolling direction is much greater than the change in loss. Furthermore, the permeability is apt to vary far more from batch to batch of core material and is more sensitive to treatment during the manufacture of the core. Also, the effect of air gaps inevitably introduced during building has a much greater effect on the magnetising current of the core than on the iron loss. Thus the prediction of magnetising currents is considerably more difficult than the prediction of iron losses. For this reason it is not customary for this aspect of core performance to be the subject of a guarantee.

Figure 4.13 shows typical figures derived from actual tests on cores, which can be used to calculate approximate magnetising currents for three-phase three-limb cores built from 56 grade material.

4.5.3   Temperature rise

The various national standards covering power transformers either set no limits to core temperature rise or limit the temperature rise of the core surface to that of

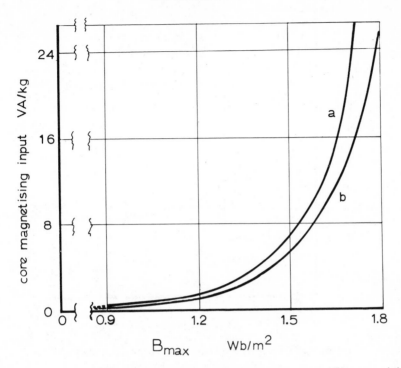

Figure 4.13   Core magnetising voltamperes at 50 Hz: curve a, cores with square joints;
curve b, cores with mitred joints

the adjacent winding but do not attempt to limit the internal temperature rise of
the core. The IEC[G1.7], for example, contents itself with recommending that no
part of the core should become hot enough to damage either itself or adjacent
parts. However, because of the risk of damage to any supplemental interlaminar
insulation, or to core bolt insulation (if bolts are used), a safe procedure is to limit
the temperature of the internal core hot spot to 120 °C.

Since it is impracticable to measure this core temperature as part of normal
acceptance tests, as is done for the windings, it is necessary to be able to calculate it
with a fair degree of assurance. One difficulty in making such a calculation is that
the thermal conductivity of a laminated core is quite different for different
directions of heat flow: it is relatively high along the laminations and relatively
low at right-angles to them. The cruciform shape of the section also adds its
complications. The simplest method of calculation is to assume that all the heat in
the largest packet of a core travels across the width of the laminations and is
dissipated at their edges, with no heat travelling at right-angles to the plane of the
laminations, or up and down the limb.

Figure 4.14 shows the relation between the thermal conductivity of
silicon – iron core material and its silicon content. Figure 4.15 shows how the
temperature rise of the hottest interior part of the core limb over the oil

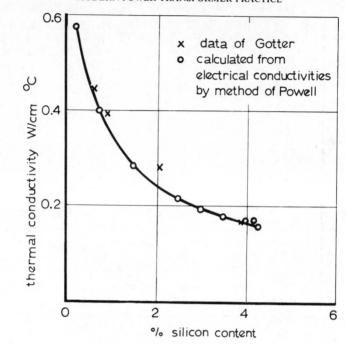

Figure 4.14   Thermal conductivity of silicon – iron alloys [8, 9]

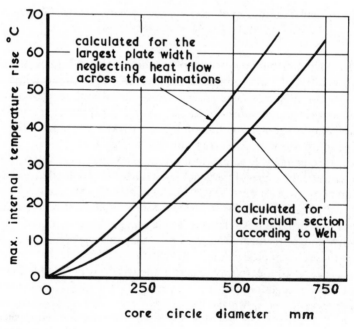

Figure 4.15   Internal temperature rise of cores without ducts, calculated[10] for 56 grade material at 1.55 T and 50 Hz

surrounding it varies with core diameter for cores without cooling ducts, calculated in the simple manner just described. From the graph it would appear unwise to use cores without ducts above a core circle diameter of 500 mm. This is, however, an unduly pessimistic approach as can be seen from the second curve in figure 4.15, which shows the results of using a more complicated form of calculation due to Weh[10]. This takes into account heat flow in a core of circular section without ducts, if we allow for conduction across the laminations and if we assume equal heat transfer coefficients and oil temperatures at all points on the outer surface. In calculations of this kind the required value of thermal conductivity normal to the plane of the laminations either can be calculated from the component conductivities of iron, insulation and oil which appear in series or can be measured[11]. Either way, the values are sensitive to the intimacy of contact between adjacent laminations and the factors such as clamping pressure which affect this.

The calculation of the maximum internal temperature of ducted cores is even more complicated, but exact solutions have been attempted by Cockcroft[12] and Higgins[13] for rectangular cores. The advent of the high-speed digital computer has allowed numerical analysis to be applied to the calculation of internal core temperatures[14], and figure 4.16 shows some isothermals for part of a ducted core obtained by this means.

The assessment of the heat transfer coefficients at the surface of the limbs and yokes is no more difficult than the estimation of similar coefficients for the windings, as considered in chapter 5. Those for the limbs will be subject to most variation, since the duct between core and winding will vary and since the vertical oil flow up the duct will also be affected by the temperature difference maintained between the inlet and outlet of the cooler. In designs with short limb lengths there is often considerable heat transfer vertically up and down the limbs into the cooler yokes, which has the effect of reducing the internal limb temperatures. However, the temperatures at the junction between limb and top yoke may be increased on this account, and, since this is surrounded by the hottest oil, they may well be the highest in the entire core.

It is not easy to confirm calculations of internal core temperatures by direct measurement, since this must involve an assessment of the likely position of the hot spot, the insertion of thermocouples or other temperature detectors and an unusual kind of heat run with the core excited. However, occasional tests of this kind are carried out by most manufacturers, particularly when extending a range of cores to larger diameters.

4.5.4 Noise and vibration

The source of transformer noise is the core, caused mostly by longitudinal magnetostrictive vibrations of the iron. This will be examined in more detail in chapter 8. Here it is sufficient to say that the magnetostrictive behaviour of grain-oriented silicon – iron is considerably affected by mechanical strain and the stress-relief anneal. Flatness of laminations is desirable so that bending strains are not

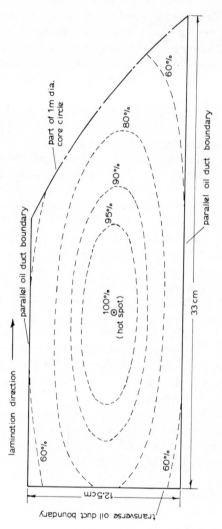

Figure 4.16 Temperature distribution in a part of ducted core

induced when the core is clamped, as well as for the reasons explained earlier in this chapter. The method of clamping is also important, as is evidenced by figure 4.17.

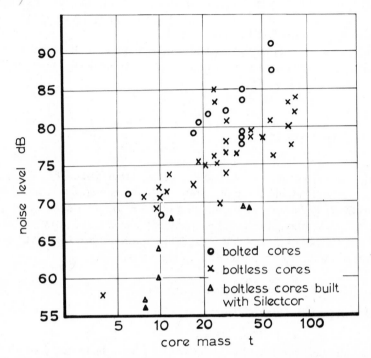

Figure 4.17   Transformer noise levels measured at 50 Hz and corrected to $B = 1.55$ T

One steel company in USA has had some success with a special grade of grain-oriented silicon – iron material which is slit to width and is subjected to a special process of annealing and insulating. This material, known as Silectcor (Allegheny – Ludlum Steel Corporation), not only has a basically low magnetostrictive constant but gives very flat laminations and is not so sensitive to core clamping pressures. Some results of noise tests on transformers whose cores were made from this material are also given in figure 4.17. Even with low basic magnetostriction, it is possible that an unsuitable choice of core dimensions will lead to mechanical resonance and to an increase in core vibration and transformer noise level[15, 16].

4.6   FUTURE DEVELOPMENTS

The most certain thing about the future of the transformer core is that any changes will be evolutionary in nature rather than revolutionary.

It is not very likely that any new material will replace grain-oriented silicon – iron in the forseeable future, but improvements may very well be made in the degree of orientation and its stress sensitivity.

Advances in automating the production of laminations and the building of cores may lead to the use of thinner material and more precise dimensions. Both factors may lead to improved core performance.

The application of faster more powerful digital computers to the calculation of temperature rise and vibration is certain to improve predictability of performance.

ACKNOWLEDGEMENTS

The author acknowledges with thanks the valuable assistance of several companies and individuals in the preparation of this chapter. Particular thanks are due to Bruce Peebles Limited (now part of the Reyrolle – Parsons Group as Parsons Peebles Limited), the Steel Company of Wales Limited (now part of the British Steel Corporation) and the Allegheny – Ludlum Steel Corporation for permission to quote from data relating to their products.

REFERENCES

(Reference numbers preceded by the letter G are listed in section 1.14.)
1. Goss, N. P., *ASM Trans. Q.*, **23** (1935) 215
2. Barton, J. P., Interlamination resistance, *Trans. Am. Inst. Electr. Eng.*, **63** (1944) 670
3. Morris, A. L., Stepped core sections for power transformers, *Engineer*, **184** (1947) 581
4. Morris, A. L., Calculation of losses in transformer cores, *Engineer*, **191** (1954) 837
5. Gordy, T. D., and Somerville, G. G., Single-phase power transformer formed cores, *Trans. Am. Inst. Electr. Eng.*, **69** (1950) 1384
6. Phillips, M., and Thomas, J. M., Opticore—a computer program to calculate optimum cross-sectional areas of transformer cores, *AEI Eng.*, January–February (1966) 40
7. Kerr, H. W., and Palmer, S., Developments in the design of large power transformers, *Proc. Inst. Electr. Eng.*, **111** (1964) 823
8. Gotter, G., *Heating and Cooling Electrical Machinery*, Springer-Verlag, Berlin, (1954)
9. Powell, R. W., Thermal conductivities of metallic conductors and their estimation, *Proc. Gen. Discuss. on Heat Transfer*, Institution of Mechanical Engineers, London (1951) 290

10. Weh, H., Die zweidimensionale Wärmeströmung im geschichteten Transformatorkern, *Arch. Elektrotech. (Berlin)*, **41** (1953) 122

11. Roberts, T. J., and Allen, P. H. G., The thermal conductivity of some electrical engineering materials, *Proc., 9th Thermal Cond. Conf.*, Iowa State University Press, Iowa (1969) 719

12. Cockcroft, J. D., Temperature distribution in transformer or other laminated core of rectangular cross-section, etc., *Proc. Cambridge Philos. Soc.*, **22** (1925) 759

13. Higgins, T. J., Formulas for calculating temperature distribution in transformer cores and other electric apparatus of rectangular cross-section, *Trans. Am. Inst. Electr. Eng.*, **64** (1945) 190

14. Rele, A., and Palmer, S., Cooling of large transformer cores, *IEEE Trans. Power Appar. Syst.*, **91** (1972) 1527

15. Matthieu, P., Calculating the natural vibrations of transformer cores, *Bull. Oerlikon*, No. 361, February (1965) 1

16. Henshell, R. D., Bennett, P. J., McCallion, H., and Milner, M., Natural frequencies and mode shapes of vibration of transformer cores, *Proc. Inst. Electr. Eng.*, **112** (1965) 2133

5

Windings

H. W. Kerr*

5.1 REQUIREMENTS WHICH CONTROL WINDING DESIGN

The design of any transformer winding must fulfil certain basic requirements. It must withstand the electric stresses imposed on it during test. These tests are intended to ensure that a transformer passing them will give trouble-free service for many years under the conditions it is likely to meet after its installation. They are therefore of three sorts: (1) a test at power frequency for about 1 min to prove its margin over its operating voltage, (2) a surge test and (3) a switching surge test—both (2) and (3) are to prove its ability to withstand voltage surges due to atmospheric disturbance and to switching.

The transformer loss must not be higher than the quoted figure, and this loss must include not only the loss in the conductor itself but also any losses in other structures which are a function of current. The leakage impedance must be of the required value, and the winding dimensions must be such that the quoted core loss is not exceeded.

The winding temperature must not rise above an acceptable level, since overheating could damage the insulation and could shorten its life.

Inevitably electromagnetic forces of considerable magnitude can be produced in large transformers under fault conditions, and the winding must be designed to withstand these forces.

Finally the whole transformer must be capable of being built and sold at a price acceptable to the purchaser and must still provide a margin of profit for the manufacturer.

5.2 MATERIALS

5.2.1 General

In spite of developments in the production of synthetic materials there has been

* Parsons Peebles Limited.

little change in the raw materials used in windings of oil-filled transformers. The conductor is generally copper, although in some cases economics has brought about the use of aluminium.

The most common insulation is paper or paper board known as pressboard[G2.4]. In certain places wood, often laminated and varnish impregnated, is employed. One common departure from these materials is the use of synthetic enamel as conductor insulation when the voltage between conductors is not too high. Many other insulating materials have been tried but either have been found wanting in some respect or have proved too expensive for general use.

Surrounding the windings and insulation of most power transformers is a liquid insulant. This serves the double purpose of providing good insulation between parts and of removing the heat resulting from the losses. The most common such material is oil[G1.18, G2.1].

The maximum viscosity permitted is $37 \text{ mm}^2 \text{ s}^{-1}$ at $21.1°C$. This low value is necessary to allow oil circulation under thermosyphon conditions. A pour point of $-31.7°C$ is specified, but where extreme cold may be experienced even lower temperatures may be required.

The maximum permitted acidity of new oil is that which can be neutralised by 0.05 mg KOH per g. The oil must withstand for 1 min a voltage of 40 kV between spheres 13 mm in diameter and 4 mm apart, but many transformer manufacturers demand higher test levels than this.

To maintain the oil in good condition in service it is essential to keep out moisture. A very common practice is to have a conservator tank mounted above the level of the main tank. This allows changes of volume with temperature while keeping the main tank full of oil. The conservator is connected to atmosphere through a breather designed to remove moisture from the air entering. Thus the air in contact with the oil is kept dry providing the breather is working satisfactorily.

This oil is flammable. Where the use of such a material is not acceptable, an alternative is the liquid known in Britain as Pyrochlor. Its chemical composition is 60% hexachlorodiphenyl and 40% trichlorobenzene.

One property of any solid insulating material in a high-voltage transformer is that it should be fairly easily impregnated by the liquid insulant.

The presence of cavities full of gas could cause internal discharges and subsequent breakdown. It is also desirable that the various insulating materials should have as nearly as possible the same values of permittivity, since wide variation throws high stress on the material of lowest permittivity and distorts the field at the interface between the two materials.

While this completes the basic raw materials in oil- or other liquid-filled transformers, there is a very wide range of materials in dry-type transformers. These are classed in accordance with the temperature at which it is safe to operate them, and in this field new materials have been tried with success. More information about this type of transformer is given in chapter 11.

A more recent development is the use of gases, of which sulphur hexafluoride (SF_6) is the most common. Under pressure this gas has very good electric strength and also good cooling properties. While gas-filled transformers have been studied in some detail in the UK, there is much more experience in the USA where numbers of quite large units have been made[1,2].

5.2.2 Conductors

While the basic materials used as conductors are limited to copper and aluminium, these can be employed in various forms, depending on the current and voltage requirements. In common use are wires and strips, and in recent years there has been an increased use of foil and sheet. Wires are suitable only where the currents are fairly low, as in the high-voltage windings of distribution transformers. Strips, either singly or in groups in parallel, can be found over a very wide range of current from about 10 A to the highest currents in the low-voltage windings of large generator transformers.

There is a considerable overlap between wire and strip. Foil can replace wire or strip where the currents are not too large, and, beyond the thickness limit of the foil, sheet can be used. There is a limit to the width in which foil or sheet can be obtained, and clearly winding with thick sheet would be quite impracticable.

Conductor insulation for oil-immersed transformer is either paper or synthetic enamel. The latter is suitable for wire and strip where its electric strength is sufficient to withstand the voltages appearing during test and in service. For higher voltages paper insulation is almost universally used. Where necessary a number of strips in parallel, each insulated by synthetic enamel, are covered overall by paper taping. The individual strips are transposed every few centimetres so that there is complete transposition every few metres (figure 5.1). The requirement for transposition is explained in section 5.6.

The insulation of foil, especially at the edge, presents some problems. Efforts have been and are being made to coat foils with insulation, but interturn insulation is more commonly provided by winding in thin paper or plastic during the manufacture of the coil.

5.2.3 Insulation within a winding

Within a winding the insulation must be arranged not only to provide sufficient electric strength to prevent breakdown under any conditions likely to be met in service but also to allow adequate circulation of the cooling medium so that no part of the winding gets excessively hot. Various winding arrangements have been developed to meet these requirements. Some are described later in this chapter. Common to all are spacers to maintain ducts, either horizontal or vertical.

These spacers are usually cut from boards of paper-type material without any impregnation by varnish or adhesive. They are similar to cardboard, although of much higher electric strength, and the most stringent precautions are necessary to prevent conducting matter or anything which could accelerate ageing of the oil reaching the final product. The same material can be cut into washers or can be formed into cylinders.

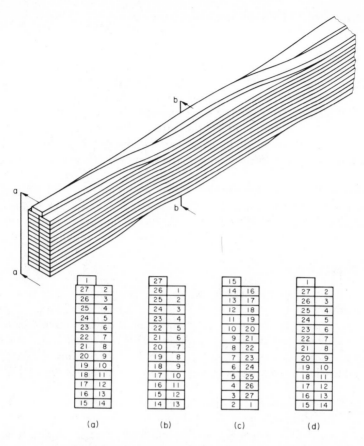

Figure 5.1 Continuously transposed multi-strip conductor: (a) at start of transposition cycle; (b) after first transposition; (c) after thirteenth transposition; (d) after twenty-seventh transposition

5.2.4 Major insulation

The major insulation is that between windings or from windings to earth. This very often consists of the same board as used for insulation within the winding. In many cases, however, more mechanical strength is required. Cylinders are wound from paper and are bonded with synthetic resin. These have good mechanical and electric properties. The puncture strength is very good, but the voltage at which failure occurs over the surface is lower than for the unbonded board. Also the permittivity is higher, resulting in a less uniform voltage distribution. Another point to be remembered is that impregnation with oil is much more difficult with the possibility of some parts being considerably weaker than others, because they are not thoroughly oil soaked.

5.3   COMMON TYPES OF WINDING

5.3.1   General

The type of winding designed for any application depends on its current and voltage requirements. It does not follow, however, that there is a unique design which meets all requirements. There is of course an overlap between the ranges covered by any two forms, but in addition external factors may have considerable influence. Obviously the manufacturing facilities must be taken into account, and the training and experience in the manufacturer's organisation cannot be neglected.

Certain winding forms may be covered by patents which, even if licences to manufacture could be obtained, would affect the economics which finally decide which winding to choose.

5.3.2   Helical windings

For low-voltage heavy-current coils the most common winding form is helical. As its name implies this is wound on a former in a helix of constant diameter, progressing turn by turn from one end to the other. It can be wound directly on a cylinder or can be separated from it by spacers to provide an oil duct. A much less common practice is to wind it on a former which is later removed. Spacers can also be arranged to provide ducts between turns if the cooling conditions warrant it.

The winding may have one or more layers. Layers must be connected in series as parallel connection would result in high circulating currents. Since this type of coil usually carries quite heavy currents, several conductors are wound in parallel. These must be transposed to keep down the stray losses. Figure 5.2 shows the low-voltage winding of a large generator transformer during manufacture.

5.3.3   Multi-layer high-voltage winding

A form of winding similar to the helical but very different in terms of the voltage range covered is the multi-layer high-voltage winding. This winding is in fairly common use for voltages above 132 kV and can be designed for a very wide range of current. Each layer is wound in the same way as a helical coil, but insulation and screening are arranged as shown in figure 5.3 to make it suitable for the highest voltages. The electric design is discussed in section 5.5.

One point which must be remembered with this type of coil is that the layer nearest the duct between high- and low-voltage windings is in a high-intensity magnetic field. This gives higher eddy current losses for this layer than for the rest of the winding and could control the thermal design. Also radial forces are highest on this layer.

Figure 5.2　Large helical coil being wound

5.3.4　Disk coils

The disk coil is commonly employed for a winding where the helical coil ceases to be economical as the voltage increases and current decreases. However, its upper limit of voltage has not been reached. It has been used up to the same level as the multi-layer high-voltage winding. Since it is only suitable for conductor in strip form, a limit is reached where the current is so low that wire is used. This is rarely if ever reached in power transformers, as the rated value tends to rise with voltage.

The disk coil consists of a number of disks wound alternately from inside to outside and outside to inside and connected in series. The series connection can be made as a separate operation, or the coil can be wound continuously from start to finish. Sometimes the coil has one terminal at the centre, and the two halves are wound in opposite directions so that top and bottom can be joined to form the other terminal. This reduces the current in each disk and may help the electric design. Figure 5.4 shows a continuous-disk coil being wound.

Special designs with screens or interwound coils have been developed to improve voltage distribution under surge conditions[3]. These will be discussed in section 5.5.

One advantage claimed for this type of winding is the ease with which ampere – turn balance between primary and secondary windings can be achieved. As will be explained in section 5.9, this keeps to a minimum the axial forces and, coupled with the inherent strength of this form of winding, makes it a very good design under short-circuit fault conditions.

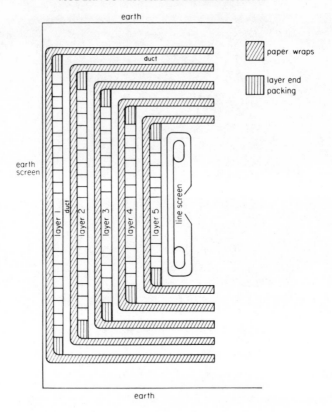

Figure 5.3   Section of multi-layer high-voltage winding

A special application of the disk winding is for very high currents such as are met in the design of furnace transformers. Then a large number of disks are connected in parallel. The low-voltage winding will be on the outside and so convenient to the connections to the terminals.

5.3.5 Coils made from wire

Power transformers rarely have currents low enough to require the use of wire instead of strip. Wire has a poorer space factor than a rectangular conductor so that it is not employed where strip is available. It is, therefore, found only in the coils of distribution transformers, which are considered in chapter 9, and in instrument transformers which form a special field.

On the rare occasions when a high voltage combined with a relatively low rated value is required, the high-voltage winding of a power transformer must be wire wound. For high-voltage windings the multi-layer type can be used, while for lower voltages similar windings to those found on distribution transformers can be employed.

Figure 5.4   Continuous-disk coil being wound

5.3.6   Coils made from foil or sheet

Coils have been manufactured from copper or aluminium foil or sheet for distribution transformers and fairly small power transformers. One of the claims made is that its space factor is better than other forms of coil and that its cooling properties are good. Both properties are dependent to some extent on the interturn insulation.

5.3.7   Tappings

The arrangement of the various coils is largely dependent on the characteristics required, the reactance being especially important. Special consideration must be given to the taps provided to adjust the voltage ratio.

Small transformers or medium-size units with small tapping range may safely be built with taps located at the mechanical centre of the coil, but, as shown in sub-section 5.9.4, axial forces increase rapidly as the out of balance increases. For transformers up to about 30 MVA taps are usually located at one-quarter and three-quarters of the distance from the top of the coil; the top and bottom halves are connected in parallel to make this the electric centre.

As units become larger even this is not satisfactory, and separate tapping coils with sections wound in parallel but connected in series must be used. Each section occupies the full winding length, and balance between primary and secondary is maintained whatever the tap position. The interconnection between sections is arranged so that the maximum voltage between adjacent conductors is that of two tap sections.

The arrangement of taps influences the variation of impedance over the tapping range. Any unbalance between windings increases reactance by producing radial flux. Normally windings are designed to balance on the mean tap with an increasing out of balance as the extremities of the range are approached.

With separate tapping coils, reactance must be calculated on normal and extreme taps. This must take into account all windings carrying current on the tap under consideration. It must be remembered that a tapping coil in the leakage flux will have losses even when it is not carrying current.

Figure 5.5 illustrates three tapping arrangements, but many more are possible.

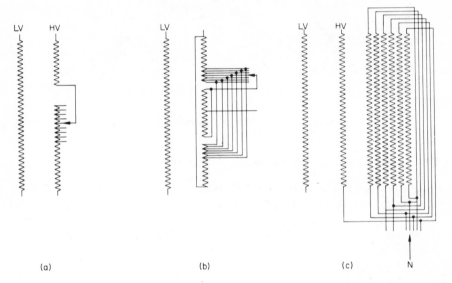

Figure 5.5   Some tapping arrangements: (a) taps in both physical and electric centre of high-voltage winding; (b) taps at one-quarter and three-quarters of distance from end but in electric centre of high-voltage winding; (c) separate tap coil connected at neutral end of high-voltage winding

5.4   WINDING ARRANGEMENTS

The simplest possible arrangement of windings has two concentric coils, the low-voltage winding next to the core and the high-voltage winding outside it. It is often necessary, however, to use a more complex design. Tappings introduce some complications, as described in sub-section 5.3.7. Special requirements on the system where the transformer will be used or limitations in size or mass may result in other variations in the winding arrangement.

While more complicated arrangements such as double concentric, where one winding is split into two parts one on either side of the other winding, present no fundamental difficulty, care must be taken in the calculation of stray loss and reactance. Calculation of surge-voltage distribution requires special care, as referred to in sub-section 5.5.2.

5.5 ELECTRIC DESIGN

5.5.1 Power-frequency voltage

The transformer windings must be designed to withstand voltages appearing in service with some margin in hand. Tests are devised accordingly. All the time it is connected to the system a transformer is subject to power-frequency voltage, varying to a small extent with load conditions. A test is therefore applied to ensure that the transformer has adequate strength to withstand indefinitely any voltage of this type to which it is subjected.

The voltages to be applied during this test are given in chapter 7. From this it can be seen that lower-voltage windings have a separate source test to prove their insulation to other windings and to earth and an induced over-voltage test to prove the insulation within the windings.

Higher-voltage star-connected windings have their neutral point insulation designed for effective earthing in service. These cannot be subjected to a sufficiently high separate source test to prove the strength between all parts of the windings and earth. The induced over-voltage test must, therefore, be modified to test the major insulation as well as that within the windings. The neutral insulation only is proved by a separate source test.

A variety of connections can be employed to obtain the voltages required, but whichever is adopted it is essential to calculate the voltages between all parts since sometimes voltages in excess of those specified may appear. Once the connection to be used has been settled, calculation of all voltages is easy since, at a given frequency, voltage is always proportional to the product of turns and flux. The voltage appearing between adjacent turns must be withstood by the conductor insulation. It must be remembered that this may be far in excess of the volts per turn since coils such as interleaved disk coils or interwound tapping coils have many electric turns between physically adjacent conductors. As a rough guide normal conductor paper 1 mm thick will break down at about 30 kV if thoroughly dried. It is of course necessary to allow a considerable safety factor, but generally this particular condition is not onerous.

Voltage appearing between parts of windings, for example between sections of a disk coil or between layers of a multi-layer helical coil, must be withstood by the conductor insulation, any cooling ducts provided and barriers of solid insulation. In power transformers the oil is always the weakest insulation, and it is subject to the highest stress because it has the lowest permittivity. Pure oil will withstand very high stress, but this value is of no practical significance since it cannot be achieved in any transformer. Also the strength of oil varies with the size of duct, and it decreases as the duct becomes wider. The inclusion of any solid insulation along which gas conduction can take place further reduces the strength.

In oil ducts up to 20 mm and without any creep the withstand stress is about $6 \, kV \, mm^{-1}$. Small samples of pressboard tested in a laboratory give much the same value, but, with large ducts and large areas of pressboard providing a creep path, failure has been known at stresses as low as $2 \, kV \, mm^{-1}$. It is impossible to formulate rules covering the wide range of conditions and materials which may be encountered.

Common practice has been to provide such an amount of solid insulation that even with all the ducts broken down sufficient puncture strength remained to withstand the voltages applied to it. This practice is not really satisfactory because gas conduction in the oil damages the solid insulation, and failure will occur if the voltage is maintained for a long time. In any case, it is bad practice to leave damaged insulation in a transformer since it can give the effect of sharp points which produce very high stresses even at working voltages. The designer must therefore aim to have no gas conduction under power-frequency test conditions. Many specifications now call for measurement of partial gas conduction and set acceptable limits.

Major insulation, that between windings or from windings to earth, is usually made up in the same way as that described above for insulation between parts of windings. The voltages are much higher, and therefore the spacing is greater, but oil spaces can still be broken up by suitable barrier arrangements. It would be ideal to have only solid insulation between windings, but, while it is possible to devise cooling arrangements to allow this, it has so far proved impossible to make use of the intrinsic strength of the solid insulation. It is extremely difficult to eliminate all oil spaces, and these would be subject to much higher stress than the solid. Gas conduction in these spaces could finally lead to failure. An even worse situation would develop if these spaces were so surrounded that oil could not get in. Gas conduction would occur at quite a low voltage, and failure in service would be very probable.

For this reason a barrier and oil duct arrangement is usually employed with the size of the individual ducts kept to a minimum to give the benefit of higher breakdown stresses in the barrier.

5.5.2 Surge voltage

The surge voltage is an attempt to simulate disturbance on a transmission line due to lightning. Any transformer which is actually struck by lightning will suffer damage. The lightning will strike the bushings which will be broken, and fire may result from oil leaks. For this reason protection against lightning strokes is provided. However, a lightning stroke on or near a transmission line will give rise to a voltage wave along the line, and this will eventually reach the transformer terminals. The value of this voltage is limited by the insulation to earth of the line, and there are rod gaps or lightning arresters located near the transformer terminals. These must not conduct at a voltage produced by normal system disturbances such as switching but must conduct at a voltage somewhat below the test voltage of the transformer. The practical design of such protective devices therefore settles the surge level required of the transformer. The levels associated with the various system voltages and the waveshapes are given in chapter 7.

For the purpose of calculating the initial voltage distribution in a winding a surge voltage in the shape of a Heaviside unit step is assumed, that is a wave which abruptly rises at zero time from zero to its full value and remains at that level to infinity.

Any transformer winding forms a complex network of resistance, inductance and capacitance. For calculating the surge-voltage distribution the resistance can be ignored, and at the start of the wave, when very high frequencies are predominant, the inductive elements become effectively infinite impedances. The whole structure therefore reduces to a capacitive network. A very simplified form of such a network is shown in figure 5.6.

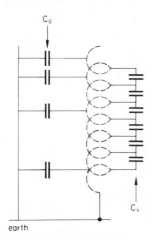

Figure 5.6 Simplified capacitance network representing transformer winding at high frequency: C_g, total capacitance to earth; C_s, total series capacitance

It can be shown that the voltage V_x at any distance x from the neutral end at an applied step voltage V is given by the expression

$$V_x = V \frac{(C_g C_s)^{1/2}\cosh(\alpha x) + C_0 \sinh(\alpha x)}{(C_g C_s)^{1/2}\cosh(\alpha) + C_0 \sinh(\alpha)} \tag{5.1}$$

where

$$\alpha = (C_g/C_s)^{1/2} \tag{5.2}$$

with C_g is the total earth capacitance, C_s the total series capacitance and C_0 the capacitance to earth at the end remote from the input.

For windings with the untested end directly earthed this expression simplifies to

$$V_x = V \frac{\sinh(\alpha x)}{\sinh(\alpha)} \qquad (5.3)$$

For windings with the untested end floating it becomes

$$V_x = V \frac{\cosh(\alpha x)}{\cosh(\alpha)} \qquad (5.4)$$

When a winding is made up of a number of parts with different values of series and shunt capacitance, the calculation must be done for each part. The value of C_0 for each part is taken as the effective capacitance to earth of the whole of the winding between that point and the untested end.

The smaller the value of α the more uniform becomes the initial distribution; α can be reduced either by reducing C_g or by increasing C_s. The final voltage distribution is controlled by the lower-frequency components of the applied wave and is therefore proportional to the turns. When a winding has one end held at earth potential while the other is tested, the final voltage distribution is uniform from the applied voltage to zero.

Between the initial and final distribution will occur an oscillation which may result in voltages in excess of the applied voltage appearing on parts of the winding (see figure 5.7). The result so obtained is pessimistic for the following reasons.

(1) The voltage rise does not take place in zero time.
(2) While the voltage is oscillating the applied voltage is reducing.
(3) There is some attenuation of voltage within the winding itself.

Clearly the desire of the designer is to obtain an initial distribution as near uniform as possible to reduce the oscillation. Also the more uniform the initial distribution the less is the voltage between parts, particularly at the ends of the winding. When it has proved necessary to use a high-voltage winding divided into two coils, the discontinuity can result in high voltages to earth appearing at the junction. This must be calculated as a winding with two parts with different capacitance networks, and allowance must be made for the oscillation about its final value.

It is now desirable to consider how the various types of winding stand up in terms of the surge requirement. The helical coil is poor because its capacitance network results in high stress, but this is not important since it is not used on high-voltage windings. At the voltage level where it is employed, mechanical and cooling considerations control the insulation arrangements so that such coils comfortably withstand the non-uniform surge voltages. When more then one layer is employed, insulation capable of withstanding the full surge level must be provided between layers.

A multi-layer high-voltage winding is illustrated in figure 5.3. The screens

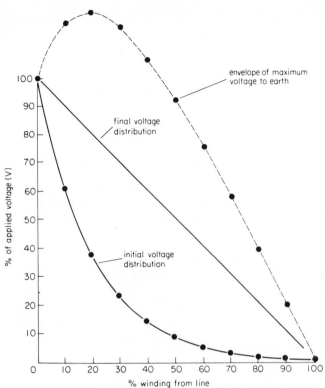

Figure 5.7   Initial surge-voltage distribution and oscillation envelope with $\alpha = 5$

provided inside and outside tend to produce equipotential lines parallel to themselves. They also cancel the interturn capacitance of the coil so that only the interlayer capacitances have to be taken into account. The series capacitance is therefore high, and, since only the layer ends are exposed to earth, the earth capacitance is low. Both factors lead to a uniform voltage distribution although at layer ends quite high interturn voltages appear with practical arrangements. In its response to surge voltages this type of winding can be said to be very good. Since the voltage to earth is graded nearly uniformly from line to earth end, the insulation to earth can be graded in proportion. This is achieved by making each layer shorter than the one inside it. The additional space at the ends is occupied by insulation barriers formed by flanging over the interlayer paper wraps. Ducts are left to allow entry and egress of oil.

The conventional type of disk winding is illustrated in figure 5.8. It tends to have a relatively low series capacitance and high capacitance to earth, but, as the size of transformer increases, the change in dimensions makes the series capacitance increase faster than the earth capacitance. Thus for a given voltage the larger the unit the better is its surge response. The total series capacitance is made up not only of the capacitance between sections but also of that between

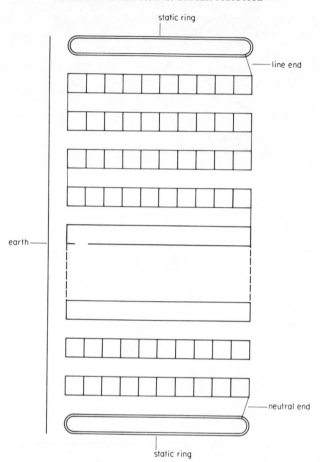

Figure 5.8   Section of conventional disk winding

conductors. For large units where one conductor is not sufficient to carry the current it is common to arrange the winding with the line end at the centre of the coil, the two halves being connected in parallel. This reduces the clearance necessary at the end of graded windings.

Where more than one conductor is required in each half the use of ordinary paper-covered strip results in a reduction in the series capacitance since each strip had to be fully insulated. This difficulty is overcome by the use of continuously transposed rectangular stranded conductor which makes a considerable improvement in the voltage distribution. Undoubtedly, however, the surge-voltage distribution in this type of winding is far from uniform, and it is common practice to have greater insulation at the line end than elsewhere in the winding. This can be done either by increasing the conductor insulation locally or by inserting extra pressboard insulation in the form of washers. This may also be required at the neutral end where there is a doubling effect due to reflection.

The static rings shown at the top and bottom of the coil in figure 5.8 increase the capacitance from the end to the inside of the first section, thus greatly reducing the stress across the section. Additional screens may be used to improve the voltage distribution near the line end. These may be similar to the line end static rings but fitted further down the winding, or they may be located outside the sections near the line end. These screens are connected to points nearer the line end than their physical position, thus increasing the series capacitance in that area. This results in a high voltage appearing between screens and adjacent sections, and care must be taken to provide adequate insulation at these points. In spite of its inherently poor voltage distribution this winding has been proved satisfactory up to the highest voltages so far manufactured.

A modification to the disk winding is the interleaved disk illustrated in figure 5.9. Two conductors starting from 1 and 9, respectively, in figure 5.9 are wound in parallel and are connected as shown in the illustration. Capacitance currents flow in opposite directions in adjacent conductors, almost eliminating inductive effects so that the apparent effect is to increase considerably the series capacitance. This arrangement undoubtedly improves the initial surge-voltage distribution considerably at the expense of a much higher voltage between adjacent turns in normal service. This form of winding does not require any special electric screens[3].

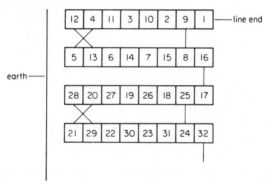

Figure 5.9   Section of part of interleaved disk winding

An extension of this winding has graded interleaving in which there is full interleaving at the line end which decreases in steps further down the winding till the main part is a convential disk.

5.6   LOSSES

5.6.1 I^2R losses

All the loss associated with the winding is classed together and called the load loss or copper loss, but in fact some of this, whilst caused by the current in the winding, does not appear in the winding itself. The I^2R loss is very easily calculated (R is the dc resistance of the winding and I the current in it). In a well-designed coil this should form the major part of the copper loss, but other losses are quite significant and must be calculated.

5.6.2   Eddy current loss in the conductor

Where alternating current passes, an alternating magnetic field is created, and, wherever flux traverses a conductor of any sort, eddy currents exist with resultant losses. Clearly the conductors of the windings must therefore carry eddy currents as well as the load current and will consequently have a loss produced by these currents.

The specific eddy current loss in thin sheets in a sinusoidal field at the maximum B_m of flux density, frequency f and sheet thickness δ with resistivity ρ and mass density ρ_d is

$$p_i = \frac{\pi^2}{6\rho\rho_d} \times B_m^2 f^2 \delta^2 \quad \text{W kg}^{-1}$$

$$= 9 B_m^2 f^2 \delta^2 \times 10^3 \quad \text{W kg}^{-1} \tag{5.5}$$

for copper at the reference temperature 75 °C. The magnetic flux density vector is assumed to be perpendicular to the sheet thickness, that is parallel to the sheet surfaces.

This formula can be applied to the axial and radial leakage flux through the conductors. It is an approximation since it takes no account of the effect of the eddy current on the flux. Also the calculation of the leakage field is inaccurate. The use of computers has greatly improved the accuracy of this calculation.

As the copper increases in section, the eddy current loss, which is proportional to the square of the conductor thickness, increases quickly until it forms a substantial part of the total load loss. To reduce the dimensions it is necessary to wind more than one conductor in parallel. In order to prevent circulating currents between parallel conductors it is essential that the conductors be transposed so that voltages induced by leakage flux cancel out. This can be achieved by arranging for each conductor to occupy every possible position for equal distances in each layer. However, for certain numbers of conductors effective arrangements involving fewer changes in the conductor positions can be devised. For example it is possible to make a perfect transposition for eight conductors with only three changes to their positions relative to each other.

For large transformers it is common to use the rectangular strand transposed conductor illustrated in figure 5.1. Since the individual strips are small and the transposition good, the eddy current losses can be kept down to about 20 % even in very large transformers.

5.6.3   Losses in clamps

The leakage flux is not confined to the windings and may traverse the metal of clamps or tanks and so cause further loss. This loss is increased by the fact that the most common material for the structural parts of the transformer is mild steel. Its magnetic properties result in a concentration of flux in it, and this results in high loss and overheating.

The flux density in any part outside the coils depends on the distance from the coil and on the magnetic properties of the material. The loss depends on the flux, on the dimensions of the clamp and on the resistivity of the material used.

Low clamp losses on large high-reactance transformers have been achieved by making clamps of insulating material, but this is not practical throughout the complete range of transformer sizes. Another fairly expensive method is the use of non-magnetic steel clamps. These costly expedients are only justified in very large units.

It is impossible to detail methods of calculating this loss, and an empirical formula is determined. Over a number of transformers ranging from 18 to 30 MVA it has been found that a reasonable approximation to the clamp loss is

$$p_{ic} = 55 B_c^5 \times 10^6 / l^{1.5} \quad \text{W m}^{-3} \tag{5.6}$$

where B_c is the flux density at the root of the clamp in teslas and l the distance of nearest horizontal part of flange from the end of winding in metres.

5.6.4 Tank losses

The tank also can be traversed by leakage flux with resulting loss. The use of the non-magnetic steels is very expensive, as also is the use of aluminium, though the latter may be used where reduction of mass is very important. Screens of either magnetic or conductive material will reduce the tank losses.

The flux density B_t in teslas in a tank with walls 12.5 mm thick can be calculated from the expression

$$B_t = 28 \frac{NI}{hb_t} \sum_{j=1}^{n} \left(\frac{b_j^2}{12} + b_{0j} \right) k_j \times 10^{-6} \tag{5.7}$$

where NI is the total ampere–turns in winding, h the height of winding in metres, b_t the distance from core to tank in metres, b_{0j} the distance from core to centre of winding j in metres, b_j the radial depth of winding j in metres and k_j the fraction of total ampere–turns in winding j, if we take into account the sign.

It is then possible to read off the tank wall loss from the curve in figure 5.10. If screens of thin magnetic sheet such as used for transformer cores are fitted, the reduction in tank flux density can be read off from figure 5.11. The new wall loss can then be found from figure 5.10[4].

It must be remembered that local high flux concentrations may cause excessive local heating although the loss is small compared with the total load loss. Local heating could damage the oil and so reduce transformer life. Care must therefore be taken to design the tank so that this does not occur.

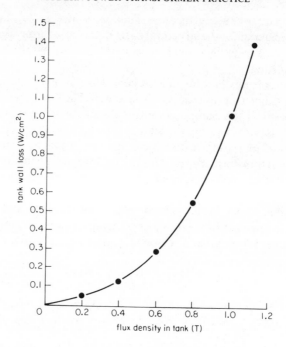

Figure 5.10   Loss in unscreened mild-steel tank wall, 12.5 mm thick

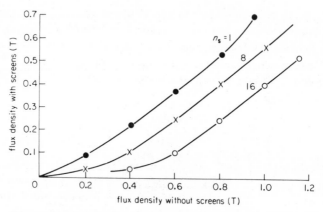

Figure 5.11   Effect of magnetic screen on tank flux density with n_s the number of plates multiplied by screen width divided by tank perimeter

5.7 COOLING

The windings are cooled by the circulation of oil over their surfaces. The oil temperature is kept to an acceptable level by coolers of various sorts outside the tank. The oil is circulated either by thermosyphon action or by pumps. Its temperature increases as it passes over the winding surface. At points it may be hotter than the oil at the top of the tank, since here it may have been mixed with oil from paths producing less heat.

The total winding temperature at any part is made up of the temperature of the oil entering the transformer, increased by the heat lost from the surface over which it has passed, plus the temperature drop through the insulation and at the interfaces between copper, paper and oil. In addition, any parts which are not directly exposed to oil will have an extra temperature sufficient to cause the heat to flow to cooling surfaces.

Complete and accurate calculation of winding temperature rise is very complex, and a number of papers have been written on the subject. It is possible, however, to make some simplification which is justified by the conditions under which transformers operate. For example, although the physical properties of the oil vary with temperature, it is sufficiently accurate in many cases to assume that these properties are constant at values corresponding[5] to 60 °C.

Hot oil naturally tends to rise so that vertical ducts provide a natural circulation path. Many of the windings described, however, depend on horizontal ducts for their cooling. This would be a disadvantage if no steps were taken to improve the situation. Figure 5.12(a) shows a disk-type coil with no means of oil direction. With equal ducts on inside and outside there is practically no tendency for oil to flow along the horizontal ducts. Unequal vertical ducts would tend to

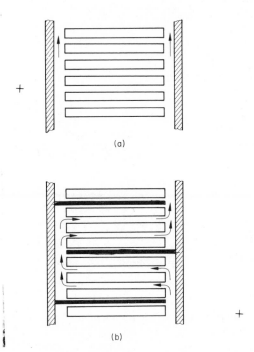

Figure 5.12   Oil flow in disk coil; (a) without oil direction; (b) with oil direction

cause some horizontal flow, but it would be fairly small. Figure 5.12(b) shows a similar coil fitted with washers to force the oil to follow a zig-zag path because its natural vertical path is blocked alternately inside and outside. Thus the oil flows over all the cooling surfaces, giving greatly improved cooling.

Generally speaking two methods of controlling temperature rise can be adopted. The designer can set limits to the heat dissipation per unit surface area for different coil designs, varying insulation thickness and different methods of oil circulation. For example a limit of 0.125 W cm^{-2} might be set for disk coils with natural non-directed oil circulation. With directed oil flow this could increase to 0.19 W cm^{-2}.

For many years this was the accepted method of calculation, but it does not allow sufficiently for all the variations which can be present in winding construction. The accuracy is, therefore, not very good, and the margins left between calculated temperatures and those guaranteed have to be considerable. The second approach is to carry out the complicated calculations necessary to allow for many variables. This is too cumbersome an operation to carry out by using only a slide rule, and so computers are employed.

When oil circulation is known, the following method of calculation gives reasonably accurate results for a disk winding.

(1)   Obtain from figure 5.13(a) and (b) the values K_λ in watts per square centimetre per kelvin of the heat transfer coefficients for vertical and horizontal surfaces.

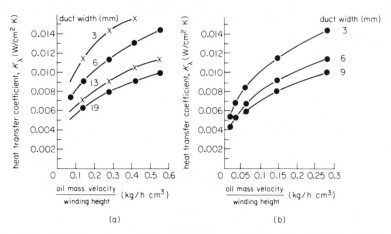

Figure 5.13   Heat transfer coefficients: (a) for vertical surfaces; (b) for horizontal surfaces

(2)   With a value of K_λ determine the associated value K'_λ of the effective heat transfer coefficient from the expression

$$1/K'_\lambda = 1/K_\lambda + \delta \times 10^6/15 \qquad (5.8)$$

where δ in metres is the value of the thickness of paper insulation whose thermal conductivity is included in the factor $10^6/15$.

(3)   The average temperature rise of a disk section over the local oil is

$$\Delta\theta_{wo} = \frac{M_e \times 10^{-4}}{K'_{\lambda h} + \frac{1}{2}(K'_{\lambda i} + K'_{\lambda o})h_W/b_W} \qquad (5.9)$$

where $K'_{\lambda h}$ is the effective heat transfer coefficient for horizontal surfaces, $K'_{\lambda i}$ the effective heat transfer coefficient for inner vertical surfaces, $K'_{\lambda o}$ the effective heat transfer coefficient for outer vertical surfaces, h_W the axial length in metres of covered disk section, b_W the radial width in metres of section winding and M_e the rate of heat transfer in watts per square metre from unit surface area of winding section considered in contact with oil.

(4)   Find the average oil temperature in the winding. It may not be the same as the average temperature of the bulk oil and should be calculated by adding to the inlet oil temperature a rise given by

$$\Delta\theta_o = 0.2P_g/hv_m b_d \qquad (5.10)$$

where P_g is the power in watts dissipated in the coil stack, h the vertical height in metres of coil stack, v_m the oil mass velocity in kilograms per hour per cubic metre in outside vertical duct and b_d the width in metres of outside vertical duct.

(5)   Add the results of (3) and (4) to get the average winding temperature rise.

Although this calculation includes a number of simplifying approximations, it is still sufficiently complicated to justify the use of a computer.

A common method of tackling the calculation of winding temperature is to calculate first of all the average watts per unit area to be dissipated and, if this is within an acceptable value, to carry out the more detailed calculation before finally deciding on the design.

5.8   MEASUREMENT OF WINDING TEMPERATURE

In most cases it is fairly simple to calculate the average temperature of a winding during tests in the factory. This is carried out by the change in resistance method (see chapter 7).

In service, however, it is quite impracticable to use this method. It is also impossible to fix thermocouples in the coils because of the voltage to earth. Instruments have therefore been devised to indicate winding temperature by using the thermal image principle. The thermal image consists of a type of thermometer immersed in the oil at the top of the tank connected to a dial-type instrument. Into this system is injected heat derived from the transformer load current through a current transformer. The temperature indicated by the instrument is the sum of the top oil temperature θ_o and $k_1 \times \Delta\theta_e$, where $\Delta\theta_e$ is the temperature difference across the insulation of the coil and k_1 is a factor that allows for local hot spots due to the fact that the spacers keep oil away from part of the surface. In Britain it is usual to take $k_1 = 1.1$.

A number of different designs have been employed[6] to indicate winding temperature. For example, in one design the temperature difference $\Delta\theta_e$ is simulated by heat from a coil wound round the thermometer bulb. In another design the extra heat is supplied directly to the instrument by placing the heating coil over a bellows which expands when heated; the movement is added to the pointer movement due to expansion of the liquid in the main bulb. A different method of converting the heat into movement uses an element, with a positive temperature coefficient of resistance, inside a tube of insulating material in which is embedded the heater coil; the resistance of the element is measured with a bridge arrangement, and a suitable detector is calibrated in temperature.

In all cases adjustment can be provided by connecting a variable resistance in parallel with the heater coil, thus varying the current passing through it. These instruments are usually provided with switches which can switch on pumps or fans or can operate alarm or trip circuits if the temperature becomes too high. With the increasing desire to obtain as much as possible from installed equipment the importance of the winding temperature indicator has greatly increased in recent years.

5.9   FORCES

5.9.1   General

Any current-carrying conductor located in and perpendicular to a magnetic field is subject to a force calculable from the formula

$$F = BIl \tag{5.11}$$

with F in newtons, B in teslas and l in metres. In this expression the current and flux density are taken at the instant under consideration. With ac, peak values of B and I give maximum force in any cycle. Since the flux density is proportional to the current in the windings, the force is proportional to the square of the current. At full load these forces are small, but under short-circuit conditions very large forces can occur to clamping structures, insulation and the conductor itself. Not only must the peak value of the short-circuit current be used in the calculation, but it must be assumed that the instant of occurrence of the fault is such that there is complete asymmetry, thus doubling the peak value. In fact, the resistance of the windings prevents complete doubling, and it is common practice to assume that the factor will be 1.8 instead of 2.

With a reactance of 10 % the force on short circuit could be 324 times that on full load. Although this occurs only on the half-cycle, with maximum asymmetry it fixes the strength for which coils must be designed.

5.9.2 Radial forces

The interaction of the main leakage flux and the winding current results in forces which tend to move the winding radially either outwards or inwards, depending on the direction of current relative to flux. The innermost winding always tends to move inwards and the outermost winding outwards, but in transformers with a number of concentric windings on each limb the forces could be in either direction and, where tap windings are involved, sometimes in one direction and sometimes in the other.

The radial force in a two-winding transformer is

$$F_r = 2\pi^2 (IN)^2 d \times 10^{-7}/h \tag{5.12}$$

where I is the current, N the number of turns, d the mean diameter of coil and h its axial length, with d and h both in metres.

More generally, this expression can be modified by replacing $(IN)^2$ by $(IN)_1 \times (IN)_2$, where $(IN)_1$ represents the ampere–turns of the winding under consideration and $(IN)_2$ the ampere–turns producing the magnetic field.

While radial forces are present in all transformers, they are generally insignificant in small units, but for large transformers it is vital to check that there is sufficient strength to withstand these forces. It may be necessary to provide special support for the inner winding or to use work-hardened copper or copper alloy to increase the tensile strength of the conductors in either winding.

5.9.3 Axial compressive forces

The leakage flux does not run straight and parallel from one end of the windings to the other. Fringing occurs towards the top and bottom of the coils, resulting in forces which tend to compress the winding axially. This force depends on the coil length, the distance between coils, the distance from coil to core and the distance from coil to tank. It varies in different parts of the same winding because the yoke causes distortion to the flux pattern.

An approximate value of the compressive force on both windings in a balanced two-winding transformer is given by the expression

$$F_c = 2\pi^2 (IN)^2 d_{\text{med}}(\tfrac{1}{3}r_1 + \tfrac{1}{3}r_2 + r_d) \times 10^{-7}/l_W^2 \tag{5.13}$$

where d_{med} is the mean diameter of coils, r_1 and r_2 the coil radii, r_d the duct radial width and l_W the winding length, all in metres.

5.9.4 Axial forces due to unbalance between windings

Additional axial forces arise when the windings of a transformer are not perfectly balanced throughout their length under all conditions. Where there is any unbalance a cross-flux, perpendicular to the main leakage flux, exists; this, in turn, reacts with the winding current to produce forces, upwards in one winding and downwards in the other.

To calculate the force a residual ampere – turn diagram is drawn, as indicated in figure 5.14 for two examples. More complicated winding arrangements result in residual ampere – turn diagrams with more triangles, but the force calculation is similar in principle.

Each triangle in a diagram gives rise to a force

$$F_u = k_2 2\pi k_3 (NI)(NI)' \times 10^{-7} \qquad (5.14)$$

where k_2 is a constant which varies from 5.5 to 6.9, depending on the winding dimensions and arrangement[4], k_3 the fractional out of balance at point considered (in figure 5.14(a), $k_3 = a$ and, in figure 5.14(b), $k_3 = \frac{1}{2}a$), NI the total ampere – turns of winding and $(NI)'$ the ampere – turns in either winding in length covered by part of diagram under consideration.

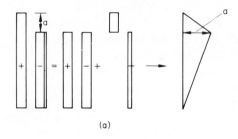

(a)

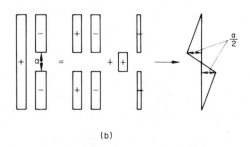

(b)

Figure 5.14   Residual ampere–turn diagrams: (a) for two-winding transformer with straight windings; (b) for two-winding transformer with one split-concentric winding

The method of calculation given in this section is approximate and usually tends to be pessimistic. The computer programmes used to calculate reactance can also be used to calculate forces, since both depend on the leakage pattern. The forces so calculated are reasonably accurate.

As transformer ratings and short circuit currents increase the out of balance which can be tolerated becomes less and less. For this reason the tapping turns of large transformers are arranged in separate coils with each tap section extending over the full winding length. With this arrangement good balance is maintained on all tap positions, and axial forces are reduced to a minimum.

It is impossible to guarantee that no unbalance will occur during manufacture because of the dimensional changes which may occur in cellulose materials. It is normal practice to assume some displacement between coil centres and to calculate forces accordingly. End clamps, winding and insulation must be designed to withstand the forces calculated.

5.10 CLAMP DESIGN

The design of winding clamps varies very widely. They must be strong enough to withstand the short-circuit forces. They must not have excessive losses due to the leakage flux, and they must not increase the possibility of electric failure. There must be a tie from top to bottom of the windings, but it is sometimes possible to use core, core clamps or tank for this purpose. Tie rods outside the windings have been commonly employed, and ties, generally strips of stainless steel, may be located between the inner winding and the core.

5.11 ECONOMICS OF WINDING DESIGN

It is important to remember that many factors other than material cost must be considered when choosing the most economical winding to use. It must pass its test and must function satisfactorily in service, but widely varying designs fulfil this requirement. The facilities available at the manufacturer's works and the experience and skill of the operators must be taken into consideration.

It is quite possible that changes in the price of raw material or advances in knowledge in the design departments could change the relative costs of different types of windings. Many new materials have become available in the last few years. Very often when first produced their price prevents their use, but as production increases price often falls so that it is necessary to examine frequently costs of using new materials which are technically suitable.

ACKNOWLEDGEMENT

The author expresses his thanks to Parsons Peebles Limited, Edinburgh, for permission to publish this chapter and for providing the illustrations used.

REFERENCES

(Reference numbers preceded by the letter G are listed in section 1.14.)

1. Camilli, G., Gas-insulated power transformers, *Proc. Inst. Electr. Eng., Part A,* **107** (1960) 375
2. Goodman, E. A., and Posner, G. C., Trends in gas-insulated transformers, *Insulation* (1965) 39
3. Chadwick, A. T., Ferguson, J. M., Ryder, D. H., and Stearn, G. F., Design of power transformers to withstand surges due to lightning, with special reference to a new type of winding, *Proc. Inst. Electr. Eng.,* **97,** Part 2 (1950) 737
4. Waters, M., *The Short-circuit Strength of Power Transformers,* MacDonald, London (1966)
5. Kerr, H. W., and Palmer, S., Developments in the design of large power transformers, *Proc. Inst. Electr. Eng.,* **111** (1964) 823
6. Brown, W. J., Kerr, H. W., Singer, D. E., and Walshe, L. C., *Accessories and Parts for Transformers, CIGRE Rep.,* No. 101 (1966)

6

On-load Tap-changing Equipment

B. C. Savage*

6.1 INTRODUCTION

On-load tap-changing equipment is required as a necessary accessory to a transformer normally employed in an electric power supply system (see chapters 10 and 12). By changing the tapping on a winding, the equipment provides the facility to vary the turns ratio of the transformer and thus the level of its output voltage.

In generator transformers where the generator voltage of about 20 kV is stepped up to a voltage in the range of hundreds of kilovolts, the tap changer is located at the neutral end of the high-voltage winding. The tapping range covered is usually about 20%, from + 5 to − 15%. Tap changers are used here for a coarse control of the power station output voltage.

Transmission auto-transformers, for example, in Great Britain link the 400, 275 and 132 kV networks. 400 to 132 kV and 275 to 132 kV auto-transformers have fully insulated tap changers which work at a voltage of 132 kV to earth, generally with a 30% tapping range. There are, however, a number of neutral-end tap changers in use, but, in general, there are various advantages for line-end units, dependent on the voltage ratio. Some of the auto-transformers in service to interconnect 400 and 275 kV systems are provided with fully insulated 275 kV tap-change units.

Transformers on distribution sub-stations may have tap changers on the star- or delta-connected high-voltage windings. Generally a tapping range of 20% in 16 steps of 1.25% or in 14 steps of 1.43% is covered.

6.2 BASIC CONDITIONS OF OPERATION

Electrically, two basic conditions have to be met in the operation of an on-load tap changer.

* Formerly Ferranti Limited.

(1) The load current in the transformer winding must not be interrupted during a tap change.

(2) The tap change must be carried out without short-circuiting a tapped section of the winding.

These conditions necessitate the introduction of some form of transition impedance during the transition stage. Such a transition impedance is provided either with a resistor or with an inductor. The amount of impedance and the method of its connection in the circuit are determined by a number of conflicting requirements. The main ones are as follows.

(1) There must be no excessive voltage fluctuations during the switching cycle.

(2) The circulating current between taps in the transition position must not be excessive.

(3) The duration of the arc, when current interruption takes place, must be kept to a minimum so as to minimise contact erosion and to reduce contamination of the oil. Contact erosion is caused when the surface of a contact piece is affected by the arc as the contact breaks to open the circuit.

The tap changer with resistor-type transition impedance has inherently the advantage that the current interrupted and the restriking voltage across the contact are in phase. This means that the arc extinguishes at a current zero and that the restriking voltage does not build up to a maximum for another quarter-cycle. The modern high-speed resistor tap changer operates so fast that with transition resistor the arc persists for only about half a cycle. Hence there is a minimum of contact erosion and minimum oil contamination.

The tap changer with inductor-type transition impedance has the advantage of absence of high-speed switch mechanisms but the disadvantage that the current on interruption and the restriking voltage across the contact at the instant of opening are in phase quadrature. Contact breaking in this condition fosters arc formation which leads to undue contact erosion and thus to contamination of the oil as an arc, on average, lasts for three to four half-cycles. This type of gear is now little used in Europe but still finds some favour in the USA, in spite of its size and cost, probably because high-speed switch mechanisms are not necessary for it.

6.3   HIGH-SPEED RESISTOR TAP CHANGER

In modern high-speed resistor tap changer the current transfer takes place in about 40 to 70 ms, depending on the type of mechanism used for switching.

The high transfer speed is obtained by using a stored-energy system, generally in the form of a bank of springs. The mechanism is such that, once the stored energy is released, the tap change is completed regardless of any external power supply failure. It is, in effect, a positive 'go – no go' mechanism with a very low risk of failure.

There are two types of resistor tap changer in common use; both use different current-transfer switching arrangements. These can be classified as the pennant switching cycle and the flag switching cycle. The terms flag and pennant are derived from the appearance of the phasor diagrams that show the change of output voltage of the transformer in moving from one tapping to the next[G1.14, G2.17].

A tap changer of the pennant switching cycle type uses a selector switch that combines the functions of tap selection and current transfer. Generally, a single resistor is used which gives an asymmetrical switching cycle. In one direction of movement, a circulating current is passing before the through-current is interrupted, whereas in the reverse direction the through-current is broken before a circulating current starts to pass.

A tap changer of the flag-switching-cycle uses a tap selector in conjunction with a separate diverter switch. Two resistors are used to give symmetrical switching. With this sequence, the through-current is broken by the main contact before a circulating current starts to pass.

In general British practice, the tap changer is metal clad and either is bolted onto the transformer or is integral with it. Tap changers for large power transformers generally have separate pockets for tap selector and diverter switch. It is now common practice to fit single-compartment combinations of diverter switches and selectors to 33 and 66 kV transformers.

In general European practice, the tap changer is suspended from the cover of the transformer tank, the whole being immersed in the main transformer oil. The tap section generally comprises a bakelised paper tube which houses the diverter switch, and the oil of this is separate from that in the main transformer. The tap selector is mounted directly below the diverter tube housing and is an open form of cage construction.

6.4 GENERAL DESIGN CONSIDERATIONS FOR A TAP CHANGER

The tap changer of a transformer has to be capable of meeting the same normal and peak rating overload conditions as the transformer itself.

Other design considerations that have to be taken into account include the following.

(1) The maximum system voltage on which the transformer has to work.
(2) Step voltage and the number of steps.
(3) The maximum rms test voltage to earth and across the tapping range.
(4) The maximum surge voltage to earth and across the tapping range.
(5) The maximum power-frequency and surge-test voltages between phases where applicable.
(6) Current rating, both for normal full-load and peak rating conditions.

(7) Type tests necessary to determine the data for a particular tap changer to ascertain whether it is capable of meeting supply authority specifications while conforming to international and national requirements.

6.5 TAPPING WINDING ARRANGEMENTS

6.5.1 General

There are a number of ways in which leads can be taken off the winding to provide the required tapping range. These are generally arranged to suit a particular design or method of construction. Considerable electrical and mechanical problems are involved with tapping lead structures on high-voltage power transformers owing to the large cross-section of conductors and the voltages that appear across the tapping range. The leads should be taken off the winding in such a way as to provide a neat connection either to a barrier panel between the switch and transformer or directly to the switch contact terminals. Three methods, in common use for providing tappings at the neutral end of a high-voltage winding, are shown in figure 6.1 and are explained below.

6.5.2 Coarse/fine-tapping winding arrangement

The arrangement shown in figure 6.1(a) has a coarse-tapping section as an extension of the main winding and is indexed by a change-over selector. The main tapping winding constitutes the fine sections.

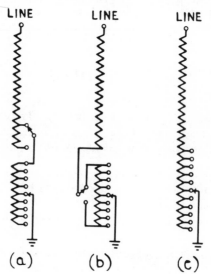

Figure 6.1   Typical tap-changer winding arrangements in common use: (a) coarse/fine tapping; (b) reverse tapping; (c) linear tapping

A typical example would have a coarse section with ten times as many turns as one fine section. The fine-tapping leads would be brought out to two rotary tap selectors, one indexing the odd tap numbers and the other the even numbers. The tap selectors, ganged for sequential operation, would make two revolutions in covering the range: one revolution with the coarse section out of service and the other with it in after the operation of the change-over selector.

Some large 20 to 400 kV generator transformer have a modified version with a linear switch arrangement for four coarse sections, each equivalent to four fine sections. Three separate fine sections are indexed in and out of series connection by means of two change-over selectors. So, starting with all the tapping windings out, 1–2–3 fine sections are added in sequence, and the first coarse section is then indexed. The fine sections are again added in sequence, and a further coarse section then brought in, and so on until the full range is spanned.

With the coarse/fine arrangement, high surge voltage can occur on the tap changer at transition.

6.5.3 Reverse-tapping winding arrangement

Figure 6.1(b) shows a typical arrangement. A separate tapping winding is arranged in conjunction with a change-over selector so that it will either buck or boost the main winding, that is the tap sections are subtracted from, or added to, the main winding polarity to vary the voltage ratio. A typical arrangement would be a tapping winding of ten sections brought out to two rotary tap selectors for odd and even tap numbers. One revolution of the double-tap-selector mechanism would be followed by an operation of the change-over reversing selector and a further tap-selector movement.

A separate tapping winding of this type generally means an additional amount of copper in the transformer with consequent high copper loss on the minimum tapping.

6.5.4 Linear-tapping winding arrangement

A typical linear-tapping winding is shown in figure 6.1(c). All the tappings are brought out in potentiometer fashion from the winding. Selection is by means of two separate selectors, one controlling the odd tap numbers and the other the even numbers. The mechanical arrangement of these tap selectors can be either rotary or in line.

This is possibly the simplest arrangement and the one in most common use, as it also means a simple mechanical drive system for the tap changer as no change-over selectors are involved.

6.5.5 Discussion of tapping winding arrangements

The three methods of tap-winding arrangement, referred to above, have their advantages according to the transformer designer's approach to the question of bringing out the tapping leads, and in respect of the surge voltages which can develop particularly across the tapping range. With the linear arrangement there is a much simpler mechanical design of tap changer as no change-over selector is required.

As a general rule, linear arrangements are used on star or delta windings of 33 and 66 kV system transformers. This method is also employed to a large extent on 132 and 275 kV two-winding transformers with the tap changer located at the earthed neutral end of the high-voltage winding. There are occasions when there are more than 16 steps, where it is more economical to use a reversing arrangement. On higher ratings, involving transformers of 400 kV and above, the tapping leads are of such dimensions that it is sometimes necessary to resort to coarse/fine or reversing arrangements with the lesser number of leads.

On 275 to 132 kV and 400 to 132 kV transmission auto-transformers, the choice of winding arrangement and the positioning of the tap section at the 132 kV point or the neutral end is influenced by such aspects as the following.

(a) Whether the voltage to be controlled is the high voltage or medium voltage.
(b) The ratio of transformation.
(c) Consideration of the use of the tertiary winding, that is possible use for power factor improvement.

It is important to realise that problems can arise when using tap changers which incorporate a change-over selector; this is especially so on transformers above 110 kV. When the change-over selector operates, the tapping winding is momentarily disconnected. Recovery voltages appear as the contacts separate, and the tapping winding assumes a voltage determined by the winding capacitance. In general, if the recovery voltages exceed that across the tapping winding, difficulties with regard to electric stress and formation of gases can arise, and special precautions must be taken to limit these recovery voltages. Limitation of recovery voltage is achieved by the following means.

(a) Capacitive control between the main winding and the tapping winding.
(b) A tie-in control resistor to locate the common of the change-over selector to a definite voltage. The value and rating of these resistors depend on the transformer design and great care must be taken when selecting them.
(c) A double change-over selector arrangement which prevents the tap winding from becoming isolated.

6.6 RESISTOR SWITCHING SEQUENCE

For the purpose of description it is necessary to consider high-speed resistor tap changers of two distinct types, both of which are operated from some form of stored energy accumulator: (1) the conventional flag-cycle double-resistor tap changer and (2) the pennant-cycle single-resistor tap changer.

6.6.1   Double-resistor (flag-cycle) tap changer

The switching sequence can be carried out by a wide variety of electrical and mechanical methods, but the basic principle remains the same. For ease of description a tap changer, fitted to the neutral end of a transformer high-voltage winding, will be considered in which all the tappings are brought out in a linear potentiometer method, such as shown in figure 6.2.

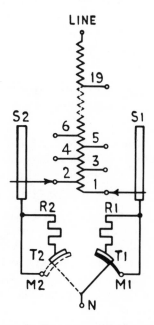

Figure 6.2   Schematic diagram for linear-type double-resistor tap changer; S1, S2, tap selectors; 1 to 19, tap selector main contacts; R1, R2, transition resistors; M1, M2, main switching contacts of diverter switch; T1, T2, transition contacts of diverter switch

The tappings, in this case, are brought out to two parallel rows of fixed tap-selector contacts, the odd numbers on one side and the even on the other. The two sliding tap-selector contacts S1 and S2 are indexed by geneva wheels and are driven by multi-start lead screws. Current is transferred via collector busbars through bushings to the sides of the diverter switch. The moving tap-selector contacts S1 and S2 are shown on taps 1 and 2, respectively. The diverter switch is shown indexed in the running position for tap 1.

The stored-energy mechanism operates, and the moving contact system of the diverter switch commences its travel as follows.

(1)   The main switching contact M1 opens, and the load current from the winding flows through the moving contact S1, the resistor R1 and the transition contact T1 to the neutral.

(2) The moving diverter contact then makes on the opposite transition contact T2. Both resistors R1 and R2 are now connected in series across the tap section with the mid-point forming the neutral. The load current is split between the two resistors, and a circulating current, limited by the ohmic value of the resistors, passes round the loop. In one resistor the phasors of half the load current and of the circulating current will be subtractive and in the other additive.

(3) As the travel of the moving contact continues, it breaks with the transition contact T1; the load current then passes through the tap-selector contact S2, resistor R2 and transition contact T2.

(4) Finally, the moving contact closes on the main arcing contact M2, and the resistor R2 is shorted out. This is the running position for tap 2.

The whole sequence from tap 1 to tap 2 involves no movement of the tap selectors. If we carry out a further tap change in the same direction from position 2 to 3, this involves a tap-selector movement as follows.

(1) The tap selector S1 moves from position 1 to 3, and, as soon as 3 has been selected, the diverter switch operates to connect it to the neutral.

(2) Continuing in sequence, the tap selector S2 moves up from postion 2 to 4, and the diverter switch operates to index tap 4 to the neutral.

Whenever a tap change is made in the same direction, the tap selector moves first, followed by a change-over of the diverter switch.

On reversal of direction, all that is necessary is for the diverter switch to transfer to the other side, for example from tap 4 to 3. During this reversal tap change, the tap-selector drive mechanism is held stationary. This is achieved by a 180° mechanical 'lost motion' device incorporated in the drive to the tap selectors. The sequence of tap selection followed by diverter switch transfer will then be picked up. The mechanical lost motion device only operates on a reversal of direction.

A tap changer of this type completes a tap change in an overall time of 3 to 8 s. The transfer of the high-speed diverter switch takes place in 45 to 70 ms dependent upon type.

The flag-cycle switching provides a universal type of tap changer for transformers in which both forward and reverse power flow directions could arise. The contact burden is equal. Figure 6.3 shows a typical selector and figure 6.4 a typical diverter switch.

6.6.2 Single-resistor (pennant-cycle) tap changer

The single-transition resistor used in this type of tap changer is carried on the moving switch arm. Very-high-speed mechanisms are employed to give rapid arc clearance.

A typical switching sequence diagram is shown in figure 6.5. The fixed tapping contacts 1 to 17 are spaced round the periphery of a circle and are indexed by a

Figure 6.3   Typical selector

Figure 6.4   Typical diverter switch

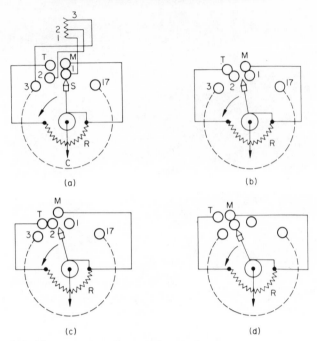

Figure 6.5   Switching sequence diagram for single-resistor tap changer: (a) running position number 1; (b) transition position, R bridging tap section 1–2 and carrying circulating current; (c) transition position, R in series with tap 2 carrying load current; (d) running position number 2: 1 to 17, fixed selector contacts; S, main moving selector contact; M, main switching contact; T, transition contact; C, switch common to neutral or continuation of winding

single rotary contact S. Its insulated support arm also carries the transition resistor R and two main switching contacts. The main switching contact M is bonded to the main current-carrying contact. The transition contact T is in series with the resistor R.

The switching sequence from position 1 to 2 is as follows.

(1)   The stored-energy mechanism is released, and the moving contact assembly begins to move. The main selector contact S opens on the fixed selector contact 1, the circuit being maintained by the main switching contact of 1 and the main switching contact M. The transition contact then makes with the main switching contact of 2, the resistor bridging the tap section and carrying the circulating current.

(2)   The main switching contact M now breaks with the contact of 1, leaving the resistor momentarily connected in series with the transformer winding and carrying the load current. The contact M then makes with the main switching contact of 2, while the transition contact T rolls round it before breaking on the final stages. The transition resistor is now shorted out, and the circuit is maintained through the contact M and the switching contact of 2.

(3) Finally, the moving contact assembly completes its travel, and the contact S indexes onto position 2.

When the direction of rotation of the contact assembly is changed, the sequence of events is reversed.

The resistor **R** in the contact assembly figure 6.5 is so connected with respect to the main winding that the phasors of circulating and load currents are subtractive during the most onerous part of the switching cycle, that is when the current is broken. Therefore, the tap changer is inherently unidirectional in respect of power flow, and it must be derated when used for power flow in the reverse direction. Figure 6.6 shows a 300 A single-resistor tap changer.

Figure 6.6   300 A single-resistor tap changer

6.7 INDUCTOR SWITCHING SEQUENCE

A typical schematic diagram for an inductor-transition tap changer suitable for use on a 33 kV star or delta winding is shown in figure 6.7. It is arranged for 15

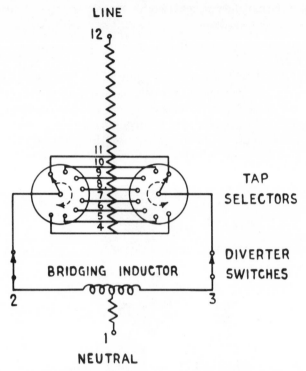

Figure 6.7   Schematic diagram for a rotary-type inductor tap changer

positions, 14 steps. Certain other methods of winding construction may be employed, but the basic principle of operation remains the same.

The tappings on the transformer winding are selected by means of two rotary tap selectors which do not make or break any current. The current is interrupted by two diverter switches, one connected to each end of the mid-point tapped inductor. This inductor is rated for continuous operation. It limits the circulating current in the transition position. Often, as in this case, the bridged position is used as a running position to give a voltage half-way between that of two main tappings, the inductor serving as an auto-transformer.

When a tap change is initiated, one of the diverter switches opens, and the corresponding selector contact is moved round onto the next fixed contact. The diverter switch then recloses, and the tap change is complete. If a further tap change is called for in the same direction, the sequence will be repeated by the other diverter switch and its corresponding tap selector.

The operation of the diverter switches and tap selectors in their correct sequence is ensured by a single actuating cam, which operates a toggle link mechanism for the diverter switches and a geneva indexing wheel to rotate the tap selectors. The use of this geneva wheel train ensures that the tap selectors are only moved one step at a time and are locked in position except during actual rotation of the drive shaft.

6.8 MOTOR DRIVE MECHANISMS

In the majority of instances either three-phase or single-phase induction motors of the capacitor start and run type are used for on-load tap-changer drive units.

It will be readily appreciated that this motor drive mechanism must be capable of operating in both directions of rotation to satisfy the normal functioning conditions of the tap changer.

The essential requirements for the control of these motors are the initiating components such as push buttons or switches and the maintaining contacts in the auxiliary equipment in order to make certain that a full tap-change sequence is completed once initiated; limit switches are provided to prevent overrun.

Further items of auxiliary equipment may have to be included in order to conform to specifications. These items could be any or all of the following.

(a) Step-by-step contacts.
(b) Tap change in progress contacts.
(c) Tap change incomplete contacts.
(d) Relay reset contacts.
(e) Sequence contact.
(f) Tap-position indication contact.
(g) Directional contacts.
(h) Certain contacts for parallel control.

Details of the motor drive mechanism are usually described in the manufacturer's maintenance manual or specific drawing.

It is usual for the motor drive mechanism to include the facility to allow the unit to be operated manually. If this facility is provided, it is essential that the electrical system is isolated automatically either before or at the time that the handle is inserted.

6.9 PROTECTIVE DEVICES

Where separate tap selectors are installed, it is usual to employ gas-actuated (Buchholz) relay-type equipment for the transformer, in order to protect them.

If the diverter switches are to be protected, then the necessary protection should be that of a low-oil-level sensing device and a device which responds to a high surge of oil or gas. Since gas is produced at each tap change, an ordinary type of gas detector element is not satisfactory.

Protection for the single-chamber-type tap changer is usually provided in a similar manner to that for the diverter switch.

It would seem reasonable to mention the following two devices.

(1) Pressure relief device.
(2) Drycol or silica-gel-breather (see figures 10.18 and 10.19) device.

Both of these items can be and are fitted when required and in their own rights are protection devices.

Attention should also be drawn to certain other protection equipment which could be fitted in associated with the tap-changing equipment. These are as follows.

(1)   Over-current equipments. These sense the load current and prevent tap changing if excessive.
(2)   Tap-change incomplete protection devices.
(3)   Parallel control systems out-of-step protection.

6.10   MAINTENANCE ASPECTS

It would be wrong to be pedantic in any way regarding tap-changer maintenance. The best advice that can be offered regarding this is to insist on reference to maintenance manuals that are usually provided by the manufacturer.

If these are not available, then contact should be made with the manufacturer in order to obtain advice.

The only advice which could be offered is that considerable care should be taken when the work is being carried out, with considerable attention being paid to cleanliness. If any form of cleaning clothes are used, it is essential that they be of the lint-free type. Particular attention must be paid to the replacement of transformer oil.

ACKNOWLEDGEMENT

The author is indebted to Ferranti Limited for permission to publish the information and illustrations contained in the text.

REFERENCE

(Reference numbers preceded by the letter G are listed in section 1.14.)
1. Savage, B. C., Testing of on-load tap changers, *Conf. on Diagnostic Testing of High-voltage Power Apparatus in Service, Inst. Electr. Eng. Conf. Publ.*, No. 94, Part 1 (1973) 99

7

Transformer Processing and Testing

H. Jackson* and K. Ripley*

7.1 INTRODUCTION

Testing is a very important stage in the manufacture of any product. Not only is the satisfactory outcome of the tests a guarantee to the customer that the equipment will meet the required specification, but it is also a confirmation to the design office of their calculations as well as providing them with valuable data for future designs.

Since the testing of transformers must be carried out in co-ordination with the works production programme, it is desirable, wherever possible, for a standard set of tests to be carried out. These must be acceptable to both the customer and the manufacturer, and an IEC publication 76[G1.7] is intended to cover this requirement. The tests can be classified as follows.

(a) Routine tests. All transformers are subjected to these.
(b) Type tests. These are carried out on the first unit only of a new design.

Occasionally, the customer may require some non-standard tests which will be arranged when placing the order, or the design office may ask for special tests for their own information.

Whilst no particular order is specified officially, the tests detailed in this chapter have been set out in the order in which they would generally be carried out in practice.

7.2 PRELIMINARY TESTS

Certain tests are carried out in the shops before the transformer is assembled in its

* Ferranti Engineering Limited.

tank, in order to ensure that any faults which may be detected at this stage can be rectified before work is continued.

7.2.1   Core insulation

As soon as the core has been assembled, an alternating voltage test at 2 kV is applied between core bolts, side plates and core, in order to prove the insulation of the magnetic circuit.

7.2.2   Ratio

The tolerance allowed for ratio is ±0.5% of the declared ratio or ±0.1% of the percentage impedance voltage, whichever is the smaller. In order to obtain the accuracy necessary for this measurement it is usual to employ a ratiometer; the basic connection is shown in figure 7.1. R consists of variable decade resistors which are adjusted during the test until a balance is obtained on the vibration galvanometer VG. The ratio of the ratiometer arms is then the same as that of the transformer under test, and this value can be read directly on the ratiometer.

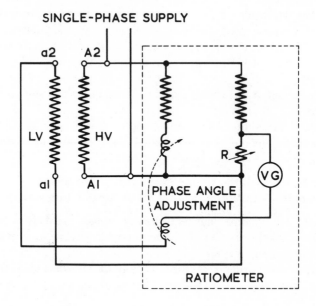

Figure 7.1   Ratio test

7.2.3   Polarity

Polarity tests are carried out to determine the phase relationship between the instantaneous induced voltages in the primary and secondary windings relative to the terminal markings.

Figure 7.2 shows the connections for carrying out this test on a single-phase transformer. It will be seen that, if like-numbered terminals A_1, a_1 are connected together and a voltage V_1 is applied across A_1 and A_2, then for standard polarity (subtractive) the voltage V_2, measured between terminals A_2 and a_2, will be less than V_1.

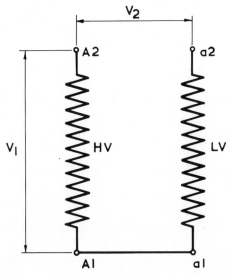

Figure 7.2   Polarity test

For three-phase transformers it is more usual to carry out a phasor reference test in which the high-voltage line terminal of phase A is connected to the low-voltage line terminal of phase a, and a three-phase supply of about 400 V is applied to the higher-voltage windings. Voltages are measured and compared with the corresponding phasor sums as indicated in figure 7.3(a) and (b) which show the conditions for star – star (phasor reference Yy0) and star – delta (phasor reference Yd1). The connections for any other groups can be checked in a similar manner.

7.2.4   Resistance

The resistance of all windings are measured at this stage 'by the voltmeter/ammeter method in order to check that there are no faulty joints or breaks in multi-stranded conductors. The official measurements are carried out during the final tests and are dealt with more fully in a later section.

7.3   PROCESSING

During storage and building the materials used in transformer insulation absorb

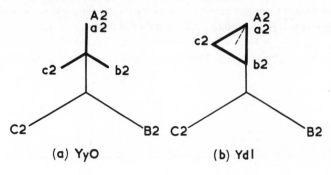

Figure 7.3   Phasor reference test (connect A2 to a2, supply A2–B2–C2, measure B2–b2, B2–C2, C2–c2, C2–b2):

(a) star – star (YyO)   B2–b2 = C2–c2 (i)   B2–C2 = C2–b2 > (i)

(b) star – delta (Ydl)   B2–b2 = C2–c2 = C2–b2 < B2–c2

moisture, the amount depending on the relative humidity of the atmosphere. This moisture, which may be as much as 6 to 8 % by mass for a relative humidity of 50 %, is fairly uniformly distributed throughout the insulation and has to be drawn to the surface and to be removed before the transformer is filled with oil. Table 7.1 gives some idea of the amount absorbed and removed during processing for different sizes of transformers, on the assumption that about 0.5 % is retained.

TABLE 7.1   *Approximate moisture content for various sizes of transformers*

(*MVA*)	(*kV*)	Mass of insulation (*kg*)	Mass of moisture 6% (*kg*)	Dry say 0.5% (*kg*)	Moisture removed (*kg*)
1	11	54.4	3.27	0.27	3.0
5	33	272	16.3	1.3	15
60	132	3180	191	15.9	175
180	275	10000	591	50.0	541
600	400	22700	1363	113	1250

The most common procedure, which may vary in detail, is to heat up the insulation to 85 to 120 °C, to circulate hot dry air round it, then to apply evacuation to complete the drying and to remove the air from the interstices of the paper and finally to soak in transformer oil.

Small transformers, either in their tanks or as core and windings, can be placed in an enclosed chamber heated with hot water or steam-heated coils (which may be around 150 to 160 °C). For large transformers, processed either in their own tanks or in an autoclave, this method is too slow; so hot air of about 110 °C is blown into the tank to circulate around the core, windings and insulation, the maximum heat being extracted from the air and introduced into the transformer. In order to prevent overheating and ageing of the insulation, careful checks must be made of winding temperature which can be determined by comparison of hot and cold resistance values. If the temperature is maintained at 95 °C throughout the processing, a good compromise is obtained between fast drying and avoidance of ageing of the insulation.

The stove heat and hot air blowing is maintained for several days in order to remove the bulk of the moisture, usually 70 to 90 %. Air, which normally has a relative humidity of 40 to 70 %, has a relative humidity of 1 to 2 % when heated to 100 °C and is then capable of absorbing 50 times as much moisture before it reaches saturation. This heated air either is blown round the transformer and out and away or, more economically, is continually circulated round the windings with a bleed-off, usually around 10 %, which is cooled and dried by passing over a chemical drier, is reheated and is then passed back into circulation. The latter method is very economical in heat energy and is particularly suitable for use with a sealed chamber or autoclave since the pipework can be left permanently installed.

Another method of drying out uses the vapour-phase heating system in which a volatile liquid, such as white spirit, is heated and introduced into an autoclave or transformer tank which has been partially evacuated. The resultant vapour condenses on the transformer windings and insulation, and the heat exchange produces a more rapid and more evenly distributed heating than that obtained by conventional methods.

The water vapour is pumped off, and this process continues until the transformer temperature has reached the required level of about 120 °C. The hot white spirit is then removed, and the final drying phase commences during which the internal pressure is reduced to a value of about 13.3 to 26.6 N m^{-2} (0.1 to 0.2 torr).

During this phase the temperature of the transformer drops owing to the latent heat of vaporisation. In order to maintain the temperature, heating coils are fitted to the sides and to the bottom of the autoclave, or, for the transformer tank, hot air is directed onto the outside of the tank.

A further advantage of this method of drying is that any impurities in the insulation are washed out by the fluid during the heating process.

Transformer insulation parameters, such as insulation resistance, power factor and the so-called dispersion value, when measured between the high- and low-voltage windings and between windings and tank, all vary with the degree of moisture content. In the dispersion method a 200 V 300 ms pulse is applied to the insulation followed by a 3 ms discharging pulse, after which it is isolated; if moisture is present, some charge will be retained and may be measured on a voltmeter which is calibrated to indicate moisture content. Monitoring any of the insulation parameters during processing will give a good indication of the degree of dryness. In practice it is usual to measure at least two of these parameters. It should be noted, however, that, since they vary with different transformers, it is not the actual values which are important but the way in which they change during the processing. For this reason results are plotted for comparison. Figure 7.4(a) and (b) show typical graphs of insulation resistance measured by means of an electronic megaohmmeter, and power factor measured by means of a Schering bridge. Drying out is continued under atmospheric conditions until the graphs show signs of levelling off, after which a vacuum is applied, whilst still maintaining the heat. The plotted characteristics then show a change, and the processing is continued until the results become stable again. At this stage the transformer, still under vacuum, is filled with first-grade transformer oil,

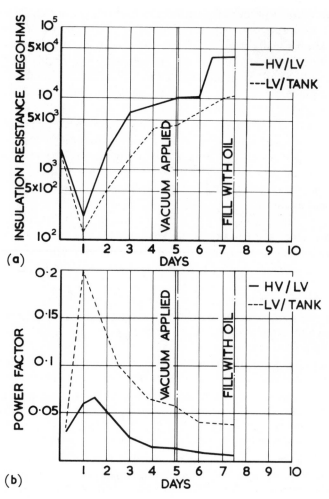

Figure 7.4   Insulation characteristics measured during processing: (a) insulation resistance; (b) power factor

preferably hot around 75 °C, which has been passed through a high-vacuum filtering plant so that it is degassed and dehydrated. The oil must be run into the tank slowly in order to maintain the vacuum already applied.

Before commencement of the final tests, the transformer is left to stand for several days under a pressure of about 35 to 70 kNm^{-2} in order to allow the oil to soak into the solid insulation and to ensure that any remaining air bubbles are absorbed by the oil.

7.4 FINAL TESTS

7.4.1 Introduction

These tests are carried out complete with all external components and fittings which are likely to affect the performance of the transformer and generally comprise the following.

(1)   Ratio and polarity.
(2)   Surge-voltage withstand test.
      (a)   Full-wave test.
      (b)   Chopped-wave test.
(3)   Separate-source-voltage withstand test.
(4)   Induced-over-voltage withstand test and internal discharge test.
(5)   Resistance of windings.
(6)   No-load loss and no-load current.
(7)   Noise test.
(8)   Load loss and impedance measurements.
(9)   Zero-sequence impedance test.
(10) Temperature rise tests.
(11) Insulation resistance.

These are all routine tests except (2)(a) and (10) which are type tests and (2)(b) and (9) which are special tests.

7.4.2 Ratio and polarity

These tests are carried out exactly as described in sub-sections 7.2.2 and 7.2.3 and are repeated at this stage in order to check that all connections to the terminals, tap-change gear, etc., have been made correctly during final assembly.

7.4.3 Surge-voltage withstand test

During service the transformer may be subjected to surges caused by lightning or by a switching operation. The surge tests are designed to prove that its insulation is capable of withstanding these conditions. The equipment to generate the surge voltages uses a voltage multiplier circuit that produces a high-voltage surge with a number of capacitors which are charged in parallel from a dc source variable from 0 to 200 kV and are subsequently discharged in series connection obtained via appropriate high-voltage spark gaps*.

Two types of test are carried out: full-wave tests and chopped-wave test.

Full-wave tests

A surge from the high-voltage surge generator simulates a travelling wave caused by a lightning discharge some distance from the transformer. For test purposes, an IEC Publication[G1.5] specifies a waveshape of 1.2/50 μs, that is the voltage wave has a front time of 1.2 μs and a time to half-value of 50 μs. A wave of this type is shown in figure 7.5.

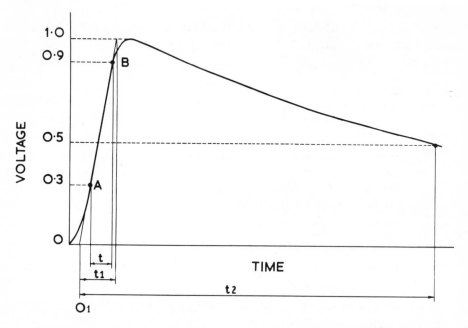

Figure 7.5   Standard 1.2/50 μs full-wave surge voltage: $t_1 = 1.2$ μs; $t_2 = 50$ μs

In figure 7.5, O_1 is the start of the wave and is a point on the time axis which is intersected by a straight line drawn through the points A at $0.3V$ and B at $0.9V$, where V is the peak value of the wave. The front time is $t_1 = 1.67 \times t = 1.2$ μs and $t_2 = 50$ μs is the time to half-value. A tolerance of $\pm 30\%$ is allowed for t_1 and $\pm 20\%$ for t_2.

The peak voltage applied for this test depends upon the system highest voltage and also on the insulation level. Values specified by the IEC for standard 1 and

---

* For a modern type of high-voltage spark gap and its application in high-voltage surge generators, see reference 1.

standard 2 insulation levels are given in tables 7.2(a) and (b) (table 7.2(b) is based on current practice in North America; see references G1.7 and G2.2).

TABLE 7.2(a)   *Insulation levels for windings and connected parts designed for surge-voltage tests series I (based on current practice other than in North America)*

System highest voltage (kV (rms))	Insulation level					
	Surge test voltage			Power-frequency test voltage		
	Standard 1	(kV (peak))	Standard 2	Standard 1	(kV (rms))	Standard 2
3.6		45			16	
7.2		60			22	
12		75			28	
17.5		95			38	
24		125			50	
36		170			70	
52		250			95	
72.5		325			140	
100	450		380	185		150
123	550		450	230		185
145	650		550	275		230
170	750		650	325		275
245	1050		900	460		395
300	—		1050	—		460
420	—		1425	—		630

TABLE 7.2 (b)   *Series II (based on current practice in North America)*

System highest voltage (kV (rms))	Insulation level		
	Surge test voltage		Power-frequency test voltage (kV (rms))
	500 kVA and below	(kV (peak))	Above 500 kVA	
2.75	45		60	15
5.5	60		75	19
9.52	75		95	26
15.5	95		110	34
25.8		150		50
38.0		200		70
48.3		250		95
72.5		350		140

Above 72.5 kV the values of series I are applicable.

Chopped-wave test (special test)

This test simulates the condition when a flashover occurs at an insulator. The chop is obtained by adjusting the setting of a rod gap or a controlled chopping gap so that the flashover occurs on the tail of the wave between 2 and 6 μs from the nominal start of the wave.

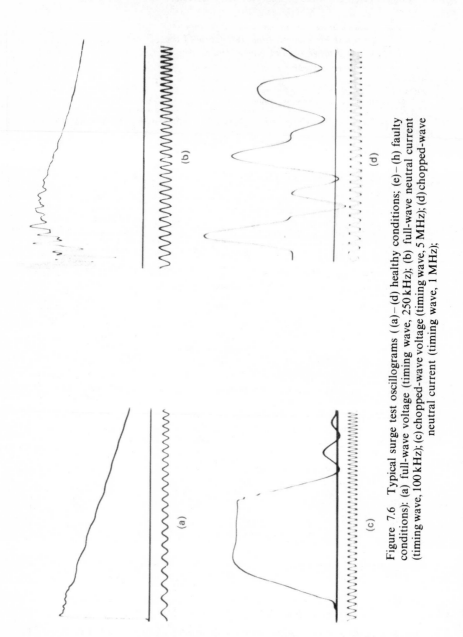

Figure 7.6 Typical surge test oscillograms ((a)–(d) healthy conditions; (e)–(h) faulty conditions): (a) full-wave voltage (timing wave, 250 kHz); (b) full-wave neutral current (timing wave, 100 kHz); (c) chopped-wave voltage (timing wave, 5 MHz); (d) chopped-wave neutral current (timing wave, 1 MHz);

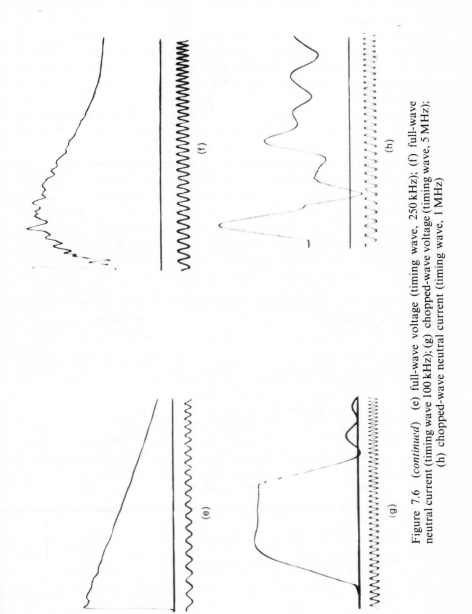

Figure 7.6 (*continued*) (e) full-wave voltage (timing wave, 250 kHz); (f) full-wave neutral current (timing wave 100 kHz); (g) chopped-wave voltage (timing wave, 5 MHz); (h) chopped-wave neutral current (timing wave, 1 MHz)

Surge testing differs from ac power testing in a number of ways. Although the intensity of current is several kiloamperes, the duration of current is so short that the thermal rating of connections need not be considered. Small-diameter flexible connections will suffice.

The connections must be as short and straight as possible in order to keep the inductance to a minimum. Care is necessary to ensure that all connections are well made, since a loose connection would result in sparking and would produce false results on the recording apparatus. Care should also be taken with earth connections which should have a large width-to-thickness ratio in order to keep the surge impedance to a minimum. They should be connected to a special earth grid which usually consists of copper strips installed under the test floor and connected to earthing rods at intervals.

Precautions must be taken to ensure that surges transferred to other windings not under test do not cause any damage to these windings. This aim is achieved by earthing their terminals either directly or through non-inductance resistors in order to limit the transferred voltage to not more than 75% of the test value.

Before the official test, it is necessary to carry out preliminary tests for calibration purposes at voltages between 50 and 75% of the full-wave value.

The final tests are of negative polarity, unless agreed otherwise. They are applied in the following order.

(1)   One full wave.
(2)   Two chopped waves.
(3)   One full wave.

During these tests, peak value and shape of the wave are recorded by means of a high-speed cathode-ray oscillograph. The wave to be recorded is supplied from the low-voltage arm of a capacitor divider. An oscillation of about 1 MHz is also recorded on the photographic plate for timing purposes.

In addition, surge currents are recorded as voltages obtained with a shunt inserted between the neutral point and earth for a star-connected transformer, or between the other two lines joined together and earth for a delta connection.

Insulation failure will be indicated by a variation in the waveshape of any of these oscillograms. As a final check they are compared with, and should be identical with, those obtained at a reduced voltage during the preliminary tests. Typical recordings are shown in figure 7.6.

7.4.4   Separate-source-voltage withstand test

This test checks the major insulation between windings and earth. The line terminals of the windings under test are connected together, and the appropriate test voltage is applied to them; the other windings are shorted and earthed during this period, as shown in figure 7.7. The value of the test voltage is approximately twice the system highest voltage for fully insulated windings, as indicated in table 7.2(a) and (b). Windings with graded insulation are subjected to a relatively low

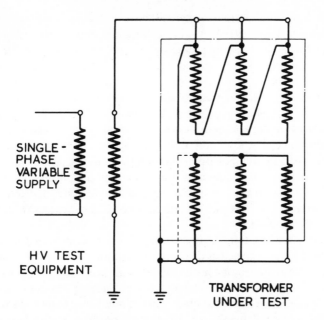

Figure 7.7 Connections for separate-source-voltage withstand test

test voltage depending on the type of earthing arrangements. The value is a minimum of 38 kV.

The specified test voltages are given as rms values on the assumption that the waveform is sinusoidal, thus producing peak voltages of $2^{1/2}$ times the rms value. It is advisable, therefore, when carrying out these tests to measure the applied high voltage by means of a sphere gap or other peak-measuring device in order to allow for any discrepancy in the relationship between rms and peak values due to possible waveform errors introduced by the generator or regulating device.

In order to prevent damage due to transients, the test is commenced at a voltage not greater than one-third of the required value. This voltage is then increased as smoothly as possible up to the test value and is applied for a period of 1 min, after which it is reduced to less than one-third of this amount before being switched off.

7.4.5 Induced over-voltage withstand test and internal discharge test

In order to prove the interturn and line end insulation it is necessary to induce a higher voltage than normal in the windings. The value depends upon the transformer insulation level as shown in table 7.2(a) and (b).

For test voltages up to 66 kV, three-phase transformers are generally supplied direct from a three-phase source, but for higher values it is usual to energise each low-voltage winding in turn from a single-phase supply.

In order to avoid saturation of the core at the test voltage it is necessary to use a supply of a higher frequency than normal, for example, 250 Hz. At this higher frequency, however, the transformer will have a capacitive reactance of between 500 to 1000 Ω (referred to the low-voltage winding), and this may require a supply of up to 3000 kVA at a leading power factor. This requirement can be reduced by means of a parallel-resonant circuit which is obtained by connecting a variable inductor across the generator terminals and by adjusting its reactance to the same value as the transformer capacitive reactance; allowance is made for the ratio of any interposing transformer which may be connected between the generator and the transformer under test. Series resonance between the capacitive load and any series inductance in the circuit must be avoided because of the danger of producing excessive voltages. Before carrying out any tests it is advisable to check from design office calculations and from the circuit parameters that this will not occur. As a further precaution, a sphere gap should be connected across the high-voltage winding so that it will spark over if any excessive voltages are inadvertently produced.

As in the preceding test, the high voltage should be measured by means of a peak-measuring device such as a sphere gap. Alternatively, a bridge rectifier and ammeter which give a reading proportional to the peak voltage may be connected in the earth end of the high-voltage capacitor bushing of the phase under test and may be calibrated against a sphere gap at a number of points up to about 70 % of the required test voltage. A graph of sphere gap voltage against bushing current can then be drawn, and the bushing current equivalent of the full test voltage may be obtained by extrapolation.

The test is maintained for 1 min for frequencies not greater than twice normal. At higher frequencies the time is reduced such that

$$\text{time of test in seconds} = \frac{60 \times \text{twice rated frequency}}{\text{test frequency}}$$

The duration of the test on a 50 Hz transformer when supplied at a frequency of 250 Hz is, therefore, 24 s.

Typical connections for single-phase tests on a three-phase three-limbed core are shown in figure 7.8. Figure 7.8(a) shows the connections required for carrying out the single-phase test on the two outer phases simultaneously. Since the voltage applied to the low-voltage winding during the test may be higher than the specified separate source test voltage, it is usual to earth the mid-point of the supply transformer in order to keep the voltage between windings and earth to a safe value. The centre phase is tested separately; the connections are as shown in figure 7.8(b). In this case, owing to the direction of the flux, a voltage of 1.5 times the test value V will appear between the high-voltage terminal of the phase under test and the adjacent terminals.

Partially graded or fully insulated windings may be tested as shown in figure 7.8(c), in which each phase is tested separately, the other two being joined together and earthed. With this connection, the voltage between adjacent terminals is limited to the test value V, but the voltage that appears at the neutral point terminal will be one-third of this amount.

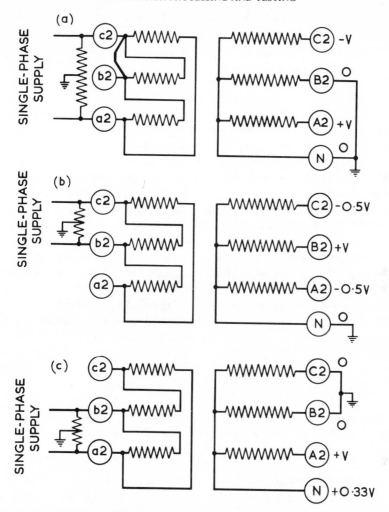

Figure 7.8 Typical single-phase connections for induced-voltage tests on three-phase transformers: (a) test on the two outer phases; (b) test on the centre phase; (c) alternative test

For a five-limbed core the tests are also carried out in single phases, each phase in turn. The flux is made to bypass the two phases not under test by connecting their line terminals to the neutral terminal and earth, thus limiting the voltage between adjacent terminals to the test value.

During the induced-voltage test the insulation is highly stressed, and measurement of internal discharges produced by such stress is becoming increasingly important.

One method of measurement[G1.16] is by a type of detector in which the discharge current over a range of 10 to 150 kHz is amplified and displayed in the form of pulses on an oscilloscope. The peak value of the discharge is measured by comparison with that of a known pulse injected into the detector.

Another method, specified by NEMA for radio interference tests[2], is often used for internal discharge detection. This instrument is tuned for a frequency of 1 MHz, and values are indicated directly on a voltmeter associated with the instrument.

In both methods, connections between the transformer under test and the measuring equipment is made via a blocking capacitor, or, alternatively, the test tapping of the transformer bushing may be used.

It is important, when carrying out the test, to ensure that the test area is free from extraneous interference and that there are no unearthed objects in the vicinity, since these are liable to produce false readings. External gas conduction at the high-voltage terminals is kept to a minimum by fitting suitable spheres or ellipsoids over the terminal leads in order to reduce the electric stress.

The measurements are carried out at various values up to the full induced test voltage and are repeated as the voltage is reduced to zero. The measured values provide a good indication of the insulation quality at the higher voltages. If, as is general, a preliminary test has been carried out at about 1.2 times the normal voltage before the commencement of any of the electric tests, comparison of the initial and final discharge measurement at that voltage will provide a good indication of whether any damage has occurred during any of the high-voltage tests.

7.4.6   Resistance

The final resistance values are measured preferably by means of a Kelvin double bridge. Certain precautions are necessary to ensure accuracy and to prevent damage to the instruments and are as follows.

(a)   Since the windings are highly inductive, time must be allowed for the current to attain a steady value before taking measurements. Failure to do so will result in false readings and may cause damage to the voltage circuit of the bridge due to the high induced voltage that is produced when switching on. The voltage leads, therefore, should not be connected until an ammeter in the circuit indicates steady reading, and, of course, they should be removed before switching off.

(b)   The temperature of the windings must be stable and known, and for this reason it is important that this test is carried out before the power tests are applied.

7.4.7   No-load loss and no-load current

The measurement of no-load loss and current is important, not only for the purpose of assessing the efficiency of the transformer but also as a check that the high-voltage tests described in the preceding sections have been carried out without damage to winding insulation. For large transformers, therefore, a no-load measurement is often carried out before and after the electric tests to ensure that there is no significant change between the two measurements.

Measurement is carried out at normal frequency supplying either the lower- or the higher-voltage winding, but it is usually more convenient to supply the former. Since the no-load current will be very small compared with the full load value, the I^2R losses in the windings will be negligible, and the measured power is assumed to represent the core and dielectric loss. The specified tolerance at normal voltage and frequency is one-seventh of the guaranteed value, provided that, when combined with the load loss, the total losses do not exceed one-tenth of the guaranteed value. Measurements are taken at a series of voltages between 85% and 110% of normal, so that curves of no-load current and power may be drawn.

For safety and convenience the voltages and currents to be measured are transformed to about 110 V and 5 A, using instrument transformers of the highest precision (class AL). The instruments also must be of precision-grade accuracy, the wattmeters being designed to produce a full-scale deflection at a power factor of 0.2.

The no-load power factor for distribution and small power transformers is generally between 0.2 and 0.5. Three-phase units of this size may be tested using the two-wattmeter method, either by two single-element wattmeters or by a polyphase instrument.

The power factor of larger units, however, especially when working at higher flux densities, is about 0.1. At this low value the two-wattmeter method results in only a small difference in the instrument readings. A small error in one instrument, therefore, will produce a large error in the result. It is preferable in this case to use the three-wattmeter method since the total loss is given by the algebraic sum of three readings and any small error in one of the instruments is not magnified to the same extent.

Since the no-load current will be non-sinusoidal, it is possible that the voltage waveform may be distorted owing to the harmonic components of the current, producing voltage drops across series impedances in the supply. This can be reduced by supplying from a large source and by keeping the series impedance to a minimum. If the distortion cannot be eliminated completely, a correction can be applied to single-phase units or to three-phase units with a delta winding.

The core loss consists of hysteresis and eddy current losses; the hysteresis loss is dependent on the peak value, and the eddy current losses depend on the rms value of flux. During the test, two voltmeters are used. One is a bridge rectifier type calibrated to indicate average voltage multiplied by 1.11 (that is rms value on a sine waveform) and the other a dynamometer type which indicates true rms values. The supply voltage is set so that the rated value is indicated on the rectifier voltmeter. Since the peak value of flux is proportional to this average voltage, the hysteresis component will be measured correctly, whilst the eddy current loss will be either lower or higher than the true value, depending upon the form factor of the supply voltage. Since the ratio between the two components is known for any particular quality of core steel, a correction can be applied by using the following formula

$$P = \frac{P_{\mathrm{m}}}{k_1 + k_2 k_3} \qquad (7.1)$$

where P is the no-load loss for a sinusoidal voltage, P_{m} the measured no-load loss, k_1 the ratio of hysteresis loss to total loss, k_2 the ratio of eddy current loss to total loss and

$$k_3 = \left(\frac{\text{rms voltage}}{\text{average voltage} \times 1.11} \right)^2 = \left(\frac{\text{actual rms voltage}}{\text{rated rms voltage}} \right)^2$$

For normal flux densities and frequencies of 50 Hz and 60 Hz, k_1 and k_2 are each taken as equal to 0.5 (for cold-rolled steel).

It is usual to analyse the no-load current waveform. This can be carried out after the measurement of the no-load loss. A non-inductive resistor of about 1 Ω may be inserted directly in the supply line (in which case the line must be earthed for safety), or the resistor may be connected to the secondary winding of a fully insulated current transformer. The resultant voltage across this resistor is then applied to a harmonic analyser which is designed to indicate harmonic components directly as a percentage of the fundamental.

7.4.8   Noise test

Whilst the transformer is still connected for energising on no load, it may be convenient at this stage to carry out the noise tests, details of which are dealt with in chapter 8.

7.4.9   Load loss and impedance voltage

The load loss (copper loss) comprises the sum of the I^2R losses in the windings and the stray losses due to eddy currents in the conductors, clamps and tank. Since the latter vary with frequency, it is important to supply the transformer at the rated frequency. The test is carried out by short-circuiting one winding and by supplying a reduced voltage (the impedance voltage) to the other winding sufficient to cause full-load current to circulate. The short-circuiting conductor consists of solid copper bars and should be of an adequate current-carrying capacity in order to keep losses in it to a minimum. All joints associated with it should be first cleaned and smeared with contact grease before being firmly bolted together. It is usually more convenient to supply the higher-voltage winding and to short-circuit the lower-voltage side since this will require less supply current. The impedance voltage varies from about 5 % of normal voltage for distribution transformers and up to about 15 % for large generator units. The measured power will include a small core loss as well as the load loss, but, since the voltage is generally only a fraction of normal, this additional loss can be ignored. If, however, as for a high-impedance transformer, the core loss is appreciable, it can be measured by supplying the impedance voltage with the short-circuiting conductor removed and by deducting the resultant loss from that measured under short-circuit conditions. The tolerance allowed for the load loss is the same as that for the no-load loss, that is one-seventh of the guaranteed value provided that, when added to the no-load loss, the total does not exceed one-tenth of the guaranteed total losses. In addition, the loss is measured on extreme tappings, and the impedance voltage on all tappings.

Since the loss varies with the temperature of the windings, it is important that the latter is stable and known. It is, therefore, preferable to limit the current to about half the rated value in order to keep the temperature rise of the windings to a minimim during the test. The measured impedance voltage can then be corrected to the rated value by multiplying by the ratio of rated current to test current, and the losses by multiplying them by the square of the ratio of rated current to test current.

The power factor of the supply is between 0.3 and 0.5 for distribution and small power transformers, and the losses can be measured by using similar instruments to those used in the no-load test. When testing very large units, however, the power factor may be as low as 0.02, in which case special reflecting-type wattmeters, with a full-scale deflection at power factors of 0.05 or less, are necessary in order to obtain a reasonable deflection. For three-phase measurements at very low power factor, the three-wattmeter method must be used in order to ensure greater accuracy as explained in sub-section 7.4.7.

The load test often requires the supply of large currents at high voltages, and instrument transformers are therefore used for safe and convenient measurements. The small phase angle errors inherent in them, however, together with that of the wattmeter voltage coil (due to its slight inductance), produce a phase angle slightly different from the true phase angle of the supply. The discrepancy may be only a few minutes. Although the effect is negligible for power factors greater than 0.05, it causes an appreciable error at lower power factors because, as the phase angle approaches 90°, a small change in the angle results in a considerable change in the cosine of the angle. The apparent phase angle ψ produced by these errors is shown in figure 7.9.

The phase angle, α or β, of the instrument transformer is assumed positive when, on reversal, the secondary phasor leads the primary phasor and negative when, on reversal, the secondary phasor lags the primary. In figure 7.9, therefore, the voltage transformer phase angle α is shown as negative and the current transformer phase angle β as positive.

The method of error correction, with k as a constant, is as follows.

$$\text{Measured power} = kI_{pc}I_s\cos\psi = V_pI_p\cos\psi \qquad (7.2)$$

Thus

$$\text{true power} = V_pI_p\cos\phi = \text{measured power} \times \frac{\cos\phi}{\cos\psi} \qquad (7.3)$$

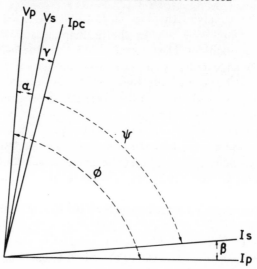

Figure 7.9   Phase angle errors due to instrument transformers: V_p, supply voltage; I_p, supply current; V_s, secondary voltage of voltage transformer; I_s, secondary current of current transformer; I_{pc}, current in wattmeter voltage coil

From figure 7.9

$$\phi = \psi + \alpha + \beta + \gamma \tag{7.4}$$

Hence

$$\cos \phi = \cos \psi \cos (\alpha + \beta + \gamma) - \sin \psi \sin (\alpha + \beta + \gamma) \tag{7.5}$$

Because $\alpha + \beta + \gamma$ is very small

$$\cos (\alpha + \beta + \gamma) = 1 \tag{7.6}$$

and

$$\sin (\alpha + \beta + \gamma) = (\alpha + \beta + \gamma) \, \text{rad} \tag{7.7}$$

At low power factors $\phi \approx \tfrac{1}{2}\pi$ rad, and hence $\psi \approx \tfrac{1}{2}\pi$ rad. Therefore $\sin \psi = 1$, and hence, from equation 7.5,

$$\cos \phi = \cos \psi - (\alpha + \beta + \gamma) \tag{7.8}$$

Thus, from equation 7.3,

$$\text{true power} = \text{measured power} \times \frac{\cos \psi - (\alpha + \beta + \gamma)}{\cos \psi} \tag{7.9}$$

With the phase angle errors given in centiradians, the percentage error can be expressed simply as

$$\text{percentage error} = \frac{(\alpha + \beta + \gamma)(\text{centiradians})}{\cos \psi} \tag{7.10}$$

The correction does not take into account any instrument transformer ratio errors. In practice they are negligible. Typical phase angle errors for precision instrument transformers and wattmeters are as follows: current transformers, $+1'$; voltage transformers, $\pm 2'$; wattmeter voltage coil, $1'$.

After correction the load losses and impedance voltages are adjusted to a reference temperature that depends on the insulation class. These values which are contained in an IEC publication[G1.7] are given in table 7.3; see also references G2.2 and G3.2.

TABLE 7.3   *Reference temperatures*

Class of temperature	Reference temperature (°C)
A E B	} 75
F H C	} 115

The calculation of the losses to the required temperature is dealt with in section 7.7. The I^2R losses are taken as increasing with the temperature and the stray losses as decreasing with temperature.

7.4.10   Zero-sequence impedance

The zero-sequence impedance Z_h is measured on all star-connected windings which have an earthed neutral in order to determine the current which will flow in the event of a line-to-earth fault.

Figure 7.10 shows the connections for carrying out this test on a delta–star transformer. The line terminals on the star-connected winding are joined together, and a single-phase supply is applied between these and the neutral point, the delta being left on open circuit during this test.

For transformers with more than one star-connected winding and neutral terminals, additional measurements of the zero-sequence impedance are made, in which the line terminals and the neutral terminal of the other star-connected winding are connected together.

Auto-transformers with a neutral terminal are treated as normal transformers with two star-connected windings. The four tests possible under these conditions are therefore as follows.

(a) Z_h for high-voltage to neutral and tertiary.
(b) Z_h for low-voltage to neutral and tertiary.
(c) Z_h for low-voltage to neutral, high-voltage and tertiary.
(d) Z_h for high-voltage to neutral, low-voltage and tertiary.

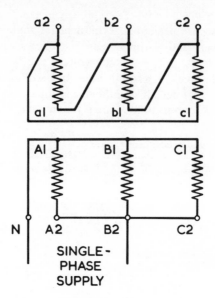

Figure 7.10   Connections for zero-sequence impedance test on a delta – star transformer

The connections for these tests are shown in figure 7.11(a), (b), (c) and (d), respectively.

The impedance decreases slightly as the current is increased. It is measured at several values up to the highest current possible within the transformer and supply limitations. The results are expressed in ohms per phase and are given by $3 \times V/I$, where V is the test voltage and I the test current.

7.4.11   Temperature rise test

The first unit of a new design is subjected to a temperature rise test in order to confirm that, under normal conditions, the temperature rise of the windings (measured by the change-of-resistance method) and the oil, where applicable, will not exceed the specified limits [G1.7, G2.2, G3.2] given in tables 7.4 and 7.5.

The temperature rises are measured above the temperature of the cooling air for all types except water-cooled transformers. In this case, the temperature rise is measured above the inlet water temperature.

The test is carried out in such a way that full-load losses, that is the sum of the no-load and the full-load copper loss, are produced in the transformer. Since it is not possible to load the transformer directly, various methods have been devised where these conditions can be achieved without using an external load.

Back-to-back or Sumpner test

This method requires basically two identical units and may be used on single-phase or three-phase transformers; the single-phase connections are shown in

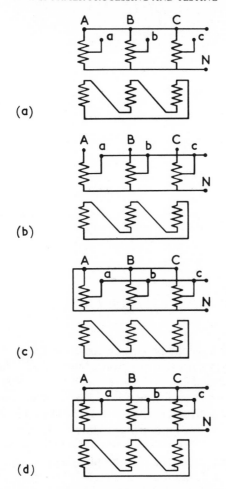

(a)

(b)

(c)

(d)

Figure 7.11 Connections for zero-sequence impedance tests on star–star auto-transformer with delta tertiary: (a) supply, ABC-N, ABC shorted, abc open; (b) supply, abc-N, ABC open, abc shorted; (c) supply, abc-N, ABCN shorted, abc shorted; (d) supply, ABC-N, ABC shorted, abcN shorted

figure 7.12(a). The two transformers are connected in parallel on both high- and low-voltage sides, and one set of windings (usually the low-voltage windings) is supplied at normal voltage and frequency, thus producing the no-load loss in both transformers. Full-load copper loss is obtained by means of an auxiliary transformer supplied from a separate source and connected between the two low-voltage windings. The auxiliary transformer must supply twice the impedance voltage of one of the transformers in order to cause full-load current to circulate in the windings as indicated by ammeter A. The current may be injected into the high-voltage windings if it is more convenient, but care must be taken to ensure that the insulation of the auxiliary transformer is capable of withstanding the

TABLE 7.4 *Temperature rise limits for dry-type transformers*

1 Part	2 Cooling method	3 Class of temperture	4 Temperature rise (°C)
windings (measured by resistance)	air, natural or forced	A E B F H	60 75 80 100 125
cores and other parts (a) adjacent to windings (b) not adjacent to windings	all		(a) same value as for windings (b) a value that will not adversely affect insulating parts that may be in contact with the windings

higher voltages which will be present on these windings.

It will be found in practice that one transformer is slightly hotter than the other because the no-load and full-load current will be nearly in phase in one and almost in phase opposition in the other. The resultant value of current will, therefore, be slightly different in the two transformers. The one with the higher temperature rise is offered for acceptance tests.

The auxiliary transformer may be omitted if suitable tappings are available on the two transformers under test. The tappings can be set on unequal tapping positions (see figure 7.12(b)), and a circulating current is produced by the unbalanced voltages.

In both cases, full-load conditions are obtained by supplying the appropriate voltages and currents. It is not necessary to measure the actual power supplied.

Delta – delta test

Three-phase transformers having delta-connected windings may be tested by the delta – delta test method, a typical connection being shown in figure 7.13(a). One winding, usually the low-voltage winding, is supplied at normal voltage and frequency to produce the core loss.

The other winding is connected in open delta by removing one of the delta links, and a single-phase auxiliary supply is injected at this point in order to circulate full-load current in the windings. If the voltage on the high-voltage side is too high for the insulation level of the auxiliary transformer, the same conditions can be produced by injecting the current in the low-voltage side, as shown in figure 7.13(b).

TABLE 7.5   *Temperature rise limits for oil-immersed type transformers*

1 Part	*2* Cooling method	*3* Oil circulation	*4* Temperature rise *(°C)*
windings temperature class A (measured by resistance)	natural air, forced air, water (internal coolers)	natural	65
	forced air, water (external coolers)	forced	65
top oil (measured by thermometer)			60, when the transformer is sealed or equipped with a conservator
			55, when the transformer is not so sealed or equipped
cores and other parts			the temperature in no case to reach a value that will injure the core itself or adjacent parts

Equivalent short-circuit run

This method is most generally used in practice, particularly on the larger units. The connections are the same as for the load test (sub-section 7.4.8), that is the low-voltage windings are shorted and a reduced supply at normal frequency is applied to the high-voltage side sufficient to produce the sum of no-load and rated load loss at the reference temperature. This total loss is measured by wattmeters in the circuit and is maintained until the top oil has reached a steady temperature.

This test could take 12, 9 or 7 h for ONAN (oil-natural air-natural) OFAF (oil-forced air-forced) or OFWF (oil-forced water-forced) transformers, respectively. After this condition has been attained, the current is reduced to its rated value and is maintained for 1 to 2 h in order to allow the windings to reach their normal temperature.

The open-circuit test

If the no-load loss is high compared with the load loss, as for earthing transformers, the equivalent short-circuit test may result in too high a current

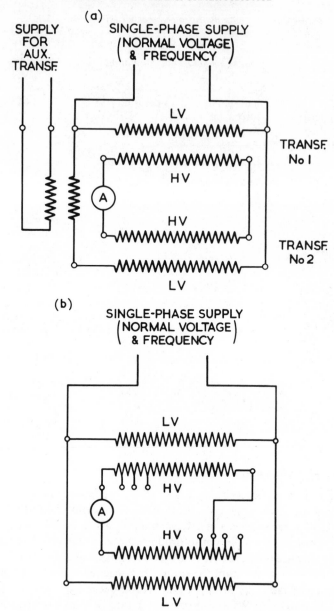

Figure 7.12   Back-to-back, or Sumpner, test on two single-phase transformers: (a) basic
arrangement; (b) arrangement with two unequal transformer tappings

density in the windings. In this case it may be preferable to supply the transformer
on open circuit at a voltage in excess of normal, possibly up to 15%, so that the
resultant core loss will be equal to the sum of the normal no-load and the load
losses.

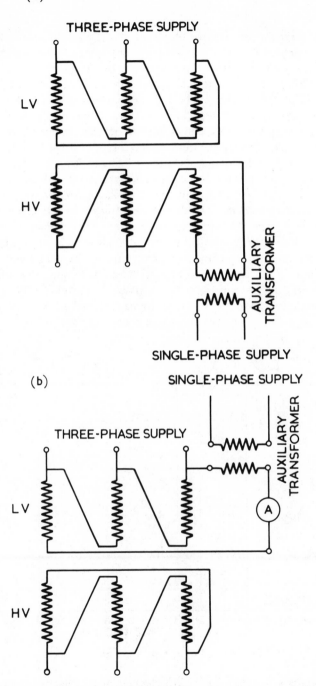

Figure 7.13 Connections for delta – delta test: (a) open-delta auxiliary supply on the high-voltage side; (b) open-delta auxiliary supply on the low-voltage side

Temperature measurements

During the temperature rise test on an oil-immersed-type transformer, hourly readings of the top oil temperatures are taken by means of a thermometer placed in a pocket in the transformer top cover. Effective average oil temperatures are determined from measurements taken at the inlet and outlet of the coolers. These are regarded as the surface at the top and bottom of a cooling tube where radiators are mounted on the tank. For separate coolers measurements are taken at thermometer pockets in the cooler top and bottom connecting pipes adjacent to the main tank. The temperature of the air is measured by taking several readings at points situated at a distance of 1 to 2 m from, and half-way up the cooling surface, care being taken to ensure that the thermometers are shielded from direct radiant heat and draughts. Time lag errors between temperature variations in the transformer and the surrounding air are reduced by placing the thermometers in special cups with a time constant similar to that of the transformer. The test is continued under full-load loss conditions until there is sufficient evidence that the final oil temperature rise would not exceed the specified limit. This can be established by extrapolating the graph of oil temperature rise against time, as shown in figure 7.14. Alternatively, the test may be continued until the oil temperature rise has reached a steady value, which in practice is regarded as an increase of not more than $1\,°\text{C}\,\text{h}^{-1}$.

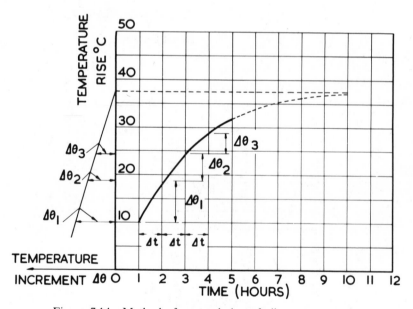

Figure 7.14   Method of extrapolation of oil temperature rise

At the end of the test the supply is switched off, and the average temperatures of the windings are determined by measuring the hot resistance values and by comparing them with the original cold values; one high-voltage and one low-voltage winding are measured simultaneously by means of two Kelvin double bridges.

Before the measurements can be made, however, the short-circuit connection has to be removed, and the dc supply used for the resistance measurements must be allowed to reach a steady value. A certain time, probably 2 to 3 min, will therefore elapse between switching off the power supply and taking the first reading, during which the resistances of the windings will be decreasing. In order to determine the temperatures at the instant of switching off the power the resistances are measured at intervals over a period of 15 min and are then converted to temperatures; allowance is made for any change in average oil temperature during this period. Graphs of winding temperatures against time are then plotted, from which the values at shutdown may be obtained by extrapolation as shown in figure 7.15.

The method of determining the temperature rise is given in section 7.7.

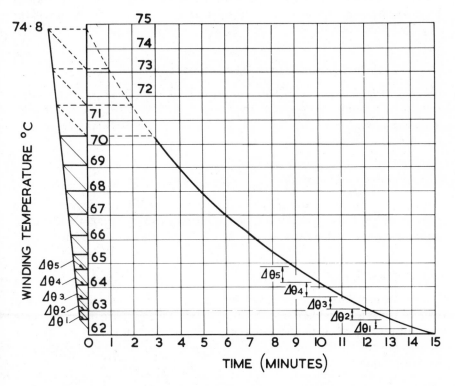

Figure 7.15   Extrapolation of winding temperature to instant of power switch-off

7.4.12   Insulation resistance

Insulation resistance is measured between all windings and between windings and tank by using a megaohmmeter insulation tester; typical values at room temperature are not less than 1000 MΩ. In addition, the insulation resistance of the magnetic circuit to the clamps and tank is checked before final despatch.

7.5 TRANSFORMER ACCESSORIES

All ancillary equipment for use with the transformer is tested separately before assembly to the main unit.

7.5.1 Capacitor bushings

Power factor and capacitance measurements are carried out at various voltages up to normal line values on all capacitor bushings using a high-voltage Schering bridge. A 1 min separate-source voltage withstand test is then applied, after which the power factor and capacitance measurements are repeated. Internal discharge tests are also carried out before and after the applied voltage test.

7.5.2 Marshalling kiosks and associated equipment

The wiring circuits are checked, and associated equipment such as relays, contactors, winding temperature indicators, etc., are tested for satisfactory operation, after which a voltage test of 2 kV is applied for 1 min between all terminals and earth.

7.5.3 On-load tap changers

This equipment is subjected to separate tests, details of which are given in chapter 6. In addition, the gear is electrically operated ten times throughout the complete range during the no-load test in order to check its satisfactory operation.

7.5.4 Oil test

Samples are tested for electric strength before any oil is accepted into storage, and again before it is introduced into the transformer. The test is carried out using a standard oil test cell with electrodes consisting of spheres 13 mm in diameter spaced 4 mm apart. The sample is required to withstand a voltage of 40 kV for a period of 1 min.

7.6 COMMISSIONING AND SITE TESTS

After installation, certain tests are required on site before the transformer is put into service. For transformers which are despatched as complete units, the following tests are carried out.

(a) Ratio and phasor reference test.
(b) Insulation resistance.
(c) Resistance of windings.
(d) Oil sample tests.
(e) All control, protective and cooling gear is checked for satisfactory operation.
(f) Insulation resistance of magnetic circuit.
(g) No-load current check at reduced voltage.

7.7 TESTS ON A LARGE POWER TRANSFORMER

The following results were obtained from tests carried out on a three-phase 750 MVA 50 Hz auto-transformer which was connected star high-voltage – star low-voltage and included a delta-connected tertiary winding.

The nominal voltage ratio was 400 to 275 to 13 kV.

Measurement of no-load loss

This test was carried out by supplying the tertiary winding at 13 kV and 50 Hz.

1.11 ×mean (kV)	(kV (rms))	(A)	(kW)
13	13.16	17	152
		17.3	
		20.3	

The waveform error was negligible.

Resistance of windings

The resistance was measured on all phases between each high-voltage terminal and the corresponding low-voltage terminal (series winding) and also between each low-voltage terminal and the neutral (common winding). The results were as follows.

Average resistance of the three series windings at 24.5°C = 0.1191 Ω

Average resistance of the three common windings at 24.5°C = 0.408 Ω

Measured load loss high-voltage to low-voltage windings

The test was carried out supplying the high-voltage terminals with the low-voltage terminals short-circuited with the following results.

(kV)	(A)	(kW)	Apparent power factor	Temperature (°C)
48.5	1083	880	0.0097	24.5

The instrument transformer errors were $+1.8'$. Therefore

$$\text{percentage error} = \frac{1.8 \times 0.0291}{0.0097} = 5.4\%$$

Thus

$$\text{corrected loss} = 0.946 \times 880 = 832\,\text{kW at } 24.5\,°\text{C}$$

From the measured values of resistance, the load loss can be calculated for the reference temperature of 75 °C as follows.

I^2R for series winding $= 1083^2 \times 0.1191 \times 3$ $= 420\,\text{kW}$
I^2R for common winding $= 492^2 \times 0.408 \times 3$ $= 297\,\text{kW}$
(Note the current in the common winding $=$ high-voltage – low-voltage current)

	Total I^2R	$= 717\,\text{kW}$
Measured loss at 24.5 °C		$= 834\,\text{kW}$
Stray loss at 24.5 °C		$= 117\,\text{kW}$
I^2R at 75 °C $= \dfrac{235+75}{235+24.5} \times 717$		$= 857\,\text{kW}$
Stray loss at 75 °C $= \dfrac{235+24.5}{235+75} \times 117$		$= 97\,\text{kW}$
Total loss at 75 °C		$= 954\,\text{kW}$

Temperature rise test

A temperature rise test was carried out at the full rated value 750 MVA (OFAF cooling) under the same short-circuit condition as for the load test. The voltage was increased, however, in order to supply the equivalent of the load loss at 75 °C plus the no-load loss. The temperature measurements were recorded, as described in sub-section 7.4.10. The actual supply to the transformer was as follows

	(kV)	(A)	(kW)	(MVA)
at normal current	48.5	1083	954	
no-load loss			152	
at full-load loss	52.2	1168	1106	106

The resistances of the series and common windings (phase B) were measured at a known temperature before the start of the test, and they were repeated at 1 min intervals over a period of 15 min after the final shutdown. These values were converted to temperature by using the formula

$$\theta_2 = \frac{R_2}{R_1}(235 + \theta_1) - 235 \qquad (7.11)$$

where θ_2 is the temperature of the winding after the test, R_2 the resistance of the winding at temperature θ_2, θ_1 the temperature of the winding before the test, R_1 the resistance of the winding at temperature θ_1 and 235 the reciprocal of the temperature coefficient for copper.

Table 7.6 shows the data recorded at the end of both the full-load loss and the full-load current tests and also the calculation for the temperature rise of the windings. The resistances of the windings at the time of shutdown were extrapolated in the manner shown in figure 7.15.

TABLE 7.6 *Temperature rise test data (750 MVA; OFAF cooling; normal position; three-phase frequency is 50 Hz)*

Load loss	954 kW	
No-load loss	152 kW	
Total rated loss	1106 kW	
Test loss	1106 kW	
Rated current	1083 A	
Test current	1083 A	

(°C)	At test load	At shutdown
Top oil temperature rise		
top oil temperature	65.1	64
average air temperature	23.7	24.6
top oil temperature rise	41.4	
Average oil temperature rise		
radiator inlet temperature	63.3	62.3
radiator outlet temperature	59.5	58.5
difference	3.8	3.8
average oil temperature	$65.1 - 3.8/2 = 63.2$	$64 - 3.8/2 = 62.1$
average oil temperature rise	$63.2 - 23.7 = 39.5$	$62.1 - 24.6 = 37.5$

(°C)	Series winding	Common winding
winding temperature at shutdown	73.5	77.7
average air temperature	24.6	24.6
winding temperature rise above ambient	48.9	53.1
average oil temperature rise at shutdown	37.5	37.5
average temperature difference between windings and oil	11.4	15.6
series winding temperature rise by resistance at rated load	$11.4 + 39.5 = 50.9$	
common winding temperature rise by resistance at rated load		$15.6 + 39.5 = 55.1$
top oil temperature rise at rated load	41.1	

7.8 FUTURE DEVELOPMENTS

Increases in both the power and the voltage of transmission systems have resulted in corresponding increases in the rating of electric plant, and it is generally accepted that this trend will continue in the future.

For transformers it is possible that the tendency will be towards single-phase units both to ease the problems of transport and also to overcome the excessive clearances between terminals which would be necessary for three-phase megavolt transformers. Even so, some units may still present despatch problems, and they may have to be built and tested at the factory, to be dismantled for transport and then to be rebuilt on site followed by proving tests. This has already been done in a number of cases where access of the complete unit to the site has not been possible.

The voltage levels of electric tests will have to be modified when dealing with higher voltage ratings since increasing the test level *pro rata* would result in excessively high voltages. The induced-voltage withstand test in its present form may have to be replaced by a lower test level but may have to be maintained for a longer period, during which time internal discharge detection tests would be carried out and would be the determining factor in the acceptance of this particular test.

Test supply requirements will be expected to increase in order to cope with the larger ratings, but to what extent it is not yet certain. It is of interest to note, however, that the largest transformers in the 1950s were only about 100 MVA. Yet today a supply of this magnitude is often required to provide the test requirements for present-day units, as can be seen from the temperature rise example in section 7.7.

ACKNOWLEDGEMENT

The authors are indebted to Ferranti Limited for permission to publish the information and diagrams contained in the test.

REFERENCES*

(Reference numbers preceded by the letter G are listed in section 1.14.)
1. Bishop, M. J., and Feinberg, R., Grundsätzliche Verbesserung des Hochspannungs-Stossgenerators—Anwendung des Polytrigatrons als Schaltgerät, *Elektrotech. Maschinenbau*, **88** (1971) 62
2. National Electrical Manufacturers' Association, *NEMA (USA) 107, Methods of Measurement of Radio Influence Voltage (RIV) of High-voltage Apparatus*

* See also references G1.4, G1.11, G2.3, G2.6 and G2.14.

8

Transformer Noise

J. Dunsbee* and M. Milner*

8.1 INTRODUCTION

Transformers emit a distinctive hum which is continuous, of constant level and irrespective of load and which consists of discrete frequency components. Although over the years the noise produced by transformers of given rated power has been reduced, the problem has become severe owing to the increasing size and number of transformers and the tendency for these to be located closer to load centres. The result has been that the likelihood of annoyance to residents has become potentially greater and that more and more precautions now need to be taken at the planning stage. Complaints can to a large extent be forestalled provided suitable estimates are made at this stage and provided measures are taken to limit the noise.

The response of the human ear to sound as a function of frequency, methods of sound level measurement and behaviour of sound in radiation, reflection, absorption and diffraction have been well investigated, as indicated by examples in references 1, 2 and 3. It is of practical interest to note that there is a difference in loudness assessment between the human ear and its simulation with an instrument and that, therefore, a distinction is made between subjective measurements based on human evaluation and objective measurements with an instrument. Such an instrument, called a sound level meter, consists of a microphone, an electronic amplifier and an attenuation filter network. The network, known as weighting, has three settings: (1) the 40 dB weighting or A scale, (2) the 70 dB weighting or B scale and (3) a C scale which is substantially unweighted.

For more detailed investigation of a noise spectrum most sound level meters can be extended by suitable analysers with either wide or narrow band. For

* Ferranti Limited.

example, an octave band analyser deals with a wide frequency band where the upper frequency is twice the lower one. A narrow band may deal with only 3 or 5% of an octave band.

8.2 TRANSFORMER VIBRATION

8.2.1 Source

The main source of transformer noise is the magnetic core which vibrates in a complex manner owing to the action of cross-fluxing and magnetostriction. It is not certain what proportion of the total noise is contributed by each of these factors. Cross-fluxing, that is flux passing between adjacent laminations, sets up magnetic forces which cause the laminations to vibrate at twice the supply frequency (that is 100 Hz for a 50 Hz supply). With modern core design cross-fluxing is probably of relatively minor importance except perhaps at the corners and at the joints where gaps intervene in the magnetic path or wherever an appreciable difference in permeability exists between points on adjacent laminations.

Magnetostriction is the relative change of dimension, due to magnetisation, of a magnetic material in a magnetic field. This change can be positive or negative, or it may be of one polarity at a low flux density and may then reverse as the flux density increases (see figure 8.1). This change not only causes longitudinal vibrations in the plane of the laminations of the core legs or yokes but also results in out-of-plane displacements of the core limbs. Magnetostriction is sensitive to mechanical stress. This is illustrated[4] in figure 8.2 which shows the effect of compressive and tensile stresses on positively magnetostrictive material. It is not sufficient, therefore, solely to aim at obtaining core material of low inherent magnetostriction but to consider also the sensitivity of the material to stress which it will be subjected to in a completed transformer core. Efforts are, therefore, directed to obtaining flat core steel as the clamping of wavy laminations in the core will result in increased stresses. Thus not only does the quality and flatness of the core material as received from the steel maker become of importance but also the method of annealing used by the transformer manufacturers must also be considered.

As the relationship between magnetostrictive strain and flux density is not linear, the magnetostrictive displacement caused by sinusoidal flux will contain harmonics of the fundamental frequency, and this again is twice the supply frequency. Although the average flux density in a core limb may be sinusoidal, the local flux density at any particular part of the core contains a large number of harmonics[5]. These, in turn, will result in an increased harmonic content of the core vibrations.

Investigations[6] have shown that the basic magnetostriction for a given flux

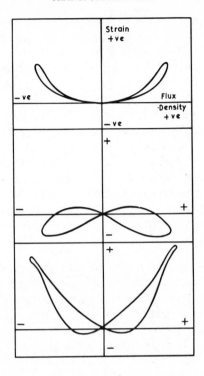

Figure 8.1 Oscillograms of magnetostrictive strain against flux density

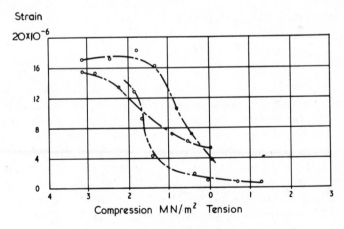

Figure 8.2 Effect of stress on magnetostriction[4]

density varies from point to point throughout the material, following an apparently random pattern. The picture can be further complicated by gaps, bolt holes and changes in permeability, causing local changes in flux density and hence additional differences in magnetostrictive displacements. Thus, when a core leg is excited it carries out a most complex motion, involving 'writhing' dishing and transverse flexural movements apart from the (expected) longitudinal displacements[7]. Both longitudinal and transverse vibrations contribute towards the noise generated by the transformer. As the legs and yokes are coupled at the corners, movement of one will affect the behaviour of the other.

As the core is a structure consisting of distributed mass and stiffness, it has an infinite number of natural frequencies, one of which may coincide with one of the harmonics of the magnetostrictive or magnetic forces, resulting in greatly magnified core vibration. Attempts[8, 9] have been made to calculate these natural frequencies for both in-plane and out-of-plane directions for different types of cores. Because of the considerable mathematical effort involved, computers are required for such calculations.

8.2.2   Transmission

Core vibrations can be transmitted to the tank through the core supports or through the oil. The transmission through the supports can be substantially reduced by resting the core on springs or other anti-vibration mountings. In any case, if possible, the core should not be rigidly tied to any part of the tank at any point. Since oil is relatively incompressible, the proportion of vibrational power transmitted through it is appreciable and is probably far more than that transmitted through the supports. If it is assumed that half the energy is transmitted through the supports then, even if highly efficient anti-vibration mounts were fitted, the transmitted energy would only be halved, resulting in a decrease in the noise level of only about 3 dB. This change in noise level is only just discernible by the human ear. For any sizable reduction it is essential that the vibrational energy transmitted along all paths is attenuated.

8.2.3   Tanks

The tank modifies the vibrational pattern in the oil and, therefore, probably also that of the core itself and adopts its own modes of vibration with the particular forcing frequencies which are imposed on it. In some cases individual panels may resonate, and this will result in increased vibrational amplitudes. As a tank possesses panels of different size and as these are neither freely supported nor rigidly fixed, the calculation of natural frequencies is very difficult and is made even more so by the various attachments to the tank such as cable boxes, tap-change gear pockets, etc. There are various ways in which tanks can be tested for resonance[10]. If such resonances are discovered, they may be remedied by altering the mass-to-stiffness ratio of the panel in question.

The amount of noise radiated depends on the vibrational pattern of the tank and tank panels, that is on the vibrational amplitudes and modes and on whether panels vibrate in or out of phase. Interference takes place between the various sound waves radiating from different parts of the tank, and this results in fairly abrupt changes in noise level around the tank.

8.3   TRANSFORMER NOISE

The sound level measured close to a transformer and around its periphery varies from point to point. For that reason it is usual to take a number of noise level readings (see section 8.5) and to average them arithmetically to obtain the arithmetic mean for the sound levels. It is often suggested that a quadratic mean is a more useful representation of a group of sound levels. The quadratic mean level L_{av} of the group of sound levels $L_1, \ldots, L_n$ is

$$L_{av} = L_{tot} - 10 \log_{10}(n) \qquad (8.1)$$

It represents the level of the average value of p_x^2, where p_x is the sound pressure in newtons per square metre at any point and is useful when sound power levels are to be calculated from the sound pressure levels, since the power is proportional to pressure squared. The quadratic mean is always greater than the arithmetic mean but, if the range of sound pressures is less than 10 dB, then the difference between both mean levels will usually be less than 1 dB.

Plotting the sound levels of transformers against the logarithm of their rated sizes shows that the points will be scattered about a straight line or a series of straight lines. Figure 8.3 gives average noise levels obtained by using measurement methods described in British specifications (see table 8.1, third column). The

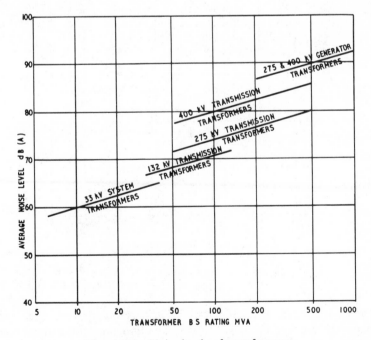

Figure 8.3   Noise levels of transformers

noise level increases with size or mass of the core and with increasing flux density. Small reductions may be made by lowering the flux density, but the resultant increase in the core mass partially offsets the improvement obtained. For a fixed core mass the relationship between the change in noise level and flux density is approximately $25\,dB(A)\,T^{-1}$.

For more detailed work the noise spectrum must be known. Figure 8.4 shows typical spectra covering different types and sizes of transformers; approximate mean levels measured 0.3 m from the transformer tank are given.

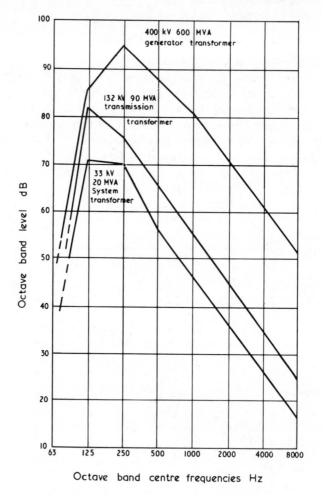

Figure 8.4   Transformer noise spectra

The effect of distance on a transformer noise harmonic and the noise level is of importance. It has been investigated by taking measurements around transformers at different distances. The harmonic readings obtained at a given distance were then averaged, and the differences between these average levels were plotted against distance. It was found[11] that the curves plotted for different transformers vary markedly even when the transformers are sited in open areas. Figure 8.5 shows a typical curve a; extreme forms are represented by curves b and c. Three

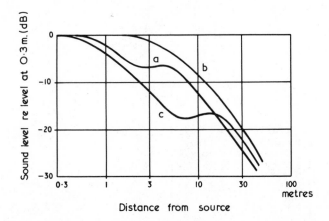

Figure 8.5   Distance attenuation: curve a, typical curve; curves b and c, extreme forms

distinct regions can be seen: (1) the near field where little attenuation occurs, (2) an intermediate region and (3) the far field. In the intermediate region the various harmonics of any one transformer may attenuate at different rates, and these rates may be completely different in other transformers. In some cases the level may even rise slightly as the distance from the transformer is increased. In the far field the different curves tend to converge, and thereafter the rate tends to stabilise at 5 to 6 dB per doubling of distance. If the A-scale sound level readings of different transformers are plotted, the curves lie very much closer together as can be seen from figure 8.6, which shows the distance attenuation of several transformers of different manufacture ranging from $7\frac{1}{2}$ to 100 MVA.

When the noise of transformers is measured on site, free-field conditions usually do not exist as switch houses, blast walls and other obstacles intervene, causing reflection, standing waves or pressure build-up. Diffractional effects are caused by walls, and this again leads to difficulties in the estimation of noise levels at some distance from the transformer. Further complications arise with wind conditions. If the air temperature changes with height from the ground, sound waves emanating from the transformers will be curved away from the warmer layer.

8.4 FAN NOISE

The noise made by the fans fitted to the cooler contributes to the total noise, but

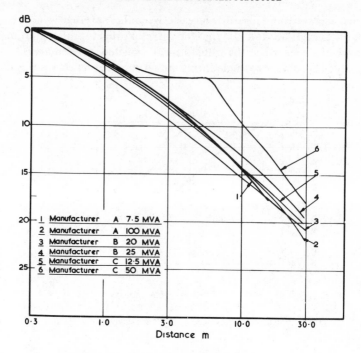

Figure 8.6   Distance attenuation, A scale

its frequency spectrum is continuous, with perhaps some isolated pure tones in it. It therefore differs from that of transformers. In many cases fan noise can be neglected, for instance when the fans are only in operation during the daytime, but in others it may be necessary to determine the spectrum of the fans. These may in fact mask the noise produced by the transformer. Typical noise spectra for two types of coolers are shown in figure 8.7[12].

8.5   MEASUREMENT SPECIFICATIONS

Specifications exist in many countries, and all stipulate the use of a sound level meter which operates on the A scale. For measurement, the microphone is held at a specified height and distance from the transformer contour, readings being taken around the periphery of the transformer with the microphone locations usually not more than 1 m apart, subject to a minimum number of readings. The contour is obtained by placing an imaginary string around the plan projection of the transformer including all pockets, boxes, stiffeners and attachments (including tank-attached radiators) below tank cover height, but excluding valves, transport lugs, projections above tank cover height and separate radiator banks. The individual A-scale readings are usually added arithmetically, and the

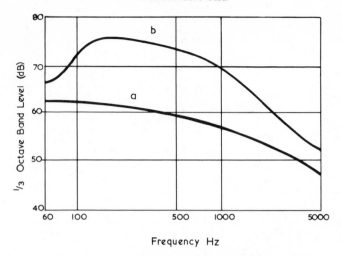

Figure 8.7   Noise spectrum of coolers, measured at 2 m distance: curve a, four propeller fans, 560 rev min⁻¹, 0.90 m in diameter, mounted under radiator bank, as used also to provide natural-air cooling; curve b, one axial-flow fan, 715 rev min⁻¹, 0.95 m in diameter, fitted to oil cooler, as used to provide only forced-air cooling

arithmetic mean value plus corrections for cable, microphone or instrument error is taken as the noise level of the transformer (in the UK this is known as the average surface noise level).

Measurements should be carried out in an environment free from reflecting surfaces and with a background level preferably more than 7 dB below the level of the transformer. To verify this, the background is measured both immediately before and after the transformer has been energised for its noise test. Table 8.1 shows the variation between national specifications. It covers the main features, but reference should be made to the original documents before attempting to work to any of these specifications.

Finally, it must be stressed that the object of a specification is to permit comparison between different transformers. However, compliance with a specification is not a guarantee that a transformer will not cause any annoyance to nearby residents.

8.6   REMEDIAL MEASURES

8.6.1   Manufacturer

The following measures can be taken by the manufacturer.

(1)   Cross-fluxing by suitable core dimensioning and design should be reduced.

TABLE 8.1 Noise level measurement specifications (1972)

Country	International	UK	USA
Issuing authority	International Electrotechnical Commission (IEC)	British Electricity Boards (BEB)	National Electrical Manufacturers' Association (NEMA)
Specification number	not yet published	T2 (1966)	TR1 (1968)
Sound level meter scale	A	A	A
Microphone height	tank < 2.5 m, one-half tank height tank ≥ 2.5 m, one-third and two-thirds tank height cooler < 4 m, one-half cooler height cooler ≥ 4 m, one-third and two-thirds cooler height	1.2 m or one-half tank height, whichever is the lesser	tank < 2.4 m, one-half tank height tank ≥ 2.4 m, one-third and two-thirds tank height
Measurement position — Minimum number	6 for transformer only; 10 when cooler involved	6	4
Maximum internal	1 m	1 m	0.9 m
Averaging method	range < 10 dB, arithmetic; range ≥ 10 dB, quadratic	arithmetic	arithmetic
Cooling equipment noise measurements	routine only if coolers < 10 m from transformer	when requested	routine (if coolers tank mounted)
Equivalent sound level of reference hemisphere	not required	not required	not required
Effective measurement surface area	must be stated	not required	not required
Harmonic analysis	not required	not required	not required
Maximum permissible sound levels	none stated	depend on megavoltamperes, voltage and cooler position	depend on megavoltamperes, voltage and cooler type

Cooler position	Cooler type	(1) (m)	(2)	(3)	(1) (m)	(2)	(3)	(1) (m)	(2)	(3)
tank mounted	natural air	0.3	TC	T	0.3	TC	T	0.3	TC	T
	forced air	2	TC	T[a], TC[b]	0.3[a,b,c]	TC	T[a], C[b]	0.3 and 1.8	TC	TC
	water	—	—	—	0.3	TC	T	0.3	TC	TC
separate, <3 m from tank	natural air	0.3	TC	T	0.3	T	T	c	c	c
	forced air	2	TC	T[a], TC[b]	0.3[a,b,c]	T[a], C[b]	T[a], C[b]			
	water	—	—	—	0.3	T	T			
separate, >3 m from tank	natural air	0.3	T	T	0.3	T	T	c	c	c
	forced air	2	T[a], C[b]	T[a], C[b]	0.3[a,b,c]	T[a], C[b]	T[a], C[b]			
	water	0.3	T	T	0.3	T	T			

TABLE 8.1 (*continued*)

Country	France	West Germany
Issuing authority	Electricite de France (EDF)	Verband Deutscher Elektrotechniker eV (VDE)
Specification number	HN52–02 (1965)	0532 Part 1 (11.71)
Sound level matter scale	A and C	A
Microphone height	tank or cooler < 2.5 m, one-half height tank or cooler ≥ 2.5 m, one-third and two-thirds height	one-half tank height
Measurement positions Minimum number	10	normally only 4 positions if rating ≤ 1.6 MVA and 8 if rating > 1.6 MVA; for comparison test minimum number is 4 at 1 m intervals
Maximum internal	1 m	
Averaging method	quadratic	arithmetic
Cooling equipment noise measurements	routine	normally routine; for comparison test, not required
Equivalent sound level of reference hemisphere	required for 3 m radius hemisphere	not required
Effective measurement surface area	must be stated	not required
Harmonic analysis	not required	required at one position
Maximum permissible sound levels	none stated	none stated (German spec. DIN 42540 does contain levels)

Cooler position	Cooler type	(1) (m)	(2)	(3)	(1)[d] (m)	(2)	(3)
tank mounted	natural air	0.3	TC	T	1, 3, 5	TC	T
	forced air	0.3[a], 1[b]	TC	T[a], C[c]	1, 3, 5	TC	TC
	water	—	—	—	1, 3, 5	TC	TC
separate < 3 m from tank	natural air	0.3	T	T	1, 3, 5	TC	T
	forced air	0.3[a], 1[b]	T[a], C[c]	T[a], C[c]	1, 3, 5	TC	TC
	water	—	—	—	1, 3, 5	TC	TC
separate, > 3 m from tank	natural air	0.3	T	T	1, 3, 5	TC	T
	forced air	0.3[a], 1[b]	T[a], C[c]	T[a], C[c]	1, 3, 5	TC	TC
	water	—	—	—	1, 3, 5	TC	TC

Column (1), microphone distance; column (2), transformer T and/or cooler C included in contour; column (3), transformer T and/or cooler fans C energised.

[a] Test to determine noise from transformer (without any cooler noise).

[b] Test to take account of cooler noise

[c] Subject to agreement.

[d] Microphone distance depends on rating: ≥ 1.6 MVA, 1 m; < 1.6 MVA, ≥ 65 MVA, 3 m; < 65 MVA, 5 m.

(2) Flat steel of low magnetostriction[13] should be used.
(3) Flux distortion should be kept to a minimum.
(4) Careful annealing.
(5) The mechanical stresses in the core[13] should be limited.
(6) Core or tank resonances should be avoided.
(7) There should be no rigid connections between core and tank.
(8) External cladding.

It is unlikely that radical improvements in core vibration and hence generated noise will occur, but research into transformer core steels is bringing about steady reductions in magnetostriction and sensitivity to mechanical stresses. Reduction of a few decibels are therefore often possible by using the latest special core steels.

8.6.2 On site

The most obvious way of reducing the effect of transformer noise is by increasing the distance between the transformer and the nearest house. Figure 8.6 could be used to estimate the reduction of noise level with increasing distances from the transformer.

Walls, anti-vibration pads

Walls may be erected, preferably at least 1 m higher than the transformer and located close to it. On average a reduction of about 10 dB(A) on the far side of the wall will be obtained, but people living on the opposite side may receive more than their share! If the wall is set too far away or not high enough, the shadow effect will be materially reduced. Shrubs are frequently planted for more psychological and decorative reasons and may thus help to reduce the likelihood of complaints. In themselves, their contribution to noise reduction is negligible.

Anti-vibration pads are used under transformers to minimise the transmission of vibration from the tank to the plinth. They are only required if the plinth is connected to any wall or building (either directly or through rocky ground) which, if subjected to forced vibration, will radiate noise. Where such connection does not exist, the fitting of anti-vibration pads will be of no value in reducing noise. The pads are made of oil-resistant elastomer and are usually designed for a low natural frequency (that is about 10 Hz) of the combination of transformer mass and pad stiffness. The dynamic stiffness[14] must be used for the calculation, which in any case is only approximate because the mass of the transformer is greater than that of the plinth, because the transformer is not a simple mass and because the compliance and effective mass of the earth are not known.

Enclosures

Enclosures[15, 16, 17] may be classified as follows.

(1) Free-standing enclosure. This is a separate building completely surrounding the transformer, fitted with roof and doors and usually with access space between the transformer and the enclosure walls. This type of

enclosure is erected after the transformer has been installed and may be built of steel, brick or concrete[18].

(2) Tank-supported enclosure (tank cladding). This is integral with the transformer tank and is supported from it. It is usually of steel and may screen the sides of the tank only or may form a hood over the whole transformer. It must be resiliently supported from the tank to minimise the transmission of vibration.

All enclosures must be adequately sealed to prevent escape of noise. The greater the desired noise insulation, the more care needs to be taken over the sealing. Where the area of gaps and cracks which are to be filled is large in relation to the surface area of the enclosure the sealing material must be applied in depth[15]. It is not sufficient to rely on the fact that the enclosure is light proof.

The doors must be made heavy enough in relation to the mass of the enclosure walls and roof and properly sealed; otherwise the sound insulation of the enclosure will be impaired.

Vents may be provided to allow oil fumes to escape, but they do not contribute any cooling for the transformer. The vent area must be limited[19] to maintain the sound insulation of the enclosure. Where large vents are needed, silencers should be fitted.

It is important to minimise the transmission of vibration from the transformer to the enclosure, for instance by pipework or electric conductor connections. Where pipes are brought through the enclosure walls, flexible sealing must be provided between them and the wall. Similar methods are used for bushings which are usually mounted on turrets, which are sealed to the enclosure roof by means of flexible sealing material.

To limit the noise emitted by pipes between the enclosure and the radiator bank, flexible bellows are inserted in the pipelines on the inside of the enclosure.

The (average surface) noise level of an enclosed transformer is most conveniently measured at a distance of 0.3 m from the enclosure. The difference between this level and the (average surface) noise level of the unenclosed transformer may be termed the average-surface-noise reduction of the enclosure.

The noise level measured inside an enclosure is usually higher than that measured at the same distance from the transformer before it was enclosed (that is under free-field conditions). This increase is called build-up. It is caused by reverberation or by sound pressure increase due to the proximity of the enclosure to the transformer, and values of 10 dB(A) and above[19, 20] have been measured. The fitting of absorbent material on the inside of the enclosure reduces the build-up, but tests[20] have shown that the decrease in sound level inside the enclosure is not always accompanied by a similar decrease on the outside.

Tank-attached enclosures give average-surface-noise reductions of about 10 to 15 dB(A), whereas with free-standing enclosures values of 20 to 25 dB(A) can be obtained. Where noise reductions greater than about 30 dB(A) are required, single enclosures are no longer sufficient. There have been built some double enclosures in the UK. The results indicate that attenuations of 40 dB(A) or more are feasible provided great attention is paid to details; in particular vibrations must be prevented from being transmitted from the transformer plinth or the inner enclosure to the outer enclosure walls.

The average-surface-noise reduction of an enclosure must not be equated with the decrease in noise level at some distance from the transformer, say, at a nearby house after an existing transformer is enclosed. This decrease, which can be termed insertion loss, depends also on the distance attenuation. With large free-standing enclosures the distance attenuation can be less than that of the unenclosed transformer[15].

8.7 PLANNING

8.7.1 Residential areas

With a knowledge of the transformer noise and its harmonics and taking account of distance attenuation, effect of reflections, walls, enclosures, etc., the noise level at the nearest house can be estimated with a reasonable degree of accuracy before the transformer is installed. The threshold of audibility of the major harmonics (say, 100 to 400 Hz) in the presence of background noise can be calculated by using the measured ambient* spectrum. Comparison between the audibility threshold values and the estimated sound pressure levels of the transformer harmonics at the house will show whether complaints are likely. Figure 8.8 shows a plot of surburban ambient noise measured on a one-third octave band analyser. The estimated[3] sound pressure levels of the pure tones of 100, 200, . . . Hz which are just masked by that ambient, that is the tones which would just be inaudible to

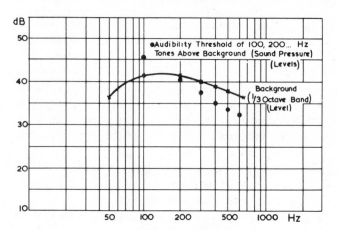

Figure 8.8   Background level and audibility threshold of major harmonics[3]

---

* The ambient is the noise existing in the area before the transformer is energised.

the average listener, are obtained by adding the values given in table 8.2 to the one-third octave band levels of the ambient read off at frequencies 100, 200, ... Hz. These sound pressure levels have been marked in on figure 8.8. Obviously large variations will occur between young and old people and between different individuals, quite apart from any psychological factors which are of great importance and may in fact determine whether the noise is annoying or not. If the estimated sound pressure levels of the transformer harmonics exceed the estimated thresholds by more than about 5 to 10 dB, it will be advisable to take some action beforehand. For a large reduction an enclosure, or in some cases even a double enclosure, will be needed. Where the difference is only 5 dB it will depend to a large extent on the policy decision of the utility which is installing the transformer, whether cladding should wait until complaints by residents are actually made.

A British standard method for predicting the likelihood of complaints of excessive noise[21] utilises measurements taken with the A weighting of the sound level meter. This has the advantage of simplicity. The use of the A scale also conforms with normal transformer noise level measurement practice. The accuracy of this method, although often adequate, will be inferior to the method involving harmonic analysis. An international standard very similar to the British standard has also been published[22].

8.7.2 Offices

Where offices are erected near a transformer the noise criterion curves[23] shown in figure 8.9 may be used, together with table 8.3, to determine whether remedial measures are necessary. The curves apply for various types of offices. The sound pressure levels of the different transformer harmonics in the offices should be estimated and should not exceed the values given by the appropriate curve. It requires some experience and knowledge to assess how much reduction to allow for intervening spaces or rooms and for walls and windows. Similarly, structure-borne vibrations transmitted into the offices must be considered.

8.8 PROSPECTS FOR THE FUTURE

It is likely that the trends which have become apparent in the late 1960s and early 1970s will continue for some time. The most striking of these is the steady reduction of transmission transformer noise (see figure 8.10(a)). This reduction can be attributed mainly to improved core steel because the 60 MVA 132 kV units, whose noise levels were used to derive figure 8.10(a), were all of similar construction. It is probable that noise levels of large generator transformers will not fall so rapidly because of the higher harmonic content of the noise due to the use of high flux densities.

When the probability of complaints of excessive transformer noise is being

TABLE 8.2 *Masking of transformer harmonics by background noise*[3]

Frequency (Hz)	100	200	300	400	500
Value to be added to background to obtain audibility threshold (dB)	+4	−1	−2½	−4	−5

Background is measured with one-third octave band analyser.

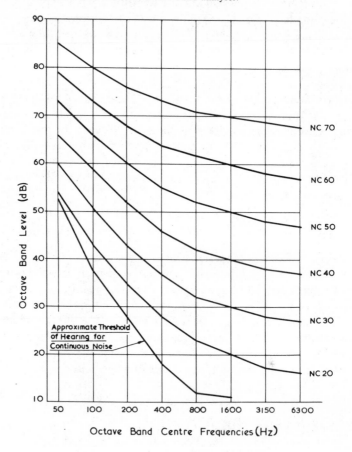

Figure 8.9   Noise criterion curves[23]

assessed, it is necessary to consider not only how the level of a particular size of transformer is falling but how the sizes of transformers in particular locations are rising. This is indicated in figure 8.10(b). The data from figure 8.10(a) and (b) have been combined to produce figure 8.10(c). It is apparent that, even though sizes of transformers are increasing, the potential nuisance could well be held at its present level.

Methods of noise measurement for transformers, well established in Europe

TABLE 8.3 *Recommended noise criteria for offices*[23] (referring to figure 8.9)

Noise criteria curve	Communication environment	Typical application
20–30	very quiet office; telephone use satisfactory; suitable for large conferences	executive offices and conference rooms for 50 people
30–35	quiet offices; satisfactory for conferences at a 5 m table; normal voice 3 to 10 m; telephone use satisfactory	private or semi-private offices, reception rooms and small conference rooms for 20 people
35–40	satisfactory for conferences at a 1.80 to 2.40 m table; telephone use satisfactory; normal voice 3 to 4 m	medium-sized offices and industrial business offices
40–50	satisfactory for conferences at a 1.20 to 1.50 m table; telephone use occasionally slightly difficult; normal voice 0.90 to 1.80 m, raised voice 1.80 to 3.60 m	large engineering and drafting rooms, etc.
50–55	unsatisfactory for conferences of more than two or three people; telephone use slightly difficult; normal voice 0.30 to 0.60 m, raised voice 0.90 to 1.80 m	secretarial areas, (typing) accounting areas (business machines), blueprint rooms, etc.
above 55	very noisy: office environment unsatisfactory; telephone use difficult	not recommended for any type of office

and the USA, are unlikely to change markedly, but there will probably be wider acceptance of international rather than national specifications. The continuing use of the A weighting network of the sound level meter is well assured, and there will be an increasing tendency to calculate sound power levels from the normal sound levels, together with the effective sound radiating area of the transformer. This will allow more valid comparisons between transformers of various sizes and constructions measured to different specifications. Noise spectrum determination will not become part of a routine test in the foreseeable future, but manufacturers and users will have to maintain their information on the harmonic content of the noise.

The future does not seem to hold any problems as far as instrumentation is concerned. Routine tests will be facilitated by the use of lightweight portable precision sound level meters, and very sophisticated equipment will be available for research.

Research and development on transformer noise reached a peak, in Europe and the USA, in the early 1960s. Since that time emphasis has continued and will continue to move steadily over to control of noise on site rather than to attempt to make very large reductions at the source.

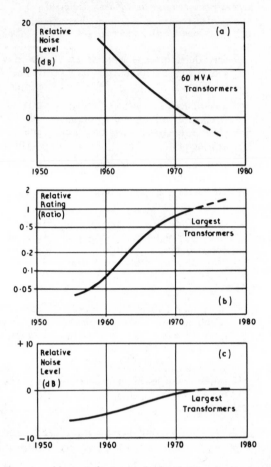

Figure 8.10 Modern trend in transformer noise level: (a) noise levels of 60 MVA 132 kV transmission transformers; (b) maximum transformer ratings; (c) noise levels of largest transformers

Legislation on maximum permissible noise levels in various localities is a possibility. Thus, electricity supply undertakings will make noise control a routine part of all sub-station and power station planning. New methods of noise control are unlikely to emerge, but existing knowledge will be more widely applied.

ACKNOWLEDGEMENT

The authors wish to thank Ferranti Limited for permission to publish this information.

REFERENCES

1. Fletcher, H., and Munson, W. A., *Equal Loudness Contours for Pure Tones, ASA Publ.*, No. Z 24.2 (1942)
2. Churcher, B. G., and King, A. J., The performance of noise meters, *J. Inst. Electr. Eng.*, **81**, (1937) 57
3. French, N. R., and Steinberg, J. C., Intelligibility of speech sounds, *J. Acoust. Soc. Am.*, **19**, (1947) 90
4. George, W. R., Holt, E., and Thompson, J. E., Magnetostriction in steel, *Proc. Inst. Electr. Eng., Part A* **109**, (1962) 101
5. Ruhlmann, R., Harmonics in the noise emitted by transformers, *Elektrotech. Z., Ausg. A,* **85**, (1964) 430
6. Whitaker, J. D., Magnetostriction in steel, *Electr. Times*, **137**, (1960) 675
7. Ferranti Limited, *Vibrations of Transformer Cores* (1960) (animated film based on measurements)
8. Jordan, H., Reinke, H., and Taegen, F., Mechanical natural frequencies of transformers, *AEG Mitt.*, **52**, (1962) 445
9. Henshall, R. D., Bennett, P. J., McCallion, H., and Milner, M., Natural frequencies and mode shapes of vibration of transformer cores, *Proc. Inst. Electr. Eng.*, **112**, (1965) 2133
10. Cervini, G. C., Variable frequency constant voltage three-phase supply for the observation of vibrations in transformers, *Elettrotecnica*, **L1-N 8 bis**, (1964) 630
11. Mattei, J., Acoustical study of electrical generating and transmitting stations, *Rev. Gen. Electr.*, **74**, (1965) 205
12. Milner, M., Problems of transformer noise, *Mach. Lloyd*, April **9**, (1963)
13. Kerr, H. W., and Palmer, S., Developments in the design of large power transformers, *Proc. Inst. Electr. Eng.*, **111**, (1964) 823
14. Brownsey, C. M., The assessment of performance of anti-vibration materials, *Proc. Inst. Electr. Eng., Part A,* **109**, (1962) Suppl. No. 3, 231
15. Keil, C., and Milner, M., *The Performance of Noise Enclosures for Transformers, CIGRE Rep.*, No. 105, No. 105A (1964)
16. Jackson, R. S., Performance of acoustic hoods, *Acustica*, **12**, (1962) 139
17. Jump, L., Causes and reduction of transformer noise, *AEI Eng.*, (1964) 200
18. Milner, M., and Bennett, P. J., Transformer noise reduction, *Electr. Times*, **25** (1963)
19. Brownsey, C. M., *The Problem of Noise with Particular Reference to Transformers, Cent. Electr. Res. Lab. (Leatherhead), Mem.*, No. 3 (1956)
20. Milner, M., and Dunsbee, J., Transformer noise harmonics, *5th Int. Cong. Acoust., Liège Pap.*, No. F. 12 (1956)
21. British Standards Institution, *BS 4142, Method of Rating Industrial Noise Affecting Mixed Residential and Industrial Areas* (1973)
22. International Organisation for Standardisation, *ISO R-1996, Assessment of*

Noise with Respect to Community Response (1971)

23. Beranek, L. L., *Noise Reduction*, McGraw-Hill, New York (1960) 519

ADDITIONAL REFERENCES

General

a. Beranek, L. L., *Acoustics*, McGraw-Hill, New York (1954)
b. King, A. J., *Measurement and Suppression of Noise (with Special Reference to Electrical Machines)*, Chapman and Hall, London (1965)

Acoustical terms

a. British Standards Institution, *BS 661, Glossary of Acoustical Terms* (1969)
b. United States of America Standards Institute, *USAS S1.1, Acoustical Terminology (Including Mechanical Shock and Vibration)* (1960)
c. International Electrotechnical Commission, *IEC 50 (08), International Electrotechnical Vocabulary, Group 08, Electro Acoustics* (1960)

Transformer noise

a. IEEE Committee Report Bibliography on transformer noise, *IEEE Trans. Power Appar. Syst.*, **87,** No. 2, February (1968)
b. British Electrical and Allied Manufacturers' Association Limited, *Guide to Transformer Noise Measurement, BEAMA Publ.*, No. 227 (1968)
c. British Electrical and Allied Manufacturers' Association Limited, *Guide to Transformer Noise Terminology, BEAMA Publ.*, No. 231 (1969)

Sound level meters

a. British Standards Institution, *BS 4197, Specification for a Precision Sound Level Meter* (1967)
b. International Electrotechnical Commission, *IEC 179, Precision Sound Level Meters* (1973)

9

Distribution Transformers

H. K. Homfray* and D. Boyle†

9.1 DEFINITION AND CLASSIFICATION

While there is no generally recognised definition, a distribution transformer may be defined as a transformer used to supply power, for general purposes, at final distribution voltage level from a higher-voltage distribution system. The minimum rating is usually regarded as 5 kVA, but the maximum is somewhat indefinite. For public supply purposes it is not common to use rating exceeding 1000 kVA, but industrial sub-stations frequently have higher ratings. 3 MVA is normally the practical limit for loading at medium voltage, as the impedance necessary to limit fault current to the usual switchgear ratings rises to impractical percentages at higher ratings.

Utilisation voltage in Britain is standardised at 240 V single-phase or 415 V three-phase. The higher-voltage system is usually 6.6 or 11 kV, but values up to 33 kV may be used, more particularly in other countries. Transformers used for conversion from distribution to lower voltages (110 V or less) for control or safety purposes are not usually regarded as distribution transformers, though similar constructions are employed for the larger sizes of such transformers. Transformers used for specific purposes such as supplying rectifiers or furnace loads would be regarded as special types, but those supplying large motors as individual loads can be regarded as distribution types, and indeed standard transformers may perform this duty. Power station auxiliary transformers, supplying medium voltage, are a special case of distribution transformers and can usefully be considered with them, as the ratings are similar, although certain special features, notably fittings, are involved. In this case the high-voltage system is commonly 3.3 kV.

* GEC Power Transformers Limited, formerly English Electric Company Limited (sections 9.1 to 9.8).
† Formerly Ferranti Limited (sections 9.9 to 9.14).

Distribution transformers may be broadly classified by type of use. Supply authority transformers are virtually all oil cooled (ON). Although they may be installed indoors or outdoors, the differences in construction for the two situations are very small. They may be further classified by type of installation, either directly on a pole or on a pole-mounted platform for rural distribution, with ratings up to 200 kVA, or for ground mounting mostly for urban distribution, from 100 kVA upwards. Consumer and industrial-use transformers are those used by large consumers who take supply at high voltage and provide their own sub-station or sub-stations. They may be indoor or outdoor.

Increasing use is being made of insulating and cooling media other than oil, either air (AN) with class C insulation or synthetic liquid (SN). With either type there is no fire risk, and fewer restrictions on siting. (AN transformers are discussed in chapter 11.) By their nature, consumer-owned transformers are usually of high rating and ground mounted, though sub-stations are not necessarily at ground level.

9.2 DESIGN CONSIDERATIONS

9.2.1 Specifications

The designer's specification (see section 3.3) is governed by standard specifications[G1.7, G2.2, G3.2], by the customer's specification (see section 3.2), for which reference 1 is a comprehensive example, and by the particular manufacturer's own standards and code of practice.

9.2.2 Economic considerations

Since the distribution transformer is in the last link of the chain from generating station to ultimate consumer, since some degree of diversity must occur between the lower- and higher-power portions of the supply system and since security of supply must be provided, the total installed capacity of distribution transformers is more than that of generating plant, the ratio being about $1\frac{1}{2}$ to 1. The chain is of course somewhat lengthy, so that perhaps a quarter of the total capacity of all transformers installed is in distribution transformers. Furthermore, since the cost per kilovoltampere increases as the rating of any plant decreases, the proportion of capital value of total installed transformer capacity represented by distribution transformers is probably nearer one-third. It is obvious that there is a strong incentive to keep capital costs down, though not at the expense of running costs, since losses at this stage of the network may represent roughly 1 % of the total power generated.

As usual in engineering, a compromise is required between the running costs of losses and capital cost which increases as loss levels decrease. Under any given set of circumstances an optimum condition can be found by applying capitalisation of loss methods. However, for the multitude of different situations in a distribution network it would not be economic to design or manufacture transformers of the same rating with many different loss levels. Therefore, for distribution purposes, an average or typical situation may be used to determine standard losses. However, for the larger transformer sizes, particularly for industrial customers, it may be worth considering individual cases. The customer will usually specify the loss level, or possibly alternative loss levels required. Where loss levels are specified, the designer's problem is basically that of finding the best compromise between manufacturing costs and technical considerations. The more transformers that can, through acceptance of standard designs, eventually be made to one design, the greater is the effort at reducing costs on that design that can be justified.

Since iron losses are always present, while copper losses are only incurred while carrying revenue earning loads, there is a tendency, especially on small units for rural distribution with low load factor, for the ratio of iron to copper losses to be very low. For example, in the specification given in reference 1 the ratio is approximately 1 to 6.5, up to 1000 kVA rating. This ratio is lower than is the practice in most other countries. However, with revised bulk tariffs there is a tendency to decrease the ratio at least on the larger sizes. The ratio to give the most economical performance in a particular situation must not be confused with the ratio that gives the highest electrical efficiency, since the assessed costs of the iron and copper losses per kilowatt are usually far from equal.

The other factor required by the designer is the percentage reactance of the transformer. Low values of reactance mean low-voltage regulation. For the range from 100 to 1000 kVA the usual reactance value is between 4 and 6 %, but on the smaller ratings lower values may be employed since an actual fault current will fall proportionally to the rating at constant reactance. The cost of a transformer is little affected by the actual reactance value over a fairly wide range of values. In making fault calculations for secondary circuits, the contribution of the primary network is rarely completely negligible and should be taken into account.

9.3   CORE CONSTRUCTION

The subject of cores is substantially dealt with in chapter 4. As described in section 4.2 the core material is invariably cold-rolled grain-oriented steel, except occasionally in the very smallest transformer sizes. This material enables low iron losses to be obtained in an economic manner. Three-phase transformers are of the core type (see section 4.3). Single-phase cores are usually of the same type rather than of shell form, as this gives better cooling of the coils, and the increased cost in winding two sets of coils is more than offset by the smaller mean turn of the coils and by smaller number of core plates that have to be cut and built. Frequently also the longer effective leg length of the single-phase core type, which is the sum of the two coil lengths, enables the required reactance to be obtained in a more economical manner. The number of steps for stepped-core sections is usually between five and seven, the choice being determined by the values of strip widths available in the manufacturer's range; the cost of the core is almost independent of the number of steps actually employed.

Mitred corner joints instead of square overlaps are commonly used to reduce the additional loss of corners (see sub-section 4.3.3). With square overlap joints it is possible to increase the section of the yokes to obtain lower losses and noise level without altering the space available for the windings, and yoke sections up to 20% greater than the leg sections are employed by some makers. On a typical 500 kVA transformer, the use of square overlap core joints with the yoke cross-section 10% greater than the leg section gives approximately the same loss as the use of mitred corner joints and equal leg and yoke sections but at the cost of using more iron. The cutting and building of mitred cores tends to require more labour and more expensive equipment, but, with reasonably large-scale production, the balance of advantage is in favour of mitred cores.

An alternative method of avoiding corner losses is to use strip-wound cores. For single-phase units these are wound from a continuous strip of grain-oriented steel, so that the direction of optimum orientation is everywhere parallel to the flux paths. By employing special winding machines, it is possible to wind coils on a closed core, the winding being supported during winding by a collapsible mandrel placed round the core legs, and driven by gears on the end of the mandrel, the mandrel being removed after the winding is completed. The losses on such an uncut core after annealing will be scarcely more than the losses of the material measured on an Epstein square test (see figure 9.1). With this method very low

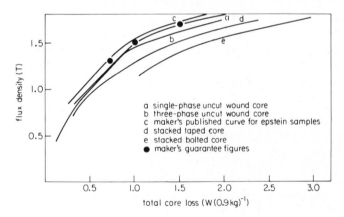

Figure 9.1   Total core loss in built cores

losses have been obtained for rural pole-mounted transformers up to 100 kVA, for example about 22 W on a 5 kVA transformer. To avoid the use of special coil-winding machines, it is possible to consolidate the core by bonding the individual turns together with epoxy resin and then to cut it into two halves through the leg portions. The cut faces are then ground and etched to obtain a close-fitting butt joint, without bridging of the interlaminar insulation by burrs, but these operations considerably increase the core loss. Various forms of 'take-apart' core have been developed and used in the USA, for transformers up to 100 to 200 kVA, which have turns cut individually, or in small groups so that the core can be inserted a portion at a time into the coils and so reassembled to give a core with the gaps usually distributed along the legs. The losses of such a core are intermediate between those of a cut and an uncut wound core. Manufacture of this type of core may have started also elsewhere.

For three-phase transformers, three wound cores can be combined to give a three-leg type core, either cut or uncut, the third core being wound over the other two placed side by side (see figure 4.7(c)). Flux redistribution between the three constituent cores takes place in the yokes, with the result that the specific iron loss is about 30% higher than that for a single-phase wound core. The uncut form is used up to 100 kVA, with advantages over a conventional core built up from plates. Theoretical considerations suggest that the ideal form of a three-phase core would have the cores disposed at the corners of an equilateral triangle, with a circular or triangular yoke, so that complete symmetry between legs is achieved. This arrangement has been used by some European manufacturers, but there are considerable mechanical difficulties in construction.

For distribution-size transformers core laminations are held together by means of tape, usually cotton webbing, wound on the legs while these are under tension. Core bolts are no longer used on this size of transformer. Yokes are usually clamped together between steel sections by bolts external to the core, though on the largest sizes of distribution transformers where core masses exceed 1 t, there may be a single bolt through each yoke on the centre line of the middle leg. The top and bottom core clamps are connected by tie rods outside the windings which serve to apply end clamping pressure onto the windings and to distribute the mass when the core and windings are lifted. With wound cores, on the small sizes, the U-shaped clamp encloses the top and bottom yokes, and with uncut cores the tie rods usually pass inside the coils in the space left by the removal of the collapsible winding mandrel. Although such tie rods are inside the coil, they do not encircle the main core flux and so do not require insulation to prevent them acting as short-circuiting turns.

Flux densities are nominally 1.65 T, but lower values are employed in order to limit noise and switching inrush currents for the larger sizes and to obtain low iron losses on the small rural types. If low noise level or iron loss is specifically required, the flux densities may be reduced to as low as 1.0 T.

9.4   WINDING CONSTRUCTION

A general treatment on windings is given in chapter 5.

9.4.1 Low-voltage windings

Low-voltage windings are placed nearest the core, since the insulation level required is low. Normal practice is to use helically wound coils. An axial cooling duct is normally provided in the middle of the radial dimension, and for the larger ratings the coil will probably be of only two layers. For the heavier currents, multiple rectangular strip conductors are wound in parallel, with transpositions at the mid-point of the windings. Although theoretically desirable if more than two conductors radially are wound in parallel, transpositions at more than one point are not always economically justified. With an even number of layers in the coil, a single transposition placed at the end of a layer—which is the most convenient position—is often sufficient.

Conductor coverings used are paper and enamel (PVA), enamel being slightly cheaper. Interlayer insulation, if required in addition to cooling ducts, is pressboard. Spiral collars of pressboard or synthetic resin-bonded paper, which has greater mechanical strength, are taped to the end turns to provide flat end surfaces for supports and to support the opposite lay of adjacent layers. If the high-voltage winding consists of more than one helical coil or is a disk winding with tappings, spacing collars may be required in the low-voltage winding to maintain the same axial ampere turns distribution as in the high-voltage winding. On single-phase transformers of core type, the coils on the two legs may be connected in series or parallel, as convenient. Where it is necessary to make a transformer to give either a voltage V in two-wire connection or the voltage $V-0-V$ in three-wire connection, a method of connection of two low-voltage half-windings is used. The necessary balance of reactance to the high-voltage winding and of resistance of the two half-windings is achieved with using four coils in total. For a small transformer, two concentric pairs are wound one on each half of the leg length and are cross-connected, as shown in figure 9.2(a), while with a core-type construction the same effect is obtained by winding two concentric coils on each leg and cross-connecting, as shown in figure 9.2(b).

9.4.2 High-voltage windings

The most common form of high-voltage winding is the helically wound winding with PVA enamelled or paper-covered round wire. By winding two wires in parallel, which are cheaper than a strip of the same area, it is possible to use this form up to 1000 kVA 6.6 kV delta-connected windings. The number of coils in series on each leg varies from one to four for voltages up to 33 kV. This type of coil has a reasonable space factor and excellent surge-voltage distribution, and it is easy to obtain good cooling by inserting axial ducts in the winding. The interlayer voltage is considerable on the larger ratings where the number of layers is few, but pressboard insulation may be used to maintain a safe value of interlayer stress. The choice of the number of coils in series is dependent on the leg voltage and allowable interlayer stresses, but single coils can be used satisfactorily up to ratings of at least 1000 kVA at 11 kV phase voltage. For lower rated voltages such

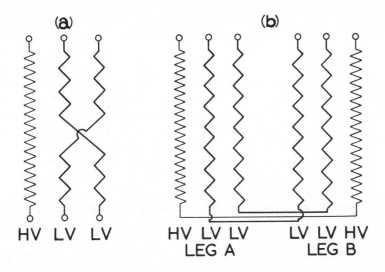

Figure 9.2   Series-parallel arrangements for low-voltage windings; (a) for a shell-type
transformer; (b) for a core-type transformer

as 3.3 or 6.6 kV in star connection, when the current in the coil is too high for
round wires, strip conductors insulated with enamel or paper can be used with a
similar type of winding. Press-paper and pressboard collars or edge strips are
employed as on low-voltage coils; they are arranged to stop short of the yokes to
allow access of oil to cooling ducts, the coils being supported by blocks from the
frames.

Multi-sectional coils are rarely found in modern designs for oil-cooled
transformers, but they may be necessary on AN class C transformers where
allowable stresses are smaller. They are more expensive to wind and have a lower
overall space factor than helical coils. Disk coils with rectangular strip conductors
are used on some larger ratings. Like multi-sectional coils they are expensive and
have a poor surge-voltage distribution.

For single-phase transformers of the core type the high-voltage coils on each
leg are connected in series for small ratings, where the current is too low to justify
parallel conductors, and in parallel on higher ratings.

9.4.3   Tappings

The normal practice is to provide high-voltage taps to allow adjustment for
voltage variations on the high-voltage side. Tapping ranges of up to $\pm 10\%$, that
is above and below rated voltage, may be required. The cost of on-load tap
changers for the ratings concerned in distribution transformers is prohibitive,

except in very special cases. Tap-adjusting switches are used with external handles operated only when the transformer is unexcited. On pole-mounted transformers, rated at 200 kVA or less, there are internal link boards in place of tap switches.

With multi-sectional or disk coils the tapping section is placed in the centre of the winding, to maintain electrical balance. This is also a point of minimum interturn surge-voltage stress. The windings are usually so arranged that the taps can be made on the outer layer of the coil, for convenience of connection. With single helical high-voltage coils it may be mechanically difficult to bring out taps from the mid-point of the coil, but it is possible to use taps on the outermost layers, that is at the finish of the winding. It must be remembered that this end of the winding can be that to which a surge voltage is applied, in which case reflections from the ends of the portions of the windings that do not carry current on minimum tap can give high values of surge voltage between taps, up to 25 % of the applied surge voltage. It is, however, quite feasible to insulate between tap connections for this value of surge stress at up to 11 kV rated voltage.

When made at the electrical centre of the legs, taps are so arranged that a portion from each half of the winding is disconnected in turn, thus giving minimum ampere – turn unbalance between the two halves. However, on core-type single-phase transformers with the high-voltage coils on the two legs connected in series it may be necessary, if the low-voltage coils are connected in parallel, to remove equal numbers of turns from both legs on every tap position; otherwise unbalanced ampere – turns will produce a circulating current between the low-voltage coils on the two legs.

9.4.4   High- to low-voltage insulation

Coils of the ratings considered are sufficiently robust to be self-supporting except at the lowest ratings and highest voltage, over 11 kV. The insulation between high- and low-voltage coils consists of a number of wraps of pressboard and usually a set of spacers. Even non-self-supporting coils can be made without individual support if wound directly onto the low-voltage coil and insulation during winding, instead of being wound and erected individually. On larger ratings a considerable space between coils is required to obtain the specified reactance; this space conveniently forms an oil-cooling duct. On the smallest sizes of transformers the clearances required to obtain the reactance may be too small to allow for a cooling duct. The axial ducts in the coils are necessary to provide cooling. However, as the leakage flux enclosed by the duct does not link with the whole of the coil, the effect of these ducts on the reactance is much less than if the same space, or even half of it, were used for a cooling duct between the windings.

9.4.5   Foil winding

If the conductor is in the form of a wide thin foil, it is possible to wind coils with only one turn per layer, with good thermal conductivity out to the edges of the coils and very low eddy current losses. For high-voltage conditions the number of turns required is such that even with the thinnest foil a number of narrow coils must be joined in series. Unfortunately this has a poor surge-voltage distribution. To obtain a reasonable space factor overall for the winding, interturn insulation can only be a small fraction of a millimetre thick. The resultant peak surge stress on the interturn insulation is high. Satisfactory thin insulation which will withstand the stress must be free from pin-holes. On low-voltage coils, however, where the number of turns is small, much thicker foil is required. Satisfactory insulation as a film applied to the foil, or as interleaved material, can be obtained. Care must be taken that the film of insulation is unbroken at the edges of the foil and that any interleaving material must extend beyond the edges.

Because each turn in low-voltage windings is the full width of the coil, unbalanced mechanical forces round the coil perimeter produced by the large helix angles of the coils are eliminated. If the winding is made longer than the high-voltage winding, the internal axial forces (balanced overall) are tensile in the foil and compressive in the high-voltage winding. Compressive forces are, of course, much more easily withstood by a coil with many turns per layer than tensile forces, whereas the full width foil has good axial tensile strength. The short-circuit strength of this construction is therefore very high. This method of construction has been extensively used in USA where reactances are much lower than in British practice (see sub-section 9.8). When the market price of copper is high, aluminium foil low-voltage windings can be justified economically.

9.5 TANK CONSTRUCTION AND COOLING

Tanks are normally rectangular in shape and are constructed from mild steel in thicknesses of 4 to 6 mm. Occasionally cylindrical shapes are employed for the smaller sizes. On the larger sizes various forms of tank stiffeners are needed, but the smaller sizes can be made adequately rigid without stiffeners. Covers for outdoor use are required to be formed so that rain cannot lodge on them. In most instances covers are bolted on, but quick-release type of fasteners may be used, particularly when it is necessary to remove the cover to obtain access to tapping-link boards.

Up to about 50 kVA the surface area of plain tanks is adequate for cooling purposes. Above this rating additional cooling surfaces are needed, and these are always integral with the tank. The extra surfaces may be provided by tubes of either circular or elliptical cross-section directly inserted into the tank walls or by radiators built from tubes set into headers or built from pressed-steel sections. Certain designs of radiators enable the space at the corners of the tank to be utilised for extra cooling surface in a way that is not practical with directly mounted tubes, and the use of radiators occupying these corner spaces is increasing.

9.6   TERMINAL ARRANGEMENTS

9.6.1   Pole-mounted outdoor transformers

Shedded porcelain bushings are employed for terminal connections on both the high- and low-voltage windings. The bushings are of the one-hole fixing demountable type. Insulating locking strips are connected between the inside ends of the conducting stems, so that in the event of damage the outer fittings can be unscrewed and one porcelain at a time can be replaced without access being required to the inside of the tank, though it is necessary to lower the oil level to replace at least the high-voltage porcelains. Low-voltage bushings are usually mounted horizontally on the side of the tank and may be mounted above oil level, since the voltage stress level on the inner end is very low. Care should be taken, however, to ensure that the terminal is not partially in and partially out of oil under any condition of service, since any contamination present is likely to float at the oil surface. High-voltage bushings are mounted at an angle of 30 to 45 ° to the vertical, in order that the rain sheds may be effective, in a pocket on the tank wall remote from the pole. On the smallest sizes it is necessary to splay the terminals apart at the tops to obtain the necessary line-to-line clearances. The top of the terminal is only just higher than the top of the tank. Double-gap arcing horns for protection against lightning are a standard fitting.

9.6.2   Ground-mounted transformers

High-voltage connections are usually made in three-core paper-insulated lead-covered cable for which a three-pole one-gland cable box is provided. If cables, which do not require sealing against ingress of moisture, are used, a recessed type of porcelain is sometimes fitted, the joint being covered with a butyl-rubber seal. Stem-type terminals may be fitted to special requirements. For indoor use these need not be shedded. Where possible, they are fitted horizontally in the tank side, but stem terminals in the cover may sometimes be required as, for instance, when the higher-voltage supply is by overhead line on an outdoor site. In this case, it is necessary to fit a conservator to maintain the oil level up to the cover or to use a special arrangement of bushing with a metallised and earthed portion reaching to below cold oil level. This avoids stressing troubles which can occur if the porcelain is partially in air and partially in oil of higher permittivity. Low-voltage connections may be either in cable boxes or as open-type stem terminals, the latter being very common for indoor use, and are normally provided on the opposite long side of the tank to the high-voltage cable box or terminals.

9.7   FITTINGS

The number of type of fittings, as would be expected, increases as the size and importance of the transformer increases. The minimum fittings, additional to tank mounting arrangements, are an oil-draining and oil-sampling device, a breather, a method of indicating correct oil level (usually an oil gauge but occasionally a dip stick), an earthing terminal and a rating plate. On the smaller transformers the oil drain, closed by a plug and with a sampling device, and on the larger transformers a drain valve are standard. Unless the cover is readily removable, as on pole-mounted types, a filling orifice is necessary. Ground-mounted transformers of ratings above 315 kVA are normally fitted with jacking lugs at the corners to facilitate movement. A usual fitting is a pocket to accept a thermometer to measure top oil temperature, which is situated at oil level or in the cover if a conservator is fitted. A thermometer, if fitted, may be a simple dial type on a short stem or a larger-scaled instrument mounted in a convenient position and operated via a capillary tube from a bulb in the thermometer pocket. Particularly in industrial sub-stations, alarm and trip contacts can be fitted to the larger thermometers to sound an alarm when an excess oil temperature is reached and, as a second stage, to trip the associated high-voltage breather if a further rise of temperature occurs. Winding temperature indicators may occasionally be fitted in industrial use of the larger sizes.

The choice of breather lies between a plain type, that is an unobstructed vent arranged to keep out rain or other forms of water, and a dehydrating breather which contains a drying agent, such as silica gel, in order to prevent atmospheric water vapour being drawn into the air space above the oil when the transformer temperature falls. The necessity for and effectiveness of this type is a matter of some debate. In indoor sub-stations where relative humidity can be high, it can be effective as long as it is properly maintained, but it has been argued that effective ventilation to keep down humidity is as effective and less likely to suffer from neglect. If ventilation is inadequate when a plain breather is used, serious trouble can be encountered with condensation on the tank and cover causing both corrosion and electric failure due to droplets of water falling into the oil.

Conservators or expansion vessels are essential for any transformer with terminals in the cover in order to maintain oil up to the underside of the tank and to enable pressure- and gas-activated (Buchholz) relays to be fitted which are one of the most effective types of overall transformer protection. The use of such a relay is, however, economically only justified in the larger sizes for key duties and is not the practice on main distribution networks. Another argument in favour of the conservator, as the name implies, is that it reduces oxidation of the oil, but the majority of distribution transformers with high voltages of 11 kV or less have not been so equipped for many years; the modern grades of oil appear to give adequate service without sludging under practical conditions. Conservators are normally fitted on higher-voltage transformers. Provision is normally made on a conservator-type transformer for oil filtration to be made via a top filter valve on the tank and the drain valve so that the oil need not be circulated through the conservator, thereby avoiding danger of spreading any contamination from deposits in the vessel. Conservators are designed with a removable end to enable them to be cleaned out. Explosion vents are not invariably fitted to conservator transformers, but they can assist to reduce potential fire risk.

9.8   A NOTE ON USA PRACTICE

American distribution practice in general is very different from British practice, and this has resulted in notable differences in distribution transformer standards. By definition, an American distribution transformer has a rating of less than 501 kVA; transformers above this rating up to 10 MVA are small power, or network, transformers.

One of the principal causes of differences is the low final distribution voltage of about 115 V single-phase (at 60 Hz). With a low housing density and the necessity to keep final distribution runs short to keep down the voltage drop, this has resulted in the use of many small final distribution transformers each supplying a small number of customers; they are fed from a main distribution voltage of from 2 to 15 kV, the older systems tending towards the lower value. With the small number of consumers the final diversity on each transformer is low. In addition, heating loads have in the past been rare, although these, together with air conditioning loads, are now increasing. The characteristic load has therefore a preponderance of motor (appliance) loads and is peaky in nature.

This has led to the adoption of lower nominal ratings for transformers for the same maximum demand than in British practice, with emphasis on low regulation and high overload capacities, particularly under summer conditions. Overload requirements can be as high as 400% for $\frac{1}{2}$h. The requirement for low regulation leads to low copper losses and low reactance, the latter often only 2% or even less. In contrast, the level of iron losses is higher than in British practice giving iron to copper loss ratios of 1 to 3 or 4, although low magnetising current is often stressed as a feature by manufacturers. These considerations in turn have led designers to place emphasis on short-circuit strength, because of the high value of fault to rated current, and on thermal distribution and reduction of temperature of hot spots, since these determine the ultimate life of the insulation. Recent results of this have been the introduction of foil windings and much development on insulation materials with good high-temperature performance, by using various forms of chemically treated cellulose. Also interest has increased in resin-encapsulated transformers partially to provide mechanical rigidity and partially to eliminate oil and reduce mass. In assessing ideas on insulation it should be borne in mind that the USA standard distribution transformer is a sealed unit, in the sense that no breathers are provided, though the sealing is by gaskets on the cover and commonly a hand hole is provided in the top to allow access to tapping links or switches.

Distribution has in the past been almost entirely by overhead lines, though underground distribution is now widespread. This has resulted in additional features such as surge arrestors and fuses or disconnecting switches which are provided on pole-mounted transformers. Other common fittings have been built-in secondary circuit breakers with tripping devices matched to the thermal time constant of the windings, as in the completely self-protected transformer. Many of these features have been incorporated in pad-mounted transformers for use with cables; these are transformers arranged with their fittings to form small kiosks, mounted on the ground and easily disguised by some form of landscaping. An alternative development has been to mount transformers in the base of street lighting fittings.

Because of the large number of small sizes of transformers employed, true mass production methods can be employed. This influences construction and details to an extent that would be impossible, for example, on the British market. Also the relative cost of materials and labour is different, so that emphasis is placed on labour saving rather than on material cost reduction. The resultant transformers bear little similarity to those of British manufacture; notable differences are the use of rectangular-section shell-type cores, with coils wound on rectangular mandrels, and the use of cylindrical tanks with cover-mounted terminals. To obtain low reactance the high-voltage winding is usually wound over one-half of the low-voltage winding and the other half of the low-voltage winding wound over the high-voltage winding. Three-phase distribution transformers as opposed to three-phase banks of single-phase units are rare, and development of these has not proceeded to the extent of single-phase units. This has resulted in some instances in the use of standard single-phase units mounted in a common tank for three-phase supplies and in the development of connections, such as the tee – tee, which allows three-phase units to be made by using only two single-phase cores.

9.9   DESIGN OF A TYPICAL DISTRIBUTION TRANSFORMER

9.9.1   Specification

In order to illustrate the problems arising in the design of distribution transformers it is useful to consider, in detail, a design based on the following typical specification which follows the example given in chapter 2.

Rating	750 kVA, three-phase ground mounting
Voltage ratio	11 000 to 433 V at no-load
Tappings	$\pm 5\%$ on high-voltage winding
Impedance	4.75 % at full-load, 75 °C
Load loss	9500 W at 75 °C
No-load loss	1420 W
Connection	high-voltage delta, low-voltage star
Maximum flux density	1.55 T
Temperature rise	55 °C top oil
	65 °C windings, measured by resistance

The transformer is to be fully in accordance with the requirements set down in reference G1.7.

9.9.2 Selection of core frame

There are many possible designs that would meet this specification, but there is a prime constraint, that of cost, which is not specifically included. The buyer will not wish to pay more than the current market price; and the transformer manufacturer, selling at the current market price, will wish to maximise his profit by producing at the lowest cost. The minimum cost constraint considerably reduces the options open to the designer. Theoretically, there will be only one design which gives the lowest cost; in practice there will be a limited number of alternatives available. This choice exists because a manufacturer will be unable to allocate marginal costs with sufficient accuracy to enable him to judge between a number of designs within a limited range. It is also felt worthwhile by manufacturers to adopt a certain amount of standardisation in design and in production method. The designer is left with a number of possibilities which may not include the design with the lowest material cost.

A design office would have a library of designs from which could be selected a core frame. Given a suitable core frame the other design features will fall readily into place. Alternatively, a design would be obtained which aimed at obtaining minimum cost within certain guidelines appropriate to the particular factory. This has been facilitated in recent years by the use of electronic computers (see chapter 3).

If the quantities required conforming with this particular specification were sufficiently large, it would probably be decided to minimise the cost of material by using parts outside the existing standard range for the particular factory. However, it will be assumed here that only a small quantity of the particular transformer is to be produced. It is considered necessary, therefore, to fit the product into an established schedule for a factory which is producing large numbers of similar, though not identical, transformers. The designers will, therefore, have a restricted range of, for example, core frames and conductor sizes available.

The standard core frames considered in chapter 2 are, therefore, considered here, with the knowledge that by limiting the choices in this way the cost of material used will be somewhat greater than the minimum cost possible.

Frame dimensions, flux density, low-voltage turns and current density, as derived in chapter 2, are summarised below.

Core area A_{Fe}	$= 0.0273\,m^2$
Half-width of widest core-plate b_{Fe}	$= 95\,mm$
Width of core window b_w	$= 152\,mm$
Height of core window h_w	$= 620\,mm$
Flux density B_m	$= 1.53\,T$
Current density J	$= 2.97\,A\,mm^{-2}$
Low-voltage turns N_1	$= 27$

There is an indication here, already, that we are paying for standardisation of the core frame. The flux density is lower than the maximum permitted; this suggests that rather more iron area is being used than is required.

9.9.3 Turns ratio

It must be noted that the voltage ratio must equal the declared ratio within the degree of accuracy laid down in reference G1.7. It is stated that the tolerance permitted on the voltage ratio is the lower of the following two values: (a) $\pm 1/200$ of the declared ratio or (b) a percentage of the declared ratio equal to one-tenth of the actual percentage impedance voltage at rated current.

Voltage ratios on tappings are subject to agreement between the purchaser and the manufacturer, and, in the absence of any specific requirement by the purchaser, it is assumed that these tolerances also apply to the tapping voltages. This is a reasonable assumption on the part of the manufacturer in this case since the number of turns on the high-voltage winding will make fine adjustment at the design stage of the voltage ratio practicable without adding to the cost. If, however, the voltages on both the high- and low-voltage windings had been low, it might be difficult to obtain an accurate voltage ratio, particularly on tappings, without excessive cost; some specific agreement might be necessary between manufacturer and purchaser. The manufacturer would be expected to advise on this at the enquiry stage.

In the case under consideration the guaranteed impedance is required to be 4.75% within a tolerance of $\pm 10\%$. The voltage ratio tolerance could, therefore, be as low as $0.1 \times 0.9 \times 4.75\% = 0.4275\%$. This must be borne in mind in the calculation of the high-voltage turns required.

The tolerances are, of course, applied to the transformer as finally produced. It is necessary for the designer to allow for possible manufacturing variations in turns wound. This is a matter of judgement, experience and production quality control.

Then

$$\text{declared phase voltage ratio} = 11000/(433/3^{1/2}) = 44$$

Therefore

$$\text{high-voltage turns on principal tapping } N_2 = 44 \times N_1 = 44 \times 27 = 1188$$

It should be noted that with a permitted tolerance of 0.4275% it is necessary for the high-voltage turns to be within ± 5 turns of this figure in the completed transformer. It can be assumed with a high degree of assurance that the low-voltage turns in the completed transformer will be as designed.

$$\text{High-voltage turns on maximum tap } N_{2m} = 1.05 \times 1188 = 1247$$
$$\text{High-voltage turns on minimum tap } N_{2min} = 0.95 \times 1188 = 1129$$

These turns must, of course, be rounded to the nearest integer, but with this number of high-voltage turns accuracy better than 0.05% can be obtained.

9.9.4 Winding current density

From sub-section 9.9.2, the current density required is $2.97\,\text{A}\,\text{mm}^{-2}$; it is necessary to ask whether this current density is acceptable. Apart from the limits imposed by the specification on load loss and in some cases by the temperature rises when carrying continuous rated current, the only other limit on current density arises from the short-circuit capability. Reference G1.7 requires that a 750 kVA transformer with an equivalent short-circuit current-limiting impedance of 5% which includes any supply impedance should withstand twenty times the rated current for 3 s. Further, the average winding temperature for an oil-immersed transformer with copper windings subjected to the over-current specified must not exceed 250 °C at the end of the short-circuit period.

The formula by which the temperature attained on short circuit may be calculated is given by

$$\theta_{wm} = \theta_{iw} + a_\theta J_k^2 t_k \times 10^{-3} \, °C \tag{9.1}$$

where θ_{wm} is the highest average temperature attained by the winding in degrees Celsius, θ_{wi} the initial average temperature of the winding in degrees Celsius, J_k the short-circuit current density in amperes per square millimetre, t_k the duration of the short circuit in seconds, a_θ a function of $\frac{1}{2}(\theta_{wm}^* + \theta_{wi})$, θ_{wm}^* the highest permissible average winding temperature attained under short-circuit conditions (this equals 250 °C for an oil-immersed transformer with copper windings) and θ_{wi} the sum of the maximum temperature of the cooling medium and the winding temperature rise measured by resistance.

For the transformer under consideration, the maximum winding temperature rise by resistance is 65 °C, and the maximum air temperature is 40 °C (reference G1.7). Thus

$$\theta_{wi} = 40 + 65 = 105 \, °C$$

With $\theta_{wm}^* = 250\,°C$ and $\theta_{wi} = 105\,°C$, the value of a_θ, from reference G1.7, is given as 8.2. The maximum current density permitted for the above conditions is 77.0 A mm^{-2} for a copper-wound transformer with a 3 s short-circuit rating. This means that, for the transformer under consideration here, the maximum working current density must not exceed $77/20 = 3.85$ A mm^{-2}. It must be noted that similar considerations apply to all the tapping positions.

The required current density of 2.97 A mm^{-2} is, therefore, seen to be comfortably within the specification short-circuit thermal requirement, even when consideration is given to the conditions on all tappings

9.9.5 Low-voltage winding inside diameter

It is necessary, now, to establish the inside diameter of the low-voltage winding which, to minimise insulation problems, will be the winding nearest the core.

As outlined in chapter 4, the core will be built of magnetic steel laminations approximately 0.3 mm thick; each lamination has a thin chemical coating which provides insulation, and the core cross-section has a stepped form. A typical space factor in core building, to allow for insulation and incomplete filling of the space due to the use of laminations, would be 0.96, that is 96% of the gross core area would be magnetic steel. It is also practicable to obtain a ratio of gross core area to core circle area of 0.94. The nett core steel area A_{Fe} contained within a given core circle diameter d, would then be

$$A_{Fe} = 0.94 \times 0.96d^2\pi/4 \approx 0.71d^2 \tag{9.2}$$

from which

$$d \approx (A_{Fe}/0.71)^{1/2} \approx 1.19A_{Fe}^{1/2} \tag{9.3}$$

Hence, with $A_{Fe} = 0.0273\,\text{m}^2$ (see sub-section 9.9.2) $d = 197$ mm.

The low-voltage winding would have as small an inside diameter as could be fitted on this core if we allow for necessary insulation between the winding and the core and for the need to hold the core leg laminations together. For a distribution class transformer it is not necessary to use a bolted construction, and it would be normal to use some form of taping on the core legs. The insulation and mechanical support requirement may be combined in one material. A typical low-voltage winding inside diameter d_{1i}, taking space factors into account, would be 6 mm greater than the core circle diameter. Thus

$$d_{1i} = 197 + 6 = 203\,\text{mm}$$

9.9.6 Formation of low-voltage winding

Rated low-voltage winding current I_1 = low-voltage line current
$$= 750 \times 10^3/433 \times 3^{1/2}$$
$$= 1000\,\text{A}$$

It is stated in chapter 2 that the minimum quantity of copper is used when the high- and low-voltage winding current densities are approximately equal. Therefore

Low-voltage winding cross-section area $a_1 = I_1/J = 1000/2.97 = 337\,\text{mm}^2$

It will be found necessary to subdivide the low-voltage conductor into a number of strands in order to reduce eddy current losses in the winding. Rectangular conductor strands would be used, and it is assumed that there are a limited number of conductor sizes available since it is required to conform to an established standard range of parts, as discussed in sub-section 9.9.2.

It will be known that a two-layer low-voltage winding with a cooling duct between layers will be necessary to meet the cooling requirements for the 750 kVA transformer of the required voltage and working at the desired current density. If this is not known at this stage, one or two trial designs will produce this knowledge. With $N_1 = 27$ (see sub-section 9.9.2) the number $N_{1\mathscr{L}}$ of turns per layer is

$$N_{1\mathscr{L}} = N_1/2 = 27/2 = 13.5$$

It should be noted that fractional turns may be wound in a layer but that the total number of turns must be an integer

The coil will be helically wound. To allow for the helix, a space equivalent to the axial dimension of one extra turn must be allowed in the winding length. In addition, it is likely that some axial packing between the turns in the centre of the winding will be desirable both to provide for the adjustment of the low-voltage winding relative to the high-voltage winding so as to achieve electromagnetic balance in order to minimise short-circuit forces and, possibly, to allow space for carrying out transpositions of the strands of the turn. This will be discussed later. It is sufficient to say here that an axial gap of 15 mm will be left at the winding centre for this purpose. The minimum end clearance for a low-voltage winding is taken as 10 mm at each end. This clearance is determined not so much by insulation requirements but to allow the provision of mechanical support at the ends of the low-voltage windings.

The winding conductor will itself be insulated. The conductor insulation could typically be either enamel or a paper wrapping. If enamel is used, it is necessary that it is hard and suitable for immersion in hot oil. It would be normal to use a polyvinyl acetate enamel. The nominal insulation strength of the conductor covering is not significant since the voltage between turns in this transformer will be only 9.4 V. It is necessary, however, that the insulation strength be preserved when the insulation is subjected to the mechanical stresses imposed during the manufacturing process or by the short-circuit forces encountered in service. If paper covering is used, it will be thicker than enamel coating of similar electric and mechanical strength. The winding space factor when using paper-covered conductor is, therefore, worse than when enamel-covered conductor is used. However, paper-covered conductors are likely to be cheaper. There is an economic assessment to be made. It is assumed that paper-covered conductor is in use in the factory where the transformer is to be made, and this is the choice here. A suitable wall thickness for the paper covering is 0.2 mm.

If we take the window height $h_w = 620$ mm from sub-section 9.9.2, then the maximum insulated axial dimension h_{t1} of one turn is given by

$$h_{t1} = (620 - 2 \times 10 - 15)/14.5 = 40.3 \text{ mm}$$

The low-voltage winding conductor will require at least four strands of typical available conductors. The bare strand axial dimension h_c would then be

$$h_c = (40.3 - \text{total conductor insulation})/4 = \tfrac{1}{4}(40.3 - 4 \times 2 \times 0.2)$$
$$= 9.7 \text{ mm}$$

Let it be assumed that conductors of width 9.5 mm are available.

Thus, the total cross-sectional area of the low-voltage conductor of 337 mm² must be obtained with a multiple of four strands in parallel. The low-voltage winding may be formed of twelve strands in parallel of 9.5 mm × 3.0 mm conductor. When allowance is made for the corner radii of rectangular conductors, this low-voltage conductor gives an area of 336 mm² which is almost the area required; a designer is not always so fortunate. The turn dimensions, including 0.2 mm wall thickness of conductor covering, are then $4 \times 9.9 = 39.6$ mm axially and $3 \times 3.4 = 10.2$ mm radially.

The vertical cooling duct between the two low-voltage layers should not be less than 5 mm wide; thus the low-voltage winding radial dimension b_1 will be

$$b_1 = 10.2 + 5 + 10.2 \, mm = 25.4 \, mm$$

The low-voltage winding outside diameter d_{1e} is then, with the inside diameter $d_{1i} = 203$ mm,

$$d_{1e} = 2 \times 25.4 + 203 = 253.8 \, mm$$

The total axial height h_1 of the low-voltage winding with the allowance of 1 turn for the helix and 15 mm for an axial centre gap is given by

$$h_1 = (13.5 + 1)39.6 + 15 = 589 \, mm$$

With the already assumed window height of 620 mm, a clearance of 15.5 mm at each end is then available. The low-voltage winding will be left at this stage with the knowledge that the end and centre gaps may be adjusted a little with the limit that the end clearance must not be less than 10 mm. This will provide some flexibility in aligning the magnetic centres of the high- and low-voltage windings to minimise short-circuit forces. It should be noted that a free choice of conductor size would have allowed a more efficient utilisation of the space available; this economy has been foregone in the interests of standardisation.

9.9.7 Formation of high-voltage winding

An insulating barrier will be required between the high- and low-voltage windings, and it is assumed that a cooling duct will be desirable under the high-voltage winding. This is normal though not obligatory. The insulation between the high- and low-voltage windings could be solid with the cooling duct provision made elsewhere. From chapter 2, a clearance of 9 mm is required between the high- and low-voltage windings with a cooling duct included. This could take the form of 3 mm of solid insulation and a 6 mm cooling duct. The solid insulation would be in the form of paper wrappings wound directly on the low-voltage winding. The high-voltage winding would then be fitted over insulation board spacer sticks, thus providing the cooling duct.

The high-voltage winding may be wound directly on the low-voltage winding plus the insulating wrap and the spacer sticks; or it may be wound separately and may be assembled over the low-voltage winding and insulation later. This is a design/production problem influenced by general design; decisions taken in the past have led to the present factory production methods. For example, it may be that it has been customary to form the high-voltage winding of a number of small coils to be assembled after winding and to be connected in series. This would mean that winding lathes for heavy low-voltage windings would be devoted solely to these windings, with a factory section of smaller winding lathes for high-voltage windings, and a separate assembly section. However, if the decision has been that the high-voltage windings would be wound directly on the low-voltage windings, this would increase the number of large lathes required but would reduce subsequent assembly time. These are marginal problems which might have a number of solutions depending on a variety of local factors, including the type of labour available.

There are also alternative types of high-voltage winding. Probably the cheapest satisfactory winding for this class of transformer is a simple multi-layer winding which uses a round conductor; a rectangular conductor may be used, and this produces a better winding space factor, although this benefit is likely to be more than counterbalanced by the extra cost of the conductor. An alternative is a disk winding (see sub-section 5.3.4). The disk winding would more commonly be found in higher-rated transformers than in the 750 kVA transformer under consideration, but there are some particular points of interest in these windings, particularly in the cooling, and a disk winding will be adopted here.

When the choice of winding type has been decided, the design can proceed.

$$\text{High-voltage phase current } I_2 = \text{high-voltage line current}/3^{1/2}$$
$$= 750 \times 10^3/11000 \times 3 = 22.7 \text{ A}$$

From the current density, derived in chapter 2, and referred to in sub-sections 9.9.2 and 9.9.4, the required cross-sectional area a_2 of the high-voltage conductor may be obtained.

$$a_2 = 22.7/2.97 = 7.64 \text{ mm}^2$$

Let it be assumed that there is a radial duct, 4 mm wide, between each pair of disk sections and an insulating washer of 0.2 mm between each section of a pair. Ideally, in this type of winding tappings would be made on the outer surface of the winding which means that, at the tapping points, the number of turns in a pair of sections should equal the number of turns between tappings. The assumption will also be made at this stage that the number of turns per section throughout the winding will be equal.

From sub-section 9.9.3 the total high-voltage turns at the principal tapping is $N_2 = 1188$. For tapping steps of $2\frac{1}{2}\%$ of the high-voltage winding, the number of turns between tapping points must be

$$\text{turns between tapping} = 1188/40$$

This equals 30 when rounded to the next highest integer, and therefore

$$\text{turns/section} = 30/2$$
$$= 15$$

$$\text{number of disk sections} = \text{total turns}/15 = 1247/15$$

This equals 84, when rounded to the next highest integer.

It may be seen that it will be necessary to omit 13 turns from 84×15 turns to give the required 1247 turns. The gaps left by these omitted turns will be distributed evenly through the winding.

The axial dimension of a paper-insulated turn can now be calculated so that the winding may be accommodated in the given window height if we allow for end clearances. The required end clearance for an 11 kV winding is $h_{o2} = 28 \text{ mm}$, from chapter 2. The insulated turn axial height h_{t2} is then given by

$$h_{t2} = (620 - 2 \times 28 - 42 \times 0.2 - 41 \times 4)/84 = 4.67 \, \text{mm}$$

Again if we take the conductor paper covering as 0.2 mm wall thickness, a bare conductor width of 4.27 mm is obtained. It is assumed that 4.2 mm wide conductors are available. An area of 7.13 mm² is obtained from a rectangular conductor 4.2 mm wide and 1.75 mm thick. This compares with the derived 7.62 mm²; it will cause higher losses but they should not be excessively high.

The axial winding length h_2 is then

$$84 \times 4.6 + 42 \times 0.2 + 41 \times 4 = 559 \, \text{mm}$$

The maximum space available is $620 - 2 \times 28 = 564$ mm. Therefore, 5 mm of extra packing is required. Any packing should be inserted in such a manner that the electromagnetic balance is maintained. It should be noted at this stage that tappings should be removed from the centre of the windings. Also, when the transformer is on the minimum turns tapping, the working steady-state voltage across the centre gap will be the voltage across the total tapping range, that is volts/turn × total turns between taps = $9.4 \times 120 = 1128$ V. The standard induced over-voltage withstand test (see sub-section 7.4.4) will be at twice this level for 1 min. It may be necessary to increase the insulation at the centre gap to deal with this electric stress.

The radial width b_2 of the high-voltage winding is given by

$$b_2 = 15 \times (1.75 + 0.4) = 32.3 \, \text{mm}$$

The inside diameter d_{2i} of the high-voltage winding is, with the low-voltage winding outside diameter $d_{1e} = 253.8$ mm and the radial width $b_0 = 9$ mm of the gap between the low- and high-voltage windings,

$$d_{2i} = d_{1e} + 2 \times b_0$$
$$= 253.8 + 18 = 272 \, \text{mm to the nearest millimetre}$$

Hence the high-voltage winding outside diameter

$$d_{2e} = 272 + 2 \times 32.3 = 336.6 \, \text{mm}$$

The required clearance between phases will be 12 mm (from chapter 2). This gives a minimim distance between leg centres of $336 + 12 = 348.6$ mm. It would be normal to round this up to 350 mm.

9.9.8 Summary of core and windings

It should be noted that the high-voltage current for the lowest voltage ratio has been calculated and included in the table below. This is simply a reminder that short-circuit requirements have to be met on all tappings. The current density obtained may be compared with the limit calculated in sub-section 9.9.4. The current density of 3.35 A mm⁻² on the minimum tapping position at 750 kVA is well below the limit set for thermal requirements in reference G1.7, provided that the short-circuit impedance on this tapping position is not unduly low.

The core and winding design can be summarised as follows.

Core
area (m^2) 0.0273
window height (mm) 620
(leg centre distance (mm) 350

	Low-voltage winding	*High-voltage winding*				
Windings						
phase voltage (V)	250	11550	11275	11000	10725	10450
Phase current (A)	1000			22.7		(23.9)
turns	27	1247		1188		(1129)
coils in series	1			84 disks (in pairs)		
turns/layer/coil	13.5			1		
layers/coil	2			15		
Conductors						
bare (mm^2)	twelve 9.5 × 3.0	4.2 × 1.75				
covered (mm^2)	twelve 9.9 × 3.4	4.6 × 2.15				
turn dimensions (mm^2)	10.2 × 39.6	4.6 × 2.15				
turn area (mm^2)	336	7.13				
current density (A mm^{-2})	2.98	3.18				(3.35)
Winding dimensions						
radial width b_1, b_2 (mm)	25.4	32.3				
inside diameter (mm)	203	272				
mean diameter (mm)	228.4	304.3				
outside diameter (mm)	253.8	336.6				
end insulation (mm)	two 10	two 28				
axial spacers (mm)	26	ducts, 41 × 4				
		washers, 42 × 0.2				
		at centre, 5				
height (mm)	600	564				

9.10 CALCULATION OF CHARACTERISTIC TRANSFORMER DATA

9.10.1 Load losses

From equation 2.1, the low-voltage winding resistance R_1 per phase at 75 °C is

$$R_1 = (21.4 \times 10^{-6} \times 27 \times \pi \times 228.4)/336 = 0.00123\,\Omega$$

Therefore

$$I_1^2 R_1 = 1000^2 \times 0.00123 = 1230\,\text{W per phase at }75\,°\text{C}$$

From equation 2.9, the percentage eddy current loss $\%P_{i1}$ is

$$\%P_{i1} = 9.5(b_{c1}IN/100hJ_1)^2 \text{ at } 75\,°\text{C and } 50\,\text{Hz}$$
$$= 9.5\{9.0 \times 1000 \times 27/(100 \times 600 \times 2.98)\}^2 = 17.5\,\%$$

where b_{c1} is the total breadth of the bare conductor, $IN = I_1 N_1 = I_2 N_2$ the ampere – turns, h the axial length of the winding and J the current density. About

15 % of 17.5 %, that is 2.5 %, is to be added to the figure of 17.5 % in order to take into account the fact that the leakage field is distorted at the ends and at tappings. Thus the revised value of $\%P'_{i1}$ is

$$\%P'_{i1} = (17.5 + 2.5)\% = 20\%$$

This value is excessive since not only do the eddy current losses increase the load loss but they are also unevenly distributed (see chapter 2) which increases the risk of winding hot spots.

The eddy current losses can be reduced at low cost by transposing the strands of the winding conductor. It should be noted that transpositions reduce that part of eddy current losses that arise from current circulating between the separate strands because of their interconnection at the winding ends.

Those transpositions that ensure that each separate strand lies in each of the three possible radial position in each layer for the same number of turns as each other strand represent the ideal condition for reducing eddy current losses to a minimim in a conductor with a given number of strands in the radial direction. The ideal transposition for the particular low-voltage winding is illustrated in figure 9.3(a). If this ideal transposition is employed the eddy current losses are those for a conductor of the radial width b_c of a single strand. Since the eddy current losses are proportional to b_c^2, they will be reduced to one-ninth of the value without transposition. This reduction applies strictly only to a uniform field in the axial direction. Empirical adjustments are usually made to allow for the non-uniformity of the field in an actual coil in a similar way to that adopted above. In practice it is unnecessary to reduce the losses in this type of transformer to the degree given by an ideal transposition. It would be more customary to use one or two transpositions, one in the centre of each layer, as in figure 9.3(b). This would reduce the eddy current loss in a uniform field to approximately one-eighth of its value with an untransposed conductor which has three strands in the radial direction instead of one-ninth with the ideal transposition. It should be noted that each transposition requires axial space equal to the width of one extra strand. This space is available in the low-voltage winding centre packing.

The eddy current losses in a uniform field would thus be reduced to $17.5/8 = 2.2\%$; with the additional allowance of 2.5 % derived above for field distortion, a total of $(2.2 + 2.5)\% \approx 5\%$ for eddy current losses is arrived at for the low-voltage winding.

Again from equation 2.1, the high-voltage winding resistance R_2 per phase at 75 °C is

$$R_2 = (21.4 \times 10^{-6} \times 1188 \times \pi \times 304.3)/7.13$$
$$= 3.4\,\Omega \text{ on principal tapping}$$

Therefore

$$I_2^2 R_2 = 22.7^2 \times 3.4$$
$$= 1752\,\text{W per phase at 75 °C on principal tapping}$$

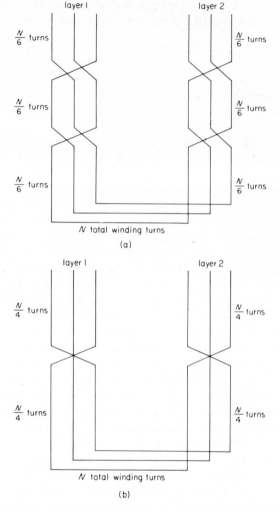

Figure 9.3   Transpositions in two-layer coil: (a) ideal transposition for three radial strands; (b) practical transposition for three radial strands

The percentage eddy current loss $\%P_{i2}$ is (see equation 2.9),

$$\%P_{i2} = 9.5(b_{c2}IN/100hJ)^2 \text{ at } 75\,°C \text{ and } 50\,Hz$$
$$= 9.5(1.75 \times 22.7 \times 1188/100 \times 564 \times 3.18)^2 = 0.7\%$$

To allow for the non-uniform field this will be increased to 1%.
   The total winding losses P_{Cu} are then given by

$$P_{Cu} = 3(I_1^2R_1 + I_2^2R_2 + P_{i1} + P_{i2})$$
$$= 3(1230 \times 1.05 + 1752 \times 1.01) = 9183\,W \tag{9.4}$$

It is necessary here to make allowance for the losses in leads and eddy current losses produced in the tank and structural steelwork due to leakage magnetic fields. The low-voltage lead length will be a significant proportion of the length of the low-voltage winding and will produce a significant extra loss. For example, if it is assumed that a 1 m length of lead per phase is required of the same cross-section as the winding conductor, this will have a resistance $R_{1\mathscr{L}}$ of

$$R_{1\mathscr{L}} = (21.4 \times 10^{-6} \times 1000)/336 = 6.4 \times 10^{-5}\,\Omega \text{ per phase}$$

Hence

$$I_1^2\,R_{1\mathscr{L}} = 1000^2 \times 6.4 \times 10^{-5} = 64\,\text{W per phase}$$

Thus approximately 200 W will be required for the three-phase low-voltage leads.

The losses produced in the tank and structural steel will be estimated on a largely empirical basis for a transformer of this type. In the original design, referred to in chapter 2, a rather conservative loss of 500 W is assumed; 200 W would be a more typical value. Therefore, the total loss $P_{\text{tot,l}}$ is

$$P_{\text{tot,l}} = 9183 + 200 + 200 = 9583\,\text{W}$$

This is within 0.9 % of the required guarantee value and is thus comfortably within the tolerance of $\pm 1/7$ (reference G1.7) of the guaranteed value of the load loss.

9.10.2 No-load losses

The usual practice with grain-oriented steel, which is in general use for these transformers, is for the cores to have mitred corner joints, as described in chapter 4.

As derived in sub-section 9.9.5 the core circle diameter is given as $d = 197$ mm. A maximum plate width (see figure 4.5(a)) appropriate to this core circle would be 190 mm.

From chapter 2, the total length of iron is

$$l_{\text{Fe}} = 3h_{\text{w}} + 4(b_{\text{cen}} + b_{\text{Fe}}) \tag{9.5}$$

where h_{w} is the height of core window, b_{cen} the distance between core leg centres and b_{Fe} the half-width of widest core-plate. Therefore

$$l_{\text{Fe}} = 3 \times 620 + 4(350 + 95) = 3640\,\text{mm} = 3.64\,\text{m}$$

The mass of iron is expressed in equation 2.64. A small additional quantity of iron is required at the end of each centre leg in order that the narrower leg plates can be tied in to the yoke plates. This extra quantity would be approximately 6 kg for this size of core. With the area $A_{\text{Fe}} = 0.0273\,\text{m}^2$, the density of iron $\rho_{\text{d,Fe}} = 7.65 \times 10^3\,\text{kg m}^{-3}$, the mass of the core iron becomes

$$m_{\text{Fe}} = 3640 \times 0.0273 \times 765 \times 10^3 + 6 = 766\,\text{kg}$$

If we take from figure 2.1 the specific iron loss in an assembled mitred core at a maximum flux density of $1.53 \, \mathrm{Wb \, m^{-2}}$ as $P_{Fe} = 1.51 \, \mathrm{W \, kg^{-1}}$, the total iron loss is

$$P_{Fe} = 1.51 \times 766 = 1157 \, \mathrm{W}$$

This is well below the guaranteed value of 1420 W. Examination of figure 2.1 indicates that the guaranteed loss could, in fact, be obtained with square-cut corners. There might be some advantage in this as the mitred-core construction tends to be rather more expensive for a given core size. It is here assumed that the particular factory is already producing mitred cores on a large scale, and consequently the mitred core will be used. However, the low loss does indicate that there is a further price that must be paid for standardisation; it suggests that a larger core with consequently less winding material will produce a cheaper transformer in terms of cost of material.

9.10.3 Reactance

In chapter 2 the percentage reactance is given from equation 2.26 as

$$\%X = 59.4IN(3b_0 s_0 + b_1 s_1 + b_2 s_2)/10^9 \times \phi_m h$$

where b_1, b_2 and b_0 are, respectively, the radial widths of the low- and high-voltage windings and of the gap between them, s_1, s_2 and s_0 the lengths of mean turn, respectively, of the low- and high-voltage windings and of the duct between them, h the axial length of the shorter winding, IN the ampere–turns and ϕ_m the maximum flux.

It may be shown that to allow for the effect of the axial duct in the low-voltage winding

$$b_1 = b_a + b_b + 3k^2 b_d$$

where b_a is the radial width of the first layer, b_b the radial width of the second layer,

$$k = b_a/(b_a + b_b)$$

and b_d the radial width of axial duct. Therefore

$$b_1 = 10.2 + 10.2 + 3(10.2/20.4)^2 \times 5 = 24.15 \, \mathrm{mm}$$

Also

$$b_2 = 32.3 \, \mathrm{mm}$$

$$b_0 = 9.0 \, \mathrm{mm}$$

$$s_1 = \pi \times 228.4 = 718 \, \mathrm{mm}$$

$$s_2 = \pi \times 304.3 = 956 \, \mathrm{mm}$$

$$s_0 = \pi \times 262.8 = 826 \, \mathrm{mm}$$

$$h = 564 \, \mathrm{mm}$$

Therefore

$$3b_0 s_0 + b_1 s_1 + b_2 s_2 = 70521 \text{ mm}$$

and

$$\%X = 59.4 \times 27000 \times 70521/(10^9 \times 1.53 \times 0.0273 \times 564) = 4.8\%$$

Also

$$\%R = 9500/750 \times 10 = 1.27\%$$

where it is assumed that the total load loss is 9500 W. Therefore

$$\%Z = (48^2 + 1.27^2)^{1/2} = 4.97\%$$

This is 4.6% above the required impedance of 4.75% and is sufficiently within the tolerance of 10% (reference G1.7) to give confidence that the impedance of the manufactured transformer will be satisfactory. The tolerance which the designer permits himself will depend upon experience, and it may be felt that some adjustment to the design would be required if confidence in the calculated reactance is to be strengthened. More complicated calculations can be performed to make allowances for gaps in the windings due to end clearances or tappings, but these are not often justified for this type of transformer.

9.11 THERMAL CALCULATIONS

9.11.1 Temperature difference between winding and surrounding oil

The average temperature difference between winding and surrounding oil in a transformer at steady load is, from chapter 2,

$$\Delta\theta_{\text{wo}} = \frac{M_e}{10^3 \lambda}\left\{\frac{A_C}{A_{\theta 1}}\delta_b f(n) + \delta_{be}\right\} + K_e M_e^{0.8} \qquad {}^\circ\text{C} \qquad (9.6)$$

with, for a coil dissipating heat only through one surface,

$$f(n) = (2n-1)(n-1)/6n \qquad (9.7)$$

and, for a coil dissipating heat through two surfaces,

$$f(n) = (n-1)(n-2)/12n \qquad (9.8)$$

where M_e is the rate of total heat transfer from unit area of the external coil cooling surface in watts per square metre, λ the thermal conductivity of copper insulation in watts per metre per kelvin, A_C the external heat transfer surface of a coil in square metres, $A_{\theta 1}$ the internal heat transfer surface between copper layers of a coil in square metres, δ_b the total horizontal thickness of insulation between copper layers in millimetres, n the number of copper layers in a coil with heat transfer only from vertical surfaces, δ_{be} the horizontal thickness of external insulation of the coil cooling surface in millimetres and K_c the empirical coefficient for heat transfer by convection from coil surfaces to oil.

For the low-voltage winding, the coil is tightly assembled on the core plus the necessary insulation which will make any duct between the core and the low-voltage winding rather ineffective for cooling. This, of course, is deliberate since the provision of an effective cooling duct would require more space and is probably unjustified. The outer diameter of the low-voltage winding is covered with a fairly heavy insulating wrap; there will be some cooling from this surface, but it will be neglected for the moment. It is, therefore, considered that all the losses generated in the low-voltage winding are dissipated into the axial duct between its two layers. Each layer then cools through one surface only.

The two layers are separated by vertical spacer sticks which would cover about 30 % of the cooling surfaces. These spacers are necessary both for support during coil winding and also to provide support when the windings carry heavy short-circuit currents which would produce compressive radial forces on the low-voltage winding. The ratio $A_C/A_{\theta 1}$ for the low-voltage winding is therefore 0.7. Thus the cooling surface A_C, of the low-voltage winding, with inside and outside diameters d_{1i} and d_{1e}, respectively, turns $N_{1\mathscr{L}}$ per layer and the turn axial dimension h_c is

$$A_C = 2 \times \tfrac{1}{2}(d_{1i} + d_{1e}) \pi \times N_{1\mathscr{L}} \times h_c A_C/A_{\theta 1}$$
$$= 2 \times 228.4 \times \pi \times 13.5 \times 39.6 \times 0.7 \, \mathrm{mm}^2 = 0.537 \, \mathrm{m}^2 \qquad (9.9)$$

It should be noted that the covered conductor axial dimensions are used. This is justifiable if the conducting path through the thin wall covering is considered.

The heat P_{Cu1} to be dissipated is

$$P_{Cu1} = 1230 \times 1.05 = 1292 \, \mathrm{W}$$

Because

$$M_e = P_{Cu1}/A_C \qquad (9.10)$$

therefore

$$M_e = 1292/0.537 = 2410 \, \mathrm{W \, m^{-2}}$$

$\lambda = 0.16 \, \mathrm{W \, m^{-1} \, K^{-1}}, \delta_b = 0.4 \, \mathrm{mm}, \delta_{be} = 0.2 \, \mathrm{mm}$, from equation 9.7 with $n = 3$, $f(n) = 10/18$, and $K_c = 0.04$ for vertical surfaces (see chapter 2, reference 1). Thus, from equation 9.6

$$\Delta\theta_{wo} = (5.4 + 20.3) \, ^{\circ}\mathrm{C} = 25.7 \, ^{\circ}\mathrm{C}$$

For the disk coils of the high-voltage winding, heat is dissipated on four surfaces for each pair of sections, two horizontal and two vertical. From chapter 2, reference 1,

$$\Delta\theta_{wo} \approx P'_{Cu2} \frac{R_{\theta b} R_{\theta h}}{R_{\theta b} + R_{\theta h}} \qquad ^{\circ}\mathrm{C} \qquad (9.11)$$

where P'_{Cu2} is the load loss per disk coil of the high-voltage winding, with

$$R_{\theta b} = \frac{1}{10^3 \lambda A_{cb}} \left\{ \frac{A_{cb}}{A_{\theta 1b}} \delta_b f(n_b) + \delta_{be} + 9\lambda \right\} \tag{9.12}$$

and

$$R_{\theta h} = \frac{1}{10^3 \lambda A_{ch}} \left\{ \frac{A_{ch}}{A_{\theta 1h}} \delta_h f(n_h) + \delta_{he} + 20\lambda \right\} \tag{9.13}$$

representing, respectively, the equivalent thermal resistances for the horizontal and vertical components of heat flow. The subscripts b and h refer, respectively, to the relevant values for horizontal and vertical direction, analogous to equation 9.6; $f(n_b)$ and $f(n_h)$ refer to equation 9.8.

On the principal tapping $P_{Cu2} = 1770$ W per phase. There are 40 coils, that is pairs of sections, in circuit; therefore $P'_{Cu2} = P_{Cu2}/40 = 44.3$ W. The inner surface is partly covered by high-voltage to low-voltage spacer sticks; a typical value of effective cooling surface A_{cb} would be 80 % of the gross surface $A_{\theta ib}$. Similarly the horizontal surfaces are partly masked, and an effective surface A_{ch} of 70 % of the gross $A_{\theta ih}$ is typical. Hence $A_{cb}/A_{\theta ib} = 0.8$ and $A_{ch}/A_{\theta ih} = 0.7$. Thus the total vertical cooling surface per coil, with the axial dimension 9.4 mm, is

$$A_{cb} = \pi \times (272 \times 0.8 + 336.6) \times 9.4 \, \text{mm}^2 = 0.01636 \, \text{m}^2$$

Because $n_b = 15$, equation 9.8 gives $f(n_b) = 1.01$. The total horizontal cooling surface per coil is

$$A_{ch} = \pi(168.3^2 - 136^2) \times 2 \times 0.7 \, \text{mm}^2 = 0.043 \, \text{m}^2$$

Because $n_h = 2$, equation 9.8 gives $f(n_h) = 0$. With $\lambda = 0.16 \, \text{W m}^{-1}\text{K}^{-1}$, $\delta_b = \delta_h = 0.4$ mm and $\delta_{be} = \delta_{he} = 0.2$ mm, equations 9.12 and 9.13 give

$$R_{\theta b} = (0.8 \times 0.4 \times 1.01 + 0.2 + 9 \times 0.16)/10^3 \times 0.16 \times 0.01636 = 1.96$$

and

$$R_{\theta h} = (0.2 + 20 \times 0.16)/10^3 \times 0.16 \times 0.043 = 0.5$$

Thus from equation 9.11 with $P'_{Cu2} = 44.3$

$$\Delta\theta_{wo} \approx 44.3 \times 1.96 \times 0.5/(1.96 + 0.5) = 17.7 \, °\text{C}$$

It would be preferable to arrange the coils such that temperature differences between winding and surrounding oil for the low- and high-voltage windings. were more nearly equal. One way to achieve this would be to reduce the number of high-voltage horizontal cooling ducts, thereby improving the high-voltage winding space factor and saving conductor material.

It is, however, necessary to pause here and to ask whether the temperature differences for both windings are acceptable. The higher value, for the low-voltage winding, is 25.7 °C. In the reference G1.20 the recommendations on transformer loading are based on the assumption, among others, that the average temperature difference is 21 °C. The low-voltage winding temperature difference is therefore higher than it should be. Since the high-voltage winding temperature difference is low, some redistribution of the heat losses to reduce the low-voltage winding temperature difference is to be made. This requires appropriate changes in the design to accomplish this desirable aim. However, the further calculations are with the windings already derived.

9.11.2   Core temperature

Reference G1.7 specifies rather vaguely that the core and other parts must not attain a temperature that will injure the core or any adjacent materials. The temperature is not measured during the standard temperature rise tests, nor are there any agreed methods of measurement.

For oil-immersed distribution transformers up to 1600 kVA it is not the practice to provide cooling ducts especially for the core, and this may be taken as satisfactory in general. However, larger transformers may require core cooling ducts.

9.11.3   Oil temperature

The tank and the external cooling arrangements will now be considered.

The tank itself must, of course, be of sufficient size to contain the core and windings with the necessary connections and terminals and also any fittings such as tapping-selector switch. Adequate insulation clearances must be provided.

It will be necessary to allow at least 120 mm in the tank width and 100 mm in the tank length in excess of the core and winding overall plan dimensions. These clearances are necessary to allow insulation clearances for the winding and leads, to allow free circulation of oil and to permit the assembly of the core and winding in the tank.

The minimum oil level must be such as to cover the core and winding and any connections and fittings which require oil for their insulation. The minimum oil level should be at a level such that required insulation is provided at the lowest average oil temperature at which the transformer is expected to operate. This is normally assumed to be $-25\,°C$ in accordance with reference G1.7 and in the absence of any other specific minimum temperature. In setting the minimum oil level it must also be borne in mind that external cooling radiators will be necessary on this transformer and that these should have a top connection to the tank sufficiently high to produce a good oil circulation.

The tank dimensions can now be derived. The length b_{T1} of the tank will be, with b_{cen} for the distance between leg centres and d_{2e} for the high-voltage winding outside diameter,

$$b_{T1} = 2b_{cen} + d_{2e} + 100$$
$$= 2 \times 350 + 336.6 + 100 \approx 1140\,mm \tag{9.14}$$

The tank width b_{T2} is

$$b_{T2} = d_{2e} + 120$$
$$= 336.6 + 120 \approx 460 \, \text{mm} \tag{9.15}$$

To obtain h_{omin} for the minimum oil level, a reasonable value for the height h_{omin} required in excess of the core height h_c will be 200 mm, with the oil at an average temperature of 15 °C. With h_w for the height of window and b_{Fe} for half the width of widest core-plate

$$h_c = h_w + 4b_{Fe} \tag{9.16}$$

for a core-type transformer with the top yoke of equal width to the leg. Therefore

$$h_{omin} = h_c + 200$$
$$= 620 + 4 \times 95 + 200 = 1200 \, \text{mm} \tag{9.17}$$

It is worth noting here that there is little to be gained in this type of transformer by economising on tank height since, if the height is lower, more radiators will be required for cooling purposes. This is because it is economical in these transformers to fit the cooling radiators directly on the tank and not to accommodate them in a separate cooling bank.

The type of radiator to be used here will be the flat-plate radiator assembled in groups; each group is connected to the tank by a common pipe. In order to assess a desirable height for the radiator it is necessary to consider the conditions for satisfactory oil circulation. Figure 9.4 shows a section through a typical transformer. To obtain good oil circulation it is desirable that the ratio h_θ/c should be low, where h_θ is the height of the centre of heating above the radiator bottom entry pipe and c is the distance between top and bottom entry pipes; h_θ/c should be less than 0.4.

If the height of the centre of the bottom entry pipe is 50 mm above the tank base and the centre of the top entry pipe is 50 mm below minimum oil level at 15 °C, the ratio h_θ/c for the tank so far derived can be calculated.

$$h_\theta = \text{height of core centre} - 50$$
$$= 1000/2 - 50 = 450 \, \text{mm}$$
$$c = 1200 - (50 + 50) = 1100 \, \text{mm}$$

Therefore

$$h_\theta/c = 0.41$$

This is rather high and should, preferably, be reduced. The minimum oil level will therefore be increased to 1300 mm, giving $h_\theta/c = 0.375$.

The heat dissipation for a given average temperature rise of the oil is a function of the total cooling surface and the proportion of radiator to convection surface. In contrast, the difference between top oil and average oil temperature is a function of the rate of oil circulation which is a function of the inverse of h_θ/c.

The limit of temperature rise of the top oil is 55 °C (reference G1.7) for a transformer not equipped with an oil conservator. Sufficient cooling surface must

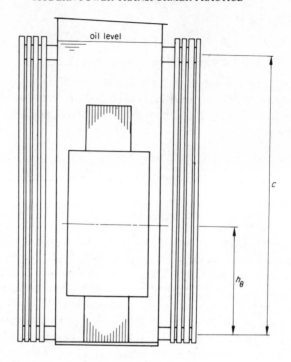

Figure 9.4   Section through typical distribution-type power transformer; c, distance between top and bottom entry pipes; h_θ, height of centre of transformer heating

be provided to meet this limitation; it must also meet the limitation that the average winding temperature rise must not exceed 65 °C. Since the low-voltage winding has an average winding temperature difference of 25.7 °C to the surrounding oil, the average oil temperature rise $\Delta\theta_{om}$ must, therefore, not exceed $65 - 25.7 = 39.3$ °C.

From chapter 2, $\Delta\theta_o = 0.7M_{eT}$ and $\Delta\theta_{om} = 0.85 \times 0.7M_{eT}^{0.8}$ for a tank with radiator-type cooling, where M_{eT} is the average thermal power dissipated from unit tank surface area in watts per square metre. Because

$$\Delta\theta_{om} = 0.85 \times 0.7M_{eT}^{0.8} = 39.3\,°C$$

Therefore

$$M_{eT} = (\Delta\theta_{om}/0.85 \times 0.7)^{1.25} = 188.3\,W\,m^{-2}$$

The top oil temperature rise is

$$\Delta\theta_o = \Delta\theta_{om}/0.85 = 39.3/0.85 = 46.2\,°C$$

This is well below the specification limit of 55 °C which reinforces the suggestion made in sub-section 9.11.1 that the low-voltage winding temperature difference to oil should be reduced. This would have the effect of increasing the acceptable average oil temperature rise and would reduce the cost of the cooling radiators. The possibility of economies of this type would certainly be explored if the design was for a transformer to be produced in large quantities.

The total power to be dissipated is the sum of the no-load and load losses at continuous maximum rating. Unless the purchaser states otherwise, the temperature rise limits, for a transformer with a tapping range not exceeding $\pm 5\%$, apply to the principal tapping only. Reference G1.7 recognises that the temperature rise limits may be exceeded in other tappings if operating at continuous maximum rating, but this is deemed acceptable for a transformer with a tapping range not exceeding $\pm 5\%$. Account may be taken of the fact that the losses may be lower than the guaranteed losses, and the cooling provision may be reduced accordingly. However, losses may be higher than the guaranteed value, but they must not exceed it by more than 10%.

The temperature rise limits, of course, apply to the finished transformer. The designer must consider the possibilities of departures from the guaranteed losses. However, it is assumed here that, after all design and manufacturing tolerances have been taken into account, the total losses P_{tot} will be the sum of the guaranteed no-load and load losses, P_{Fe} and $P_{Cu,t}$ at the continuous maximum rating:

$$P_{tot} = P_{Fe} + P_{Cu,t}$$
$$= 1420 + 9500 \text{ W} = 10920 \text{ W} \tag{9.18}$$

Therefore

$$\text{surface area required} = 10\,920/M_{eT} = 10\,920/188.3 = 58.0 \text{ m}^2$$

It is quite practicable to obtain a cooling surface of about 0.48 m^2 for every metre of height of a radiator which would occupy a plan area of $0.02 \text{ m} \times 0.045 \text{ m}$ with allowance made for adequate spacing between each radiator. if we refer to figure 9.4, where c is 1.2 m, the total radiator height could be 1.3 m. This will give a possible surface of $1.3 \times 0.48 = 0.624 \text{ m}^2$, for each single radiator.

The tank itself will have a surface area A_T of

$$A_T = 2(\text{length} + \text{width}) \times \text{oil level}$$
$$= 2(1.14 + 0.46) \times 1.3 = 4.16 \text{ m}^2$$

Some allowance should be made from this area for the effective area lost by the fitting of cable boxes and the radiator pipes; a deduction of 0.5 m^2 will be made for this. Therefore

$$A_T = 4.16 - 0.5 = 3.66 \text{ m}^2$$

$$\text{Total area required} = 58.0 \text{ m}^2$$

and thus

$$\text{area of radiator required} = 58.0 - 3.66 = 54.34 \text{ m}^2$$

$$\text{number of radiators required} = 54.34/0.624 = 88$$

These may be arranged in a number of groups such as shown in figure 9.5. It should be noted that provision has to be made for fittings such as external terminals,tap-selector control, temperature indicator and oil level indicator.

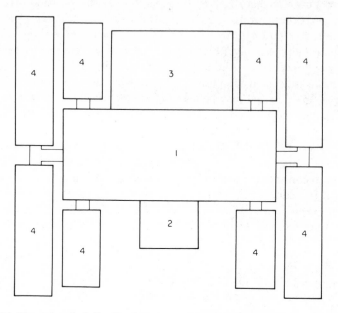

Figure 9.5   Plan of typical distribution-type power transformer: 1, tank; 2, high-voltage cable box; 3, low-voltage cable box; 4, radiators

9.12   SHORT-CIRCUIT REQUIREMENTS

Mention has been made in sub-section 9.9.4 of the thermal limitations imposed by short-circuit requirements. The transformer must also be built to withstand the mechanical forces produced under short-circuit conditions. The methods described in chapter 5 may be used. The type of transformer designed here is fairly straightforward in this respect, and the application of well-established standards is normally sufficient to ensure that the windings and connection are mechanically satisfactory. It could normally be sufficient, for example, to ensure that approximately 30% of the circumference of winding is supported. A point of weakness often lies at the ends of the helical low-voltage winding where trouble is occasionally experienced from lack of adequate support of the connections as they enter the coil proper.

9.13   NOISE LEVELS

Chapter 8 deals with questions of transformer noise. There is no requirement relating to noise levels in reference G1.7. Reference 1 lays down limits for noise level for the range of distribution transformers covered. These levels would be considered as typical, and it would be expected that a distribution transformer would conform, although it may not be strictly in accordance with that specification.

ACKNOWLEDGEMENT

H. K. Homfray wishes to thank GEC Power Engineering Limited for permission to publish the information contained in sections 9.1 to 9.8.

REFERENCE*

(Reference numbers preceded by the letter G are listed in section 1.14.)
1. British Electricity Supply Industry, *ESI* 35–1, *Distribution Transformers* (*from* 16 *to* 1000 *kVA*), The Electricity Council, London (1971)

* See also chapter 12, references 1 to 8.

10

Power System Transformers and Inductors

R. Feinberg* and R. J. Gresley*

10.1 INTRODUCTION—THE GENERAL POWER SYSTEM

Modern power supply networks are usually very complex, with many generating stations, often a variety of transmission voltages and a large number of interconnection and distribution points. Quantities of energy transmitted are large, and short-circuit fault levels are high.

At any point where there is a change of voltage a transformer is required. Usually the larger transformers are equipped with on-load tap-changing equipment for voltage control purposes. Quadrature boosters may be needed for the control of load flow, whilst, to control reactive power, other devices such as shunt inductors or synchronous compensators may be connected to the system. Short-circuit levels, if not sufficiently limited by the inherent impedance of the system, may need to be further reduced by the use of series inductors. All these items of equipment except generators and synchronous compensators are dealt with in this chapter.

10.2 POWER STATION TRANSFORMERS

10.2.1 General

Transformers for use in power stations comprise two basic types, those for transforming the generated power to a voltage suitable for the associated transmission system and those for supplying the station with the power necessary for auxiliary supplies. The latter are similar to the transformers described in other chapters.

---

* Consultants, both formerly Ferranti Limited.

10.2.2 Generator transformers

The function of these transformers is to step up from the voltage of the generators to that of the transmission system. It is almost always the practice for such transformers to be connected in delta on the low-voltage side and in star on the high-voltage side. The connection of the latter winding permits the tap changer, if one is fitted, to be at relatively low voltage, electrically adjacent to the neutral point which is usually directly earthed. When a three-phase on-load tap changer is fitted, the neutral point is often made inside the tap changer and is brought out through a bushing for earthing. Because of the heavy current, it is frequently the practice for the delta connection to be made external to the tank, each end of each phase winding being brought out through a separate bushing.

Voltage adjustment provided on the generator may be of relatively small range, as in UK practice. Alternatively, as in USA, sufficient variation may be provided to be able to dispense with tappings on the transformer. Typical voltage variation limits for UK machines are $\pm 5\%$; thus the flux density of the generator transformer lies within close limits, and the variation of the voltage supplied to the station auxiliary plant via the station unit transformer is minimised.

The defined rating of a transformer is the product of no-load voltage and full-load current; thus the no-load voltage of the high-voltage winding must be calculated from a knowledge of the required full-load voltage, throughput power, generator power factor and the transformer resistance and reactance. The no-load voltage of a 400 kV generator transformer is typically about 430 kV. Tapping ranges are chosen to permit variation of the full-load voltage within desired limits, and the range may be chosen to extend generator performance under leading power factor conditions. As tapping ranges are expressed relative to no-load voltage, the latter action results in an off-set tapping range, usually $+2$ to -16% of the no-load voltage on normal tap.

The choice of reactance in a generator transformer is to some extent dictated by generator stability requirements. The generator and transformer must be regarded as an entity. A typical large generator will have a transient reactance of about 25 to 32%. The maximum permissible reactance of generator and transformer, for stability reasons, is often about 50%. As a result, the generator transformer reactance must not exceed about 18% for use with higher-reactance machines. Reactance variation over the tapping range, provided that the maximum permissible value is not exceeded, is not of great consequence.

The choice of winding arrangement on the core depends on the reactance required and the variation that can be tolerated. In addition, some winding arrangements have higher losses than others. One arrangement is the so-called double-concentric winding (figure 10.1(a)). Here, the high-voltage winding is in two parts, an inner and an outer, with the low-voltage winding between. The percentage of high-voltage winding in each part may be apportioned to give the desired reactance. However, when this type of winding is used, extra care must be taken in the provision of the correct insulation for the various

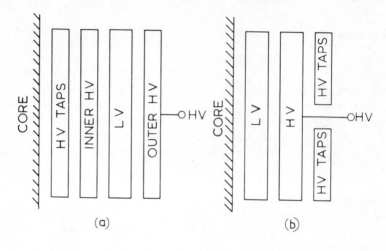

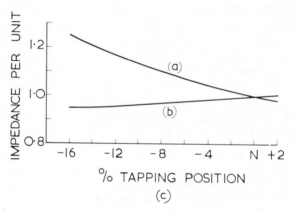

Figure 10.1   Winding arrangements: (a) double-concentric winding; (b) conventional
arrangement; (c) impedance variation for (a) and (b) over tapping range

windings, particularly in relation to the stresses resulting from atmospheric over-
voltages or surge-voltage tests.

The double-concentric construction tends to have higher winding losses than
the conventional arrangement of windings shown in figure 10.1(b), but it greatly
reduces short-circuit mechanical forces. It has a further advantage in that the
radial stray flux entering the core is reduced, thereby lessening the circulating
eddy currents in the outer laminations of the core and the extra heating which
they cause.

Typical reactance variations over the tapping range are shown in figure 10.1(c)
for both types of windings.

Since the tapping windings are electrically adjacent to the neutral point, they
may be placed near the lower-voltage parts of the high-voltage windings or near

the low-voltage windings. However, attention must be paid to the method of surge-voltage testing to be employed because a directly applied surge voltage on the low-voltage winding may result in an overstressing of the tapping winding as can never be met in service (see chapter 7).

It is usual to use three-phase generator transformers transportable in one piece. Without oil, a transport mass of 224 t nett has been achieved for a 600 MVA 400 kV generator transformer. Where transport facilities have permitted heavier units with consequent lower losses, transport masses of 305 t have been found possible. Transport is dealt with more fully in section 10.8, but it may be remarked in passing that the above transport masses have only been possible by using high core flux densities.

Future units are likely to be of even larger ratings, for example 800 MVA, and in suitable instances three-phase construction may still show economic advantages, in first cost, over single-phase units. However, single-phase units offer the possibility of a significant reduction of outage time in the event of a failure, since a spare can be kept readily available for quick replacement of a damaged transformer.

Tanks for some large generator transformers have been constructed from aluminium to reduce transport mass and stray losses. Where tanks are made from aluminium, care must be taken to prevent corrosion at junctions with electrochemically dissimilar metals, for example the copper earth electrode. It is also essential to design the tank to limit stresses due to transport, lifting, etc., since aluminium is a much weaker and softer material than steel (see sub-sections 10.7.2 and 10.8.2).

A generator transformer is rarely run at part load; thus two-stage cooling, for example ONAN – OFAF, is not required. It is usual to employ OFAF or OFWF cooling; the latter method has advantages because it requires a smaller space for the oil – water heat exchangers compared with those for oil – air units and also because of the general availability of cooling water in the power station. Further details of oil – water heat exchangers are given in sub-section 10.7.1.

As mentioned earlier, the low-voltage current of a large generator transformer is very high. If the delta connection is made external to the transformer, this means that smaller bushings can be used, although six are then required compared with three if the delta connection had been internal. The strong magnetic field of a heavy-current bushing may give rise to large circulating currents in the tank which may cause localised overheating. A method of reducing this is to mount the bushings in inserts of non-magnetic steel fastened into the tank cover or, if bushing turrets are employed, their top-plates may be of aluminium, brass or non-magnetic steel. The use of an external delta connection and the resultant need for two bushings per phase gives some degree of magnetic field cancellation since the current directions in the two bushings are opposed to each other.

Whilst it is usual to operate on the basis of one generator, one transformer, there are some installations, notably in hydroelectric power stations, where the output of two generators is fed into one transformer. Such an arrangement is shown in figure 10.2. It has the advantage of costing less and of having higher efficiency than if two transformers were used. Special winding arrangements must be employed to achieve the required reactances, and provision is usually needed to permit operation of one generator whilst the other is disconnected for maintenance or repairs. Thus a switch is provided between each generator and its associated low-voltage windings. These switches are not normally capable of breaking a fault current but are of the load – make type and are used for isolation purposes and during synchronisation of the two machines.

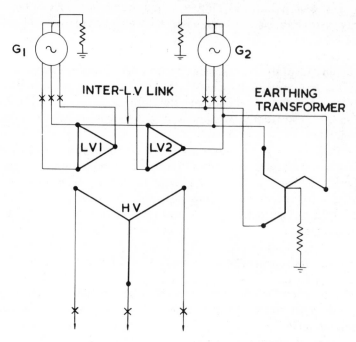

Figure 10.2   Generator transformer with two input windings

To ensure the correct voltage with respect to earth of a winding when separated from its machine, it is necessary to use an earthing transformer permanently connected to the winding. Only one earthing transformer is needed for the two low-voltage windings, since correct voltages to earth are maintained on both if an interconnection is made from one corner of one delta connection to the like corner of the other.

10.3   TRANSMISSION TRANSFORMERS

10.3.1   General

Transmission transformers may be auto- or two-winding transformers depending on a number of factors, some of which relate to the characteristics of the systems being coupled and some to economic considerations. The main factors affecting the choice are given in sub-sections 10.3.2 and 10.3.4.

Apart from the larger sub-stations, many sub-stations are of the two-transformer type, where the transformers are of such rating that each can support the entire sub-station load if the other is out of circuit for maintenance or repairs. In the larger sub-station, and depending on their importance, even higher security standards are adopted, for example by the use of more than two transformers.

Since a load is almost always shared by at least two transformers, not run at full rating, and since the load often varies considerably during a 24 h period, it is common practice to install transformers with dual rating, that is with natural and forced cooling, thus avoiding the running of cooling plant at times of light load and reducing any possible nuisance value due to noise from the cooling plant.

If voltages of the high- and low-voltage networks are in phase and the networks solidly earthed at their neutral points, it is possible to use auto-transformers with graded insulation. For larger transformation ratios than about 3 to 1, the saving of size and cost by using auto-transformers becomes marginal when compared with two-winding transformers (see sub-section 10.3.2).

10.3.2 Auto-transformers

The general principles of auto-transformers are covered in chapter 1. The type of auto-transformer almost always used on large power systems employs star connection, although other connections are possible[1].

With the conventional star-connected auto-transformer, there is no phase shift between primary and secondary voltages. Therefore its use is restricted to coupling two networks whose voltages are in phase. This can also be accomplished by using a two-winding transformer connected star – star, in which case there is no electric interconnection between the networks. No hard and fast rule can be laid down for when an auto-transformer should be used in preference to a two-winding unit. Various considerations must be taken into account, for example system earthing, ratio, tapping range, availability of tap changers, permissible transport masses and dimensions, etc.

An auto-transformer has two ratings: (1) the throughput rating and (2) the frame rating. The former is the megavoltamperes flowing into or out of the auto-transformer, whilst the latter is the rating of the auto-transformer if it were reconnected as a two-winding transformer, winding currents and voltages remaining constant, that is the frame rating is the transformed megavoltamperes.

If we take the transformation ratio as $n = $ high voltage/low voltage, the relation between frame and throughput ratings is given by

$$\text{frame rating} = \text{throughput rating} \times (1 - 1/n) \qquad (10.1)$$

The quantity $1 - 1/n$ is sometimes known as the auto factor. Thus, the more closely the low voltage approaches the high voltage, the smaller becomes the frame rating for a given throughput rating. Conversely, the smaller the low voltage, the nearer does the frame rating approach the throughput rating and the lower are the economic advantages of the auto-transformer compared with the two-winding unit. In practice, the frame rating is usually larger than that calculated by the above method, since allowance must be made for tapping windings, delta-connected stabilising windings, etc.

In Great Britain, it has been found feasible to use auto-transformers of up to 1000 MVA throughput rating between the 400 and 275 kV networks. Between the 400 and 132 kV networks, smaller units are required and auto-transformers have been found economically practicable even at this ratio. Particulars of auto-transformers used in Great Britain are given in table 10.1. The most appropriate values for other systems may be different from these.

Tappings in auto-transformers may be located in various positions relative to the other windings; examples are shown in figure 10.3. The choice of tapping position depends to a major extent on the voltage which it is desired to control, but other factors must be taken into account such as the availability of suitable tap changers. Whilst tap changers have been made with insulation for $275/3^{1/2}$ kV to earth, the provision of this insulation makes the tap changer very bulky.

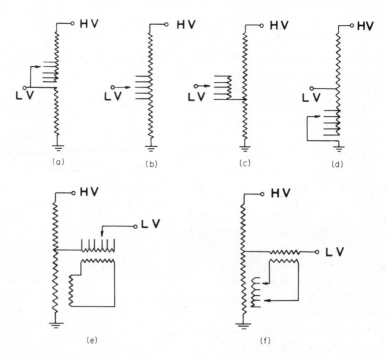

Figure 10.3   Location of tappings in auto-transformers: (a) tappings in series winding; (b) tappings to provide voltage variation on low-voltage side; (c) "fork" connection to function as for (b); (d) tappings in common winding; (e) tappings in secondary winding of separate booster transformer; (f) tappings to provide variable voltage to energise separate booster transformer

TABLE 10.1 *Characteristics of auto-transformers used on the British Central Electricity Generating Board System*

Voltage transformation of high voltage to low voltage (kV)	Throughput rating (MVA)	Reactance on throughput rating (%)	Tapping range of low voltage (%)	Position of tappings[a]
400 to 275	1000	16	no taps	—
400 to 275	750	12	no taps	—
400 to 275	500	12	no taps	—
400 to 132	240	20	+15 to −5	(a) or (b)
275 to 132	240	20	±15	(a)
275 to 132	180	15	±15	(a)
275 to 132	120	15	±15	(a)

(a) Tappings located adjacent to 132 kV terminals; (b) tappings located adjacent to neutral point.
[a]All units have 13 kV delta-connected stabilising windings suitable for supplying an external load of 60 MVA.

International practice seems to have taken $220/3^{1/2}$ kV to earth as about the maximum voltage at which tap changers can be economic.

In figure 10.3(a) the tappings are shown in the series winding and thus are principally of use when the low voltage is constant and when provision is to be made for varying the high voltage.

The arrangement in figure 10.3(b) is frequently used on auto-transformers on the British network and is mainly of use where the high voltage is fairly constant and where the tappings are required to provide low voltage variation to compensate for varying loads, etc. The arrangement in figure 10.3(c) is also used and performs similar functions. It is sometimes known as the fork connection, whilst that shown in figure 10.3(b) is known as the linear or potentiometer connection. In both these connections, the number of turns between high-voltage terminal and neutral is constant irrespective of tapping position, and thus the core flux density and the voltage of the stabilising winding remain constant if the high voltage is constant.

In figure 10.3(d), the tappings are in the common winding, and it will be seen that the alteration of tapping alters the total turns between the high-voltage terminal and neutral as well as between the low-voltage terminal and neutral. If either the high voltage or the low voltage is kept constant, the core flux density will vary with tapping position. This would also be the case with the arrangement in figure 10.3(a) if it were used to vary the low voltage when the high voltage was constant. Because of the variation of core flux density, the core is not so fully utilised at one end of the tapping range as the other, resulting in the need for an increased frame size for a given throughput rating.

To calculate the variation of flux density and the relative turns in the respective windings of an auto-transformer with neutral-end tappings, the following formulae may be used.

$$N_s = N_c(n_{max} - 1) \tag{10.2}$$

$$N_t = N_c \frac{n_{max} - n_{min}}{n_{min} - 1} \tag{10.3}$$

$$\frac{B_{max}}{B_{min}} = \frac{n_{min}(n_{max} - 1)}{n_{max}(n_{min} - 1)} \tag{10.4}$$

where N_s, N_c and N_t are the full numbers of turns in series, common and tapping windings, respectively, and n_{max} and n_{min} are the maximum and minimum ratios. Tapping turns and flux density at any point on the tapping range may be determined by substituting the appropriate ratio in the formulae.

To illustrate the increase in frame size, a comparison is made below of two designs of auto-transformer, each for 400 to 132 kV $+ 15\%$, $- 5\%$, one having tappings as in figure 10.3(b) or (c) and the other neutral-end tappings as in figure 10.3(d). It is assumed that the high-voltage terminal is at a constant 400 kV. If, for either design, the limiting flux density is taken as 1.7 T, the former unit may be designed for this flux density since it does not vary with tapping position. For the latter design, the flux density on the maximum positive tapping can only be 1.54 T in order that the limiting flux density is not exceeded at the other extreme tapping. Thus this unit must have a core about 10% larger than that of the former unit. Consequential savings in other direction may make neutral-end tappings economic for this ratio, but they become less attractive on lower-ratio auto-transformers with larger tapping ranges when a much greater derating of the core becomes necessary.

An advantage of tappings in the common winding is that they can be located close to the neutral point and can therefore be protected, to some degree, from surge voltages. The current in the common winding is less than the line current at the low-voltage terminal; thus, compared with the arrangements in figure 10.3(b) and (c), a smaller less highly insulated tap changer may be used.

When neutral-end tappings are employed, it is sometimes required that the voltage of the auxiliary winding be kept constant despite flux density variations, for example when an external load is to be supplied. This can be achieved with a booster transformer fed from fixed points on the tapping windings and with its secondary windings connected in series with the auxiliary winding in such a manner that the phase of the voltage output is unaffected by the positive or negative boost applied. One way of achieving this is shown in figure 10.4. It is the method found on some of the British 400 to 132 kV auto-transformers listed in table 10.1.

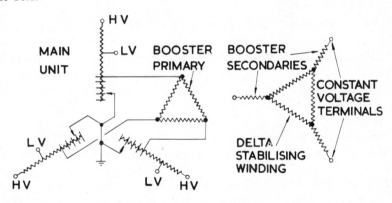

Figure 10.4   Neutral-end tappings to energise separate booster transformer

The arrangements in figure 10.3(a), (b), (c) and (d) are known as direct regulation units, the tapping windings are linked by the same core flux as the main windings. The arrangements in figure 10.3(e) and (f) employ separate booster transformers; the main core flux does not link with the regulating windings. These units are called indirectly regulated units. The booster transformer is energised from an auxiliary winding on the main transformer which is usually, but not always, the stabilising winding.

The secondary windings of the booster transformer are in series with either the low- or high-voltage line and may be arranged to boost or buck the line voltage. The employment of separate booster transformers is popular in Europe, and, by suitable choice of interconnection between main unit and booster, a degree of quadrature as well as in-phase voltage control can be provided. As the main and booster units can be transported separately, the method lends itself to constructions for sites where limited transport masses are permissible.

Because of the interconnection of windings, both directly and indirectly regulated auto-transformers are more prone to damage by surge voltages and short-circuits than are two-winding transformers. In designing to withstand surge voltages, particular care must be taken when the tapping windings are near a high-voltage terminal, for example as in figure 10.3(b) and (c), since higher voltages may be developed between tapping sections.

Another point of danger from surge voltages is when the ratio of the auto-transformer is small, that is the series winding is of relatively low voltage and few turns. In this case a surge arriving at the high-voltage terminal may be distributed almost entirely across the series winding because the low-voltage terminal is kept at more or less constant voltage by the connected network. As a result, high electric stresses can be set up in the series winding. Limitation of the surge voltage across the series winding may be effected by a surge arrester, provided the normal voltage across this winding is not large. Tapped windings of or associated with booster transformers (figure 10.3(e) and (f)) are also liable to be subjected to high surge voltages and frequently are protected by surge arresters connected across them.

Short circuits are particularly onerous to the series windings because full high or low phase voltage may be impressed across them in the event of a low- or high-voltage fault. Apart from the mechanical stresses due to the over-current, there will be electric stresses because of the over-voltage on the winding.

Stabilising windings, when not used for an auxiliary load, need only be designed for the forces due to the circulating current which arise from a single line fault to earth on the high- or low-voltage system. When the windings are also intended to supply an auxiliary load, the risk increases greatly owing to the possibility of a three-phase short circuit. A balance must be found, when choosing values for reactance between the stabilising windings and other windings, between limiting the severity of a three-phase fault and yet ensuring sufficiently small regulation of output voltage. The latter is particularly important when supplying capacitive or inductive compensation equipment, since the load taken by such equipment is proportional to the square of the voltage. (For further consideration of stabilising windings, see section 10.6.)

10.3.3 Quadrature boosters

A quadrature booster transformer is a special case of auto-transformer in that the boost voltage is arranged to be in quadrature to the supply voltage instead of in phase with it, as is the case with the conventional auto-transformer. This assists with the control of load flow between interconnected circuits.

Ideally, a quadrature booster should alter the phase difference between the phasors of incoming and outgoing voltage without affecting the magnitude of these phasors. As the boosting voltage is injected with its phasor at right-angles to the phasor of the incoming voltage for any value of boost except zero, the above requirement can only be met approximately for small values of boost. To compensate for phase difference would require excessive complications in a device already far from simple.

Figure 10.5 illustrates a typical quadrature booster design which will be seen to consist of two transformers, a star – star-connected excitation unit which feeds a booster unit with delta-connected primary and independent secondary windings for each phase. In this design, the variation of boost voltage is effected by tappings at high voltage on the independent secondary windings of the booster transfor-

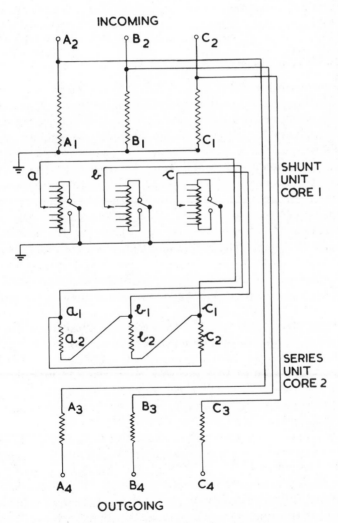

Figure 10.5   Typical quadrature booster design

mer, but such a method may not be practicable at very high voltages owing to insulation limitations of available tap changers. Many other methods of controlling the boost voltage are available, limited only by the ingenuity of the designer and the characteristics of the tap changers available. It should be noted that the boost angle of the design shown in figure 10.5 may be altered so that the outgoing voltage lags or leads the incoming voltage simply by reversing the phase of the boost winding, facilities for which are provided in many tap changers. Problems in control of surge voltages are similar to those described for conventional auto-transformers, and quadrature boosters are very prone to difficulties under short-circuit conditions, particularly when the amount of booster winding in circuit is small. For this reason, it is often practice to connect an inductor in series with the booster in order to limit the short-circuit level.

10.3.4 Two-winding transformers

As explained in sub-sections 10.3.1 and 10.3.2, if the ratio of transformation is large, the advantages of the reduced cost and losses of an auto-transformer may be small when compared with a two-winding transformer with electric separation between its windings. If a change of phasor relationship is required between the high- and low-voltage networks, then a two-winding transformer is necessary.

The electric separation of the windings reduces the design limitations, since to some extent, but not entirely, each winding may be treated independently. Care must still be taken, however, in ampere – turn balance and in allowing for voltage stresses due to inductive and capacitive coupling between windings.

Low-voltage connections of two-winding transformers may be delta or star, depending on the phasor relationship required. When a delta-connected low-voltage winding is employed, links may be incorporated within the transformer to enable the windings to be connected for alternative phasor groups. This is usually advisable on a large network to secure a common design for most interconnection points.

If the low-voltage winding is star connected, it is usual, for the sake of standardisation, to specify an additional delta-connected stabilising winding, although this may not necessarily be required with certain system conditions (see section 10.6). On the larger sizes of two-winding star – star transformers, the delta-connected stabilising winding may provide a supply to reactive power compensation equipment in the same manner as on the large auto-transformers described in the previous section.

Impedances must be chosen to prevent exceeding the switchgear rating. In the larger sub-stations, when more than two transformers may be connected to a busbar and when there is often a further infeed due to local generation, it is frequently the case that impedances would need to be so high that voltage regulation would become excessive. When this happens, the busbars must be split into sections, thus permitting lower impedances without excessive short-circuit levels. Depending on the short-circuit levels when part of the transformer complement is out of circuit for maintenance or repair, the sections may be joined to give increased security of supply.

A useful transformer arrangement, the dual-low-voltage winding, permits two supplies to each half of a split busbar and thus limits short-circuit level whilst ensuring continuity of supply in the event of a transformer failure. Figure 10.6 shows two such units feeding a split busbar. To achieve the same degree of security with conventional two-winding transformers would require two transformers per busbar, each rated at half the primary rating of the dual-low-voltage units, and would necessitate extra space on the sub-station site.

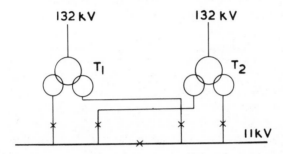

Figure 10.6   Two dual-low-voltage winding transformers feeding a split busbar

Dual-low-voltage transformers have an equivalent impedance network, as shown in figure 10.7(a), in which the apportioning of impedances between the various branches may be altered by varying the degree of interleaving between the two low-voltage windings. A possible arrangement of windings in a dual-low-voltage unit is shown in figure 10.7(b). The high-voltage main and tapping

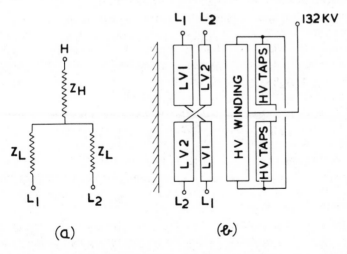

Figure 10.7   Dual-low-voltage winding transformer: (a) equivalent impedance network; (b) winding arrangement

windings are in two parallel halves, thus ensuring correct ampere–turn balance between these and the low-voltage windings under differing low-voltage loads.

Impedances of dual-low-voltage units must be chosen with care since a design with a high value of Z_H and low values of Z_L will give high regulation owing to the large common impedance and may necessitate an extension of tapping range. Conversely, if Z_L is high, regulation drops will be large in the low-voltage windings, and, with differing low-voltage loadings, control of the voltages of both windings may become difficult with the single tap changer which can be fitted to the high-voltage winding.

When the dual-low-voltage windings are delta connected, it is possible to use a single earthing transformer, as in figure 10.2.

10.3.5 Earthing transformers

If an earth fault occurs on one line of an isolated system, usually one fed by a delta-connected main transformer winding, no return path exists for the earth fault current, and hence no current flows. The system will continue to operate, but the other two lines will rise in voltage from $V/3^{1/2}$ above earth to V above earth, where V is the system voltage. Overstressing of insulation will result both in the transformer and on the system.

To prevent such an occurrence, a return path must be provided for the earth current, and this is usually done by connecting an earthing transformer to the system; the earthing transformer has an earthed neutral point through which the fault current may flow.

The simplest form of earthing transformer, in this case more strictly speaking an earthing inductor, is a star-connected three-phase winding. However, such a device, if it is not to take an excessive magnetising current from the system during normal operation, must have a high impedance. It thus considerably restricts the current at the time of fault. Insufficient current may pass to operate the protective equipment, for example earth fault relay, particularly if the earth fault is remote from the earthing device. To reduce the impedance to fault current, a closed delta winding may be fitted to the inductor, making it into an earthing transformer.

A low-voltage supply is often required in a sub-station for operation of auxiliary equipment, and, since this supply must also be earthed, it is conveniently provided by an extra star-connected winding on the earthing transformer. Thus a feasible combined earthing and auxiliary transformer would have three windings connected star–delta–star.

A simpler arrangement (figure 10.8) may be achieved by the use of an interconnected-star winding in place of the star and delta windings described above. The interconnected-star winding has a much lower impedance to zero-sequence currents, owing to the cancellation of the fluxes set up by the individual half-windings. A delta winding is thus no longer required, although it is common practice to fit a star-connected auxiliary winding to supply auxiliary loads.

An added advantage of the interconnected-star connection compared with the star connection is that, when used with a main transformer connected in

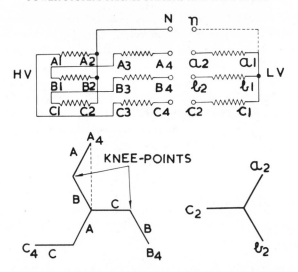

Figure 10.8   Earthing transformer with interconnected-star winding

star – delta, the auxiliary winding voltage of the earthing transformer is in phase with the primary voltage of the main transformer. This is useful in checking sub-station phase relationships during commissioning and subsequent operation. To enable phase relationships to be maintained if provision for alternative phasor group connection is made on the main transformer, provision for alternative phasor group connection must also be provided on the earthing transformer. This may take the form of links to enable reconnection of the parts of the interconnected-star winding, or an equivalent can be obtained by reconnection of the star low-voltage windings.

The latter method is preferable because of over-voltages which can occur within the interconnected-star windings as a result of system transients. These over-voltages are between the 'knee-points' and may considerably exceed the insulation levels of the winding even though the transient voltages at the terminals are within these limits.

Whatever connection is used for the high-voltage winding, the conductor cross-section must be chosen so that the temperature rise of the windings, when carrying short-circuit current, is not excessive. The maximum value of earth fault current is usually given in the purchasing specification; hence the value of zero-sequence impedance required may easily be calculated.

When the earthing transformer is operating as an auxiliary transformer, supplying a load on the low-voltage winding, the current density in the primary winding is usually very low in comparison with that when fault current is passing. Hence the load loss of an earthing transformer tends to be small, and it is generally possible to employ a plain tank. Typical ratings of earthing transformer auxiliary loads are 120 to 150 kVA. There is a tendency for these to be increased for supplies to the larger sub-stations, and cooling tubes or radiators may then be required. Common practice is for the low-voltage side to be connected to the system via a switch-fuse unit which provides a termination for the low-voltage cables. High-voltage terminations may be bushings or cable boxes, the latter being normally of the three-phase type. However, on earthing transformers connected to the delta stabilising windings of large two-winding or auto-transformers, when these windings are externally terminated for supplying compensation equipment, separate cable boxes or ones employing interphase shielding are fitted. This reduces the possibility of phase faults and consequent onerous mechanical stresses on the stabilising windings.

10.4   SHUNT INDUCTORS

10.4.1   General

In urban areas, power distribution is frequently by underground cables. Owing to the large increase in consumption in recent years, a need has arisen to site primary sub-stations as close as possible to the load centres with the result that extensive high-voltage cable networks are built up. The charging currents of such networks can be very considerable, for example 22.5 to 23 A per circuit kilometre of 275 kV 2000 mm^2 cable, equivalent to 17 to 17.5 Mvar. It is apparent that, if uncompensated, only a few kilometres of cable would take a considerable charging current, resulting in increased heating and a significant decrease in the effective cable throughput capacity. Furthermore, the effect of these capacitive currents is to produce a rise of system voltage under light load conditions. Another difficulty with large capacitive currents is that, being out of phase with the voltage, the interruption duties placed on the circuit breakers are made much more onerous.

To counter the above effects, shunt inductors may be installed at intervals along the cable network. A shunt inductor is an inductive device which, if we neglect the slight effect of losses, takes a current which lags the applied voltage by exactly $\pi/2$ rad. Thus the lagging current of the shunt inductor may be chosen to compensate for the leading current taken by the cable. It is usually unnecessary to have complete compensation from the switchgear performance considerations.

It should be noted that the shunt-inductor current is proportional to the system voltage as is that of the cable; thus a combination of cable and shunt inductor is self-compensating, a rise of leading current in one being counterbalanced by a rise of lagging current in the other.

Shunt inductors may also be used in association with long-distance overhead lines, where again capacitive currents may reach quite high values.

On very-high-voltage systems, shunt inductors are invariably star connected, the star points being earthed. The basis of a shunt inductor is a coil connected between line and earth; the required impedance is achieved by the diameter of the coil former, the number of turns and the proportions of the flux path in iron and air.

10.4.2 Gapped-core inductors

The cores of these inductors are divided into packets separated by insulating material since, if a complete core were used, the inductance of the coil would be too great, and insufficient magnetising current would be taken. Figure 10.9 indicates the basic assembly of one limb of a gapped-core inductor.

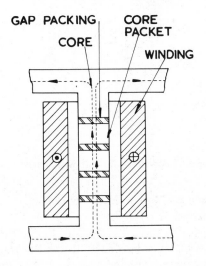

Figure 10.9   Basic assembly of one limb of a gapped-core inductor

The disadvantage of gapped-core inductors is that they tend to be noisy unless extensive measures are taken to counteract the causes which are principally the vibration, at twice the supply frequency, of the packets due to the attractive forces between them and the fringing of the flux at the gaps between the packets.

10.4.3 Core-less inductors

A core-less inductor may be considered as a gapped-core inductor in which the gaps are expanded to comprise the whole leg. In a gapped-core inductor the return limbs guide the coil flux from its point of exit from the coil back to the point of entry, thus preventing flux from entering the tank and other metal parts where it could cause overheating and increased losses. In a core-less inductor, similar provision for flux control must be made and various possibilities exist.

(i)   Each coil may be surrounded by a core-like screen with yokes and return limbs but no centre limb (figure 10.10). To form a three-phase inductor, three such units are mounted in one tank. A variant on this method is that the screen can have a four-limb construction, one phase winding being inserted into each window. Another variant is that each phase winding can be enclosed in an annular screen consisting of radial core-plate segments

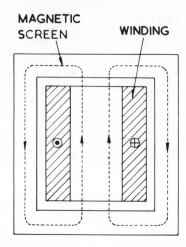

Figure 10.10   A coil of a core-less inductor with an individual magnetic screen

and with top and bottom mosaics of core-plate to collect the flux leaving the ends of the coil.

(ii)   The screening may be applied to the tank. For a single-phase inductor, the construction would thus be similar to the last variant mentioned in (i) above, but, for a three-phase inductor, a construction is possible in which three unscreened phase windings are enclosed in a tank with screening on its inside surface. In order to maintain symmetry, it follows that the phase windings are arranged in trefoil pattern; thus each winding is affected equally by mutual coupling to the two adjacent windings and unbalance is prevented.

Such a construction is illustrated in figure 10.11. It will be seen to have aluminium flux screens at top and bottom instead of the core-plate screens used in other types of inductors. Aluminium screens have the advantage of lightness and, in the construction illustrated, avoid the provision of heavy structures to support the complicated arrangement of core-plate screens which would otherwise be necessary.

As they are in shunt with the system, shunt inductors do not have to sustain forces due to through short circuit. The winding design need cater only for sufficient mechanical strength to withstand the electromechanical forces involved in normal service.

Two principal types of winding design exist: (1) the disk type and (2) the layer type. In the disk-type design, it is common practice to place the high-voltage end of the coil at the centre and to wind outwards to the ends which, as they are at earth voltage, need have only nominal insulation.

When layer windings are employed, because the outside of the coil is adjacent to earthed metal, it is usual to make the neutral on the outside and for the coil to

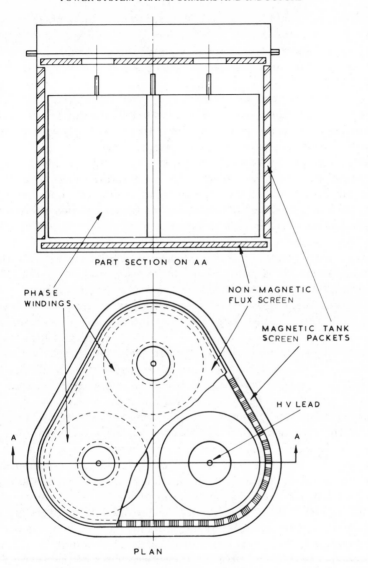

PART SECTION ON AA

PHASE
WINDINGS

NON-MAGNETIC
FLUX SCREEN

MAGNETIC TANK
SCREEN PACKETS

HV LEAD

A   A

PLAN

Figure 10.11   Three-phase arrangement of a core-less inductor with the tank as a
common magnetic screen

have decreasing lengths of layers towards the inside where the high-voltage end is
located. With this arrangement, the lead from the bushing passes down the bore
of the coil.

Unlike transformers, where a high loading usually is required only at peak
periods when increased noise due to forced-cooling plant is unlikely to be a
nuisance, shunt inductors take an almost constant load depending only on system
voltage and frequency. Thus, if forced cooling were used, it would run

continuously and could be a noise nuisance at times of low ambient noise level. Because of this, many inductors are designed for ONAN cooling despite the encumbrance of the very large cooling banks of radiators required. The inductors may be designed for the fitting of noise enclosures, as further described in subsection 10.7.2.

An important point to be remembered in connection with shunt-inductor cooling is that the winding loss is proportional to the square of the system voltage. Therefore, shunt inductors may need to be designed to comply with temperature rise guarantees at higher than nominal system voltage.

Shunt inductors may be subjected to switching surges or surge voltages of atmospheric origin and should therefore be designed to have the same insulation levels as other equipment on the system. Surge-voltage tests of the usual type may be applied to verify adequacy of insulation; full-wave surge-voltage tests may also be applied in lieu of induced-over-voltage tests which are almost impossible to arrange on this type of device.

Measurement of losses on shunt inductors is very difficult since, in contrast with transformers, it is not possible to measure no-load and load losses separately. All that can be measured is a total loss comprising winding, screen and stray losses; furthermore, some of these losses may be non-linear with voltage. Thus a curve of loss against voltage, up to rated voltage, should be established for an inductor before relatively low-voltage tests can be accepted for subsequent inductors. This is the ideal which in practice may not be achieved owing to limitations of factory power supplies. It may be possible to reach full voltage for a brief period to measure full loss, but rarely can temperature rise tests, which require an extended test period, be carried out at full voltage.

A further difficulty in loss measurement is caused by the very low power factor of the current which results in gross inaccuracies even if special low-power-factor wattmeters are used. A power factor of 0.003 is a typical value for a very-high-voltage inductor. Frequent recourse has been made to bridge measurements and calorimetric methods, both of which, if carefully carried out, can give results certainly no less inaccurate than methods involving the use of wattmeters[2, 3, 4].

Several 100 Mvar three-phase shunt inductors are already operating on the British 275 kV system, whilst future inductors for the 400 kV system may be of 200 Mvar rating. In other countries, probably the highest-voltage shunt inductors in service are the 735 kV 110 Mvar single-phase units of the Hydro-Electric Commission of Quebec, Canada, installed as 330 Mvar banks.

10.5 SERIES INDUCTORS

10.5.1 General

At each voltage level of a transmission system the load is supplied from a number of generating sources through varying degrees of transformation via interconnected circuits. Under conditions of healthy system operation, the size and direction of the power flow is determined by the value and location of the loads on these circuits. The connections between the points of generation and supply are chosen to maintain a balanced system. Care is also taken to ensure that, in the event of a fault on the system, the power flowing to the fault can be limited so as not to exceed the rupturing capacity of the circuit breakers. The permissible duration of a full short circuit upon any part of the system is dictated by the short-circuit withstand capability of the system plant. In practice, it is taken, for example as 1 s in a 400 kV or 275 kV system and as 3 s in a 132 kV system; the system protection is set to clear a fault normally in much less time than this.

The degree of fault infeed can be calculated by theoretical analysis of the system when operating under various fault conditions, thus permitting a comparison to be made with the known fault withstand capability of the system plant. For any given theoretical system fault condition such comparisons may indicate that the system impedance alone is not sufficient to limit the fault infeed and that there exists a need for added fault-limiting impedance. Other similar situations occur on a system owing to load growth or as a result of additional system interconnections to provide greater load sharing by plant. In either case it may be found necessary to reinforce the system impedance at selected points to limit the increased fault current or alternatively to provide a balanced network of system impedance to maintain a uniform sharing of load.

It is common for such situations to arise. Analysis of the problem often indicates that it is not practical to obtain the required degree of impedance reinforcement simply by increasing the impedance of system equipment. Additional system impedance may be obtained by increasing the impedance of transformer and generator plant when the system is at the design stage, but there is a limit beyond which this becomes impracticable or uneconomic. It is in these circumstances and those arising on an established system that any additional impedance is provided by installing series inductors.

These are employed at all system voltage levels and are usually located close to the equipment they protect. Methods of design and construction vary, but in nearly all cases series inductors are designed on the basis of a core-less inductor. Together with additional constructional features, series inductors can be divided into three fundamental types: (1) air-cooled cast-in-concrete, (2) oil-immersed magnetically screened and (3) oil-immersed non-magnetically screened.

A fourth type known as an oil-immersed screened magnetic-cored inductor may also be used, but the types listed above are the most commonly found in practice. Of these, the oil-immersed screened inductor is widely used since its design and construction is adaptable to almost any application except perhaps for low-impedance low-voltage heavy-current applications when the cast-in-concrete type may be favoured. As will be seen later, all three types can be manufactured for indoor or outdoor use irrespective of their cooling arrangement.

To be successful as a current-limiting device, a series inductor must be designed so that it does not decrease in effectiveness as the current through it increases to the value of maximum fault current. This means the relationship between through-current and reactance should be as shown by figure 10.12, curve a. This relationship is representative of a core-less inductor without any form of magnetic screening and situated away from any metallic structure or other apparatus which may be affected by its high stray field. Only the cast-in-concrete and non-magnetically screened inductors have a similar characteristic of current against reactance since neither design incorporates a magnetic circuit outside the winding to provide a return path for the main flux. Such a path is provided in the magnetically screened type in order to restrain the flux from entering the tank wall and from inducing eddy currents with consequent tank heating and increased losses. The effect of this screen is to produce a non-linear characteristic of current against reactance, as shown by figure 10.12, curve b. This is due to the changing permeability of the screen laminated-core-plate material as the flux density in the screen increases with increasing current.

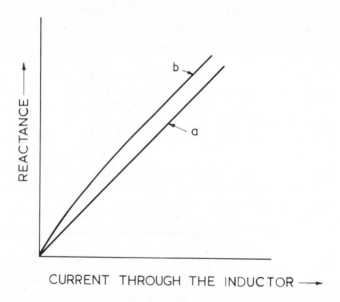

Figure 10.12   A characteristic of reactance against current of core-less inductor: curve a, without magnetic screen; curve b, with magnetic screen

A feature that is common to all inductors and must be given full consideration in all designs concerns the mechanical strength and thermal performance of the winding. The forces acting upon the winding are increased in proportion to $(I_k/I_m)^2$, where I_k is the current under fault conditions and I_m the normal full-load current. They may be greater with non-magnetically screened inductors if any magnetic unbalance exists between the winding and the screen. To prevent possible damage to the inductor under fault conditions the windings and screen must be securely clamped. In addition the winding conductor must be of sufficient cross-sectional area to give it the mechanical strength to withstand the compressive forces existing during short circuit. These forces act axially upon the conductors, interturn and interdisk insulation, their cumulative effect being a maximum at the centre of the winding. To prevent movement of the major insulation and turns, the windings are securely clamped axially between top and bottom non-magnetic frames by means of brass or other non-magnetic tie rods.

A further important consideration affecting the cross-sectional area of the conductor is the thermal performance of the winding. The losses under normal load conditions are usually small and therefore ON cooling is generally sufficient to limit the winding temperature rise to within the permissible limits. Under short-circuit conditions, however, these losses are increased in proportion by the square of the short-circuit ratio, that is $(I_k/I_m)^2$. The winding is unable to dissipate these quickly enough to prevent the winding temperature from rising at a very rapid rate. The maximum permissible winding temperature at the end of the short circuit is taken as $250\,°C$, whilst the duration of the short circuit is not to exceed 5 s. On the assumption that all losses in the inductor during short circuit are stored wholly in the winding copper, the winding final temperature can be calculated from the following formula.

$$\theta_f = P_{Cu}t_k\left(\frac{b}{2\Theta_i} + \frac{620a}{b}\right) + \theta_i \tag{10.5}$$

where θ_f is the final temperature in degrees Celsius, θ_i the initial temperature, in degrees Celsius, Θ_i the absolute initial temperature in kelvins, t_k the short-circuit duration in seconds, a the ratio of eddy current to I^2R loss at $75\,°C$,

$$P_{Cu} = \frac{1}{82} \times \text{watts loss per kilogram of copper at } \theta_i$$

$$= 1.1\left(\frac{\text{short-circuit fault current in amperes}}{\text{conductor cross-section area in square millimetres}}\right)^2\Theta_i \times 10^{-16}$$

and

$$b = 2\Theta_i + P_{Cu}t_k$$

In general, inductor windings are designed to have small losses at normal load for economic reasons. In consequence they have a substantial copper section which tends naturally to improve the thermal capacity of the winding under short circuit.

10.5.2 Cast-in-concrete inductors

The term used to describe this type of inductor arises from the method of construction employed. Basically it is an air-cooled core-less coil whose turns are supported and restrained from any mechanical displacement by being held solidly in position with concrete piers (figure 10.13). Usually the winding is of disk-type construction with the appropriate number of solid or stranded bare copper conductors for the current rating. The clearances between turns and between disks are greater than those employed in oil-immersed units because of the low voltage withstand strength of the concrete and insulation. If intended for outdoor

Figure 10.13   Three-phase air-cooled core-less cast-in-concrete inductor

use, the winding conductors may be insulated with a class B insulation consisting of bitumen-impregnated asbestos or similar material, though this is not always considered to be essential. Sometimes insulated cable is used as the conductor.

One method of manufacturing this type of inductor is to build the winding on a former which also serves as a mould for casting the specially prepared concrete. The completed casting consists of solid concrete piers which hold the turns in place. The size, number and spacing of those piers is chosen so that sufficient strength is available to restrain the turns from any movement which might be caused by the compressive stresses during short circuit.

The completed windings may be mounted horizontally or vertically. For three-phase inductors the individual phases can be arranged side by side or in tiers. Whichever method is adopted, the insulation of the winding either from earth or between phases when necessary is, in general, maintained by means of porcelain insulators.

This method of construction is extremely robust, but because of its reliance upon air as a coolant and the low insulation strength it is restricted in size and rating to low-voltage network applications. This, together with its inability to compete with the economy of space afforded by oil-immersed inductors, has led to more general use of oil-immersed inductors.

10.5.3 Oil-immersed screened inductors

General

This type of inductor can be made suitable for indoor or outdoor use at all levels of system voltage. It is in many respects similar in external appearance to the conventional transformer in that it is contained in a steel tank which may be fitted with bushings or cable boxes, oil-cooling tubes or radiators, winding temperature indicator, gas-actuated device (Buchholz relay) protection, oil-pressure relief device and oil breather. In almost all cases the inductor cooling is of the ONAN class, and the inductor has a thermal performance in accordance with reference G1.7.

When compared with cast-in-concrete inductors, the oil-immersed inductor can be made physically smaller for a given rating because of the greater voltage withstand strength and better cooling properties of the oil. Conventional types of windings can be used, and internal clearances can be reduced. For the same reasons, this type of inductor can be designed for much higher ratings and the highest system voltages.

Internally the construction consists of a core-less paper-insulated disk- or layer-type winding with diameter, length, radial depth and number of turns of the required magnitude to give the desired value of reactance. The design of the winding may be varied as to conductor size, number of turns in parallel and quantity of insulation according to the voltage and current applications envisaged. However, it must ultimately result in a fully insulated winding capable of withstanding surge voltages appropriate to the designated voltage class at each of its terminals. The size of the winding conductor is chosen to maintain a low current density at normal current and to have a high thermal capacity under short circuit, as described in sub-section 10.5.1.

The winding is securely clamped axially by means of non-magnetic tie rods and clamp-plates. The assembly, together with the inductor, is firmly located within the tank to prevent any possible movement under fault operation. Adequate electric clearances must be provided between the winding and all metal parts within the inductor, as these are solidly earthed to prevent any induced voltages arising owing to stray flux.

The air-cooled inductor has basically an unrestricted flux path, but special treatment has in some instances been necessary to prevent overheating in surrounding safety metal screens and constructional steelwork. A serious consideration in the design of any oil-immersed inductor is to prevent this flux from passing into the steel tank where it induces eddy currents which cause local heating and increased load losses. To overcome this problem, two different methods of construction are to be found in common use, each designed to restrain the flux from entering the tank. The first employs a magnetic circuit about the winding to contain the flux, and the second prevents the flux from entering the tank by establishing an opposing flux in a non-magnetic screen surrounding the winding. To differentiate one method of construction from the other, oil-immersed inductors are described as either of the magnetically screened type if the former construction is employed or of the non-magnetically screened type if the latter construction is adopted.

Magnetic screening

This consists of a laminated core-plate screen arranged about the winding. This screen forms a low-reluctance path for the flux outside the winding, thus preventing it from entering the tank. The core-plate at the top and bottom of the winding is usually supported on stout non-metallic platforms, whilst that surrounding the winding may take the form of a cylindrical screen or alternatively packets of core-plate can be bolted to the inside tank wall. The screen is designed to have a low flux density of about 0.4 to 0.5 T at normal current. Under fault conditions it becomes saturated, and the inductor then functions as a coreless inductor. As explained in sub-section 10.5.1, owing to the varying permeability of the screen material with increasing current, the characteristic of current against reactance of the magnetically screened inductor is thereby non-linear.

Depending upon size, three-phase inductors may have separate phase coils arranged side by side in a common tank or mounted vertically on a common winding mandrel. In both cases, each phase can be separately screened, or in the latter a common screen can be provided for all three phases.

Non-magnetic screening

In this design, a non-magnetic screen of solid copper or aluminium is arranged as a short-circuited turn about the inside surface of the tank. This arrangement causes the path of the flux linking the winding to lie wholly within the screen, thus preventing it from entering the tank wall. A separate non-magnetic screen is also fitted at each end of the winding. An alternative method of mounting the main screen is sometimes used where a separate screen is securely clamped about the circumference of each phase winding. With either method of construction, care must be taken to ensure that the magnetic centres of the main screen and the winding are coincident; otherwise any unbalance will greatly increase the mechanical forces acting upon the winding and the screen under short-circuit operation.

Apart from the considerations enumerated above, the design, testing, operation and maintenance aspects of series inductors are similar to those for power transformers. From the point of view of testing, however, it is not possible to apply an induced-voltage test, as for transformer testing. To ensure that a satisfactory voltage withstand strength has been attained, one method is to subject each series inductor to a series of surge tests at a level appropriate to the system working voltage. This applies equally to inductors with cable box or bushing terminal arrangements.

10.6 STABILISING WINDINGS

10.6.1 General

The use of stabilising windings in transformers as a means both of harmonic suppression and of fault control is discussed in this section. Before considering these two aspects in detail it may be useful to note that it is usual to include stabilising windings on auto-transformers and on star – star-connected transformers. Some star – star-connected three-phase three-limb transformers without stabilising windings have been used on distribution networks. It appears likely that increasing use will be made of this type of unit in the future. However, general practice at present is to use stabilising windings. Various arguments have been considered questioning the need for stabilising windings, particularly with regard to three-phase three-limb transformers, but the factors requiring consideration relate to particular situations where the characteristics of the circuit involved and the type of transformer to be used are known. The effect of eliminating the stabilising winding can then be anticipated. The purpose of fitting stabilising windings to power system plant without discrimination is, therefore, to avoid the need for such considerations and to achieve standardisation and inter-changeability of the equipment.

10.6.2 Harmonic suppression

The two fundamental characteristics of core steel, permeability and hysteresis, produce appreciable distortion in the waveform of the transformer magnetising current, as flux or induced voltages depend upon the waveform of the applied voltage.

When the applied voltage is sinusoidal, as it usually is in practice, the resultant flux is sinusoidal, but the magnetising current is non-sinusoidal. Analysis of the magnetising current waveform shows that it consists of a fundamental plus a series of odd harmonics, of which the third harmonic is the principal. The presence of these harmonics is due to the behaviour of the core steel under ac excitation. The third and other triplen harmonics are generated as a result of the non-linear $B - H$ characteristic of the core steel, and the remaining odd harmonics are due to the hysteresis effect. If the magnitude and waveform of the flux and the corresponding hysteresis waveform are known, the magnetising current waveform can be represented by a Fourier series consisting of a fundamental sinusoidal current plus the sinusoidal odd harmonics. The effect of hysteresis is to cause the no-load current to lead the true magnetising current by an angle depending upon the magnitude of the hysteresis present, but the triplen harmonics remain substantially in phase with the fundamental.

The amplitude of the harmonics and hence the degree of distortion of the magnetising current waveform is dependent upon the peak flux density in the core, the type of core steel employed and the method of core construction. The introduction of grain-oriented core steel, together with various methods of improved core design, such as mitred corners, matched leg and yoke sections and boltless construction (see chapter 4), has led to considerably reduced values of magnetising current compared with previous transformers of similar rating using non-oriented core steel. These reductions are due to the better characteristics of the grain-oriented core steel. However, the non-linear magnetic effects are still present, and the magnetising current must still contain the odd-harmonic components if the induced voltages and flux are to remain sinusoidal.

If we neglect the higher-order triplen harmonics, the effect of the third-harmonic currents lies principally in the fact that in a three-phase star-connected transformer the third-harmonic currents in each phase are in phase with each other at the neutral, the fundamental and other odd-harmonic components summating to zero at the same point. Unless a path is provided for these currents from the transformer neutral to the neutral of the supply and hence back into the system, the current in each phase will be sinusoidal, and the line-to-neutral voltages at the transformer will in consequence contain a third-harmonic component. These will be in phase with each other at the neutral point, causing it to shift and thereby increasing the voltage stress on the windings. Under these conditions the magnitude of the third-harmonic voltage component can be 50% of the phase voltage.

Such a path as that required may be provided for the third-harmonic current by earthing the primary or secondary winding. If the latter method is used, the secondary system must be earthed elsewhere by means of an earthed star – delta transformer or, alternatively, of an earthing transformer (see sub-section 10.3.5). Whichever neutral is used, an earth return is provided for the third-harmonic current, and the problem of third-harmonic voltages on the transformer and the system is resolved. Two complications, however, can arise as a result of earthing the transformer neutral. The triple-frequency current that passes in the circuit consisting of the earth path, source neutral and transmission lines may inductively affect parallel communication circuits and may result in interference. Control of this effect can be very difficult to achieve in practice since it is unlikely that a system will have only a single source earth. In this case the third-harmonic current path can be complex.

Depending upon the system characteristics, a second complication can occur as a result of earthing the transformer neutral when no source earth is available or, to a lesser extent, when the earth return path presents a high impedance to the flow of third-harmonic current. In these circumstances the third-harmonic voltage at the neutral is impressed on a circuit comprising the line capacitance to earth in series with the open-circuit third-harmonic inductance of the transformer winding. The condition can become very dangerous when the values of capacitive and inductive reactance of the circuit are almost equal. In these circumstances resonance can occur, causing the third-harmonic voltage at the neutral, and hence the system voltage, to increase to a peak value such that the line-to-earth voltage may be high enough to cause a flashover at the transformer terminal and possible breakdown of the winding. Intermediate values of the circuit parameters give rise to conditions which are less onerous but which require equal consideration.

The magnitude of the effects described varies according to the type of core construction employed. With the transformer neutral isolated the third-harmonic fluxes, established in each phase by their associated voltage and current, are in phase with one another. With three-phase core constructions, such as the shell type and the five-limb core, a high-permeability path is provided for the third-harmonic fluxes. Because of the low magnetising impedance to the third-harmonic current, the third-harmonic voltages may be about 30 to 70% of the applied phase voltage. Where star-connected five-limb or shell-type core constructions are used or where single-phase transformers are star connected to form a three-phase bank, particular attention must be paid to the method of earthing the neutral point if the effects described are to be minimised.

For three-phase three-limb cores, the third-harmonic flux paths are through the high-reluctance air path outside the core. The flux is therefore much smaller compared with other three-phase core constructions, and the resulting third-harmonic voltages are only about a few per cent of the applied voltage.

These effects, together with the differences due to core construction, are minimised if the transformers have delta-connected stabilising windings. These provide a low-impedance circuit to the third-harmonic voltages, and the foregoing third-harmonic effects, which otherwise appear at the transformer terminals, neutral point or on the system, become negligible. In general, the advantage of the stabilising winding is to achieve for the auto-transformer and star – star-connected transformer a stabilised neutral point, freedom from excessive transformer and system third-harmonic voltages and freedom from causing interference with communication circuits. Their use is of greatest advantage in transformers with five-limb or shell-type cores and least in the case of three-limb cores.

10.7 AUXILIARY ASPECTS

10.7.1 Cooling arrangements

Introduction

In respect of the type of cooling used, power system transformers may be divided into two groups: (1) transmission transformers and (2) generator transformers. Generator transformers operate primarily at full load and therefore have small need for cooling plant with an intermediate rating. Also, their location at a power station where there is usually an abundant and reliable supply of cooling water provides the facility to use oil–water heat exchangers with the consequent economic savings in space over the conventional oil–air cooling equipment. The distinction between the two groups is, therefore, that generator transformers may, if required, employ OFW cooling, whereas transmission transformers are cooled by any of the alternative standard methods.

Generator transformer cooling arrangements

Generator transformers, if equipped with OFW cooling, may have three oil–water heat exchangers connected in parallel and each rated at 50 % of the full-load losses of the transformer. This arrangement permits the transformer to operate at its rated output with only two heat exchangers in service, the third heat exchanger serving as a standby to maintain security of supply in the event that one heat exchanger fails or is taken out of service temporarily for periodic maintenance.

The heat exchangers are designed to have a specific heat dissipating capability for defined oil and water rates of flow and temperature. The velocity and quantity of oil required to maintain the maximum transformer operating temperature within the permissible limits is a function of the average winding-to-oil temperature difference and the dynamic head of the oil circuit, as measured across the inlet and outlet of the oil pump. This is the total pressure drop in the oil circuit when the pumps are operating and includes the pressure drops across the windings, pipework, heat exchangers and pumps. Similarly the velocity and quantity of water required through a given heat exchanger in order to obtain the desired transformer oil inlet temperature, are a function of the exchanger oil inlet and outlet temperature difference, the water temperature and the dynamic head across the water side of the heat exchanger. In practice, the rate of flow of oil and the pump capacity required to circulate this through the oil circuit are determined from empirical formulae obtained from studies made on completed cooling systems where oil flow patterns, particularly through the windings, are known. The water side of the problem is usually considered, firstly, in terms of utilising standard heat exchangers of known thermal performance under specified hydrodynamic and temperature conditions on site or, secondly, by designing a heat exchanger to meet the particular operating performance required.

The type of heat exchanger commonly used is known as a two-pass cross-flow heat exchanger, since the water passes twice through the cooling tubes across which the oil is circulated. Each heat exchanger is mounted in a vertical position and is equipped with a removable header at the upper end to allow the tube nest to be withdrawn as required. Whilst every precaution is taken to ensure that the oil and water systems are kept separate, failure of the oil–water separation must be anticipated, and adequate safeguards must be provided. The equipment is designed so that the oil pressure is greater than the water pressure, thus ensuring that any leakage between the two systems will be from oil to water. Pressure gauges are fitted on the oil and water inlet branches to maintain a visual check on the cooling system pressures, in addition to the automatic safeguard provided by a differential pressure gauge connected across the oil outlet and water inlet branches. This incorporates electric-alarm contacts which are set to initiate an alarm when the heat exchanger oil outlet pressure does not exceed the water inlet pressure by at least $20 \, kN \, m^{-2}$. This requirement of a $20 \, kN \, m^{-2}$ positive oil-to-water differential pressure applies to both static and dynamic oil pressures.

Further precautionary measures can be taken to maintain oil–water separation by adopting alternative and more sophisticated types of heat exchanger. Two types of heat exchanger, designed specifically to achieve this, are the intercooler and the double water-tube heat exchanger. The intercooler consists of a two-stage heat exchanger comprising an oil-to-water heat exchanger and a water-to-water heat exchanger, the dynamic oil and water systems thus being separated by an intermediate water system. The second type, the double water-tube heat exchanger, is similar in construction to the conventional cross-flow oil-to-water heat exchanger but has an additional tube over each water tube, with a clearance between both tubes. The cooler water is circulated through the inner tube in the normal manner. When puncture of this tube occurs, the cooler water passes into the outer tube where it is vented to atmosphere. Both these alternatives are more expensive than the conventional heat exchanger. Experience of the British practice indicates that the latter is eminently successful with added safeguards outlined above.

Adequate precautions must be taken against the possibility of corrosion. The composition of the cooling water will determine which materials will be suitable for the cooling tubes and, since it may change, must be checked regularly. The overall design of the system, including the pipework, must take into consideration incoming and outgoing water speeds and pressures so that these are suitable for the heat exchanger. Careful installation and maintenance procedures are also required, especially to ensure cleanliness in the heat exchanger tubes.

Transmission transformer cooling arrangements

In practice, these units often have a mixed cooling facility that permits operation on an ONAN rating and on a forced-oil rating as either OFAN or OFAF.

Transformers which have ONAN cooling must be capable of dissipating the whole of the full-load losses. Where forced-oil cooling is used, the cooling plant not only must be designed to cater for the full-load losses under the forced-oil condition but must also be capable of dissipating not less than 50% of these losses when the forced-oil circulation is not in use.

The methods adopted to fit the required amount and type of cooling to the transformer vary considerably. There are, however, certain fundamental common factors which permit some general conclusions to be drawn.

Depending upon the amount of cooling involved, most transformers up to approximately $50 \, MVA$ can carry the required amount of cooling surface on the tank. This is usually provided by means of detachable radiators arranged circumferentially about the tank, as shown in figure 10.14. The arrangement shown is typical for an ONAN transformer whereby the addition of air-blast fans mounted either beneath each pair of radiators or at each end of the radiator bank,

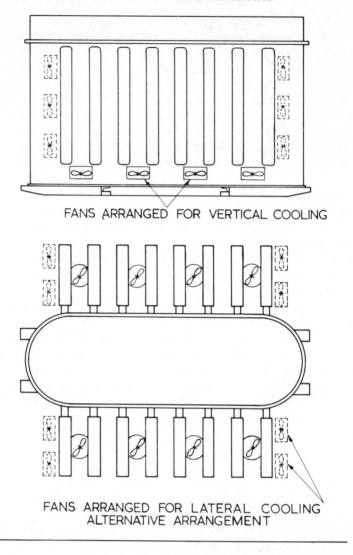

FANS ARRANGED FOR VERTICAL COOLING

FANS ARRANGED FOR LATERAL COOLING
ALTERNATIVE ARRANGEMENT

Figure 10.14 Transmission transformer cooling arrangements with radiators detachable from tank

the rating of the transformer can be increased to that of ONAF. An alternative arrangement is sometimes employed to gain a further increased rating whereby the individual radiators are connected, top and bottom, to separate headers on each side of the tank. These are arranged to feed into the tank via oil pumps. This allows the transformer to be rated OFAN or OFAF as desired. When the required area of cooling surface is greater than can be accommodated on the tank or where a transformer noise enclosure is to be used, the cooling plant is mounted on a separate structure (see figure 10.15). In place of the detachable radiators shown here, an alternative construction is sometimes used consisting of two banks of tubes between top and bottom headers and extending the length of the structure. Where oil-forced cooling is employed, it is common practice to have a double pump and pipework layout to ensure that at least 50% cooling is available in the event that one heat exchanger or pump fails in service[5].

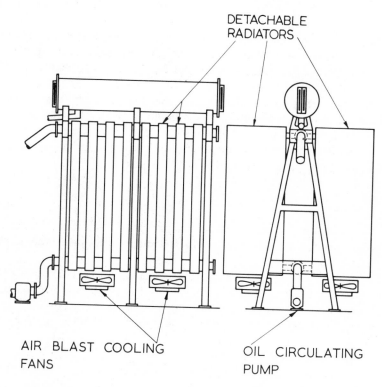

DETACHABLE RADIATORS

AIR BLAST COOLING FANS

OIL CIRCULATING PUMP

Figure 10.15   Cooling plant mounted on a separate structure

Oil-circulating pumps

The type of pump employed[5] is the totally oil-submerged impeller pump, usually with a squirrel-cage induction motor. This is fitted into the heat exchanger outlet pipe so as to be an integral part of it without recourse to exposed shafts, couplings or mounting arrangements that require independent foundations. Facilities must also be provided to permit the pump to be withdrawn from the oil circuit without the need to remove oil from either the transformer or the heat exchanger. The pump and its impeller are designed to offer a low impedance under conditions of thermosyphon oil flow as occurs when only the ONAN rating of the transformer is required.

An important consideration arises in connection with the start-up and shut-down of the pumps. Depending upon the overall pump capacity and the type of radiator used, it is possible to cause maloperation of the gas-actuated protective (Buchholz) relay when starting or stopping the pumps owing to the sudden change in oil movement. To overcome this, the heat exchanger control circuit should include a scheme for sequential start-up and shut-down of the oil-forcing equipment. Similarly, maloperation of the relay due to vibration from the heat exchangers or pumps must also be avoided.

Fans

The addition of air-blast cooling to an ONAN-class transformer can increase the rating of that transformer by about 20 to 30 %. The increased rating obtained by the addition of a number of fans is not dependent solely upon the rate of air flow but also upon the efficiency of cooling. Various methods of arranging the fans about the heat exchanger installation have been tried in order to maximise the effective cooling. However, in practice two methods are used: (1) vertical cooling or (2) lateral cooling, as shown in figure 10.14. Many of the precautions applied to the installation of oil-circulating pumps apply also to fans, particularly with regard to noise and vibration[5].

10.7.2 Tanks

With the exception of some cast-in-concrete series inductors designed for installation in enclosures, all other transformer and inductor plant is contained in some form of metal tank.

The simplest kind of tank is of mild steel construction with a separate cover bolted to the tank at a level above the top yoke of the core and any external fittings or cooling tubes. This position of connection between tank and cover is common to many transformers, irrespective of size, since it allows removal of the cover to give access to the core and windings without need to withdraw a large quantity of oil from the transformer.

As regards material, transformer tanks are fabricated in nearly all cases from mild steel plate. There are exceptions, particularly on large transformers, where aluminium is used in order to reduce mass or to avoid excessive tank heating due to induced eddy currents. Careful consideration must be given in these cases to assess the advantage to be gained from using aluminium, since the manufacture of a tank of this kind requires special techniques and skill, particularly with regard to welding, which are not met in the fabrication of steel tanks. Further alternative methods of tank design and construction are available, by using high-grade steels and pre-stressed panels to reduce the overall tank mass. In those instances, where excessive eddy current heating is anticipated, aluminium or coreplate screens can be fitted to the internal surface of a steel tank to control this.

Apart from these considerations, the use of steel or aluminium makes very little difference to the fundamental shape of the tank. A number of basic tank shapes are seen in practice. The aim is to achieve a mechanically strong vessel capable, on one hand, of supporting the mass of the core, windings and oil during transport and of lifting and, on the other hand, of containing the minimum volume of oil without detriment to the safe and reliable operation of the transformer. Figure 10.16 illustrates some common tank profiles. Modifications to these basic profiles are necessary in order to accommodate the various additional fittings that may be necessary to the operation of the transformer. These include bushings, cable boxes, radiators, tap changers, pipework, instruments and protection equipment. Most international and customer specifications include details of the type of equipment considered essential for the safe and reliable operation of the transformer and for which some provision must be made in or on the tank[G1.7, G2.2, G3.2].

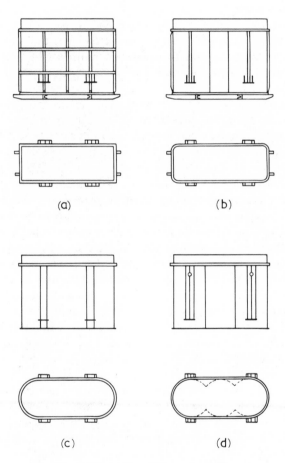

(a)   (b)

(c)   (d)

Figure 10.16   Common tank profiles

The use of round-ended or coil-contour tanks for large transformers leads to some saving of oil and at the same time permits a reduction in the quantity of stiffening that is otherwise required on a rectangular tank in order to make it a rigid vessel. However, the need for mechanical rigidity does not always dictate the quantity and position of stiffeners, since an important consideration in this respect is the flexibility of the tank panels and the adverse effect this has on limiting noise radiation. Generally, the larger transformers radiate the lower-frequency noise at a high level. This would be transmitted easily to the surrounding air if the tank panels were of the size equivalent to the wavelength of these frequencies in air. Stiffening by means of rolled channels or bars welded to the tank is essential to reduce the effective overall size of the tank to a series of panels of smaller dimensions and to provide them with a degree of boundary stiffness. A compromise is inherent, since the amount of tank stiffening required to obtain the minimum of radiated noise would be far in excess of that necessary for mechanical reasons.

A third factor concerning tank stiffening is the provision on the tank of lifting bollards, jacking pads, transport supports and draw holes, the latter being used to allow the transformer to be hauled or slewed in any direction. These are features that are to be found on all tanks in one form or another and are an integral part of the tank design and mechanical stiffening.

The problems associated with transport and installation of the transformer on a plinth must also be taken into consideration. Two methods of construction requiring specially designed tanks are discussed in section 10.8. The more usual methods of transport either by platform vehicle or by side-beam wagon are generally taken into account when designing the tank, together with any special requirements concerning design of the underbase. Frequently this is a plain metal base-plate which is of a thickness to permit movement of the complete unit by using rollers or slide rails although neither is required to be placed in a particular position under the transformer. On smaller units, an underbase of rolled-steel channel construction welded to the tank base-plate is often used. Facilities can be provided on such an underbase to assist the movement of the transformer on site and may include bi-directional plain or flanged wheels and skids, though the latter are to be found mainly on small transformers.

Finally, with regard to tank tests, it is necessary to subject the completed tanks to both routine and type tests in order to determine oil tightness and rigidity. The method of testing and measuring each of these qualities varies according to the customer's requirements. Tanks designed and constructed for transformers on the British power system are subject to a routine oil-leakage test at $35\,\mathrm{kN\,m^{-2}}$ above normal pressure for 24 h, during which time no leakage shall occur. Type tests consisting of a low-pressure test at $6.6\,\mathrm{kN\,m^{-2}}$ and an oil-pressure test corresponding to twice the normal head of oil, or to the normal pressure plus $35\,\mathrm{kN\,m^{-2}}$, whichever is the lower, are mandatory on one transformer of each size and manufacture. In each case, after completion of either test, the permanent deflection of flat plates is not to exceed a specified value expressed as a dimension according to the horizontal length of flat plate[5].

10.7.3 Oil preservation

Of the many problems associated with transformer engineering practice perhaps none has been subjected to greater investigation or discussion than that concerning the preservation of the insulating properties of transformer insulation and insulating oil. Without a liquid insulant, a transformer of given physical size is limited in both megavoltampere rating and voltage withstand strength, the former because of inadequate cooling and the latter because of the low electric strength of air and of non-impregnated paper insulation. The use of a mineral or synthetic liquid insulant provides improved thermal performance and enhances the electric properties of transformer insulation. At the same time, however, they introduce other problems because of their own inherent characteristics and the effect these have on other materials, particularly paper insulation. The use of synthetic insulants is not discussed here, since they are not usually employed in transformer plant installed on a power system. The present discussions, therefore, are concerned only with mineral-oil-filled transformers.

There are two aspects arising from the use of mineral oil as an insulant that necessitate the maintenance of the oil above a minimum quality. These occur as a result of the deterioration which the oil undergoes during the life of the transformer, and the effect this deterioration has on the life of paper insulation. Paper is a fibrous material whose main constituent is cellulose. It has a complex molecular structure which can be affected by a number of factors, of which temperature, oxygen and humidity are the most important[6].

The temperature of the insulation in a particular transformer depends on the loading history of the transformer and on the ambient conditions. Most international standards relate the continuous maximum rating of a transformer to a specific thermal performance and expected life. It is generally concluded that this corresponds to an insulation hot-spot temperature of 98°C (see reference G1.7). The effect of exceeding this temperature is cumulative, the rate of ageing of the insulation being doubled for every 6°C increase in excess of 98°C (see reference G1.7), with consequent reduction in transformer life. The temperature at which the transformer operates also affects the life of the oil, especially with prolonged oil temperatures in excess of 75°C when the general cause of deterioration is oxidation. The effect of temperature on both the life of the insulation and the oil is therefore dependent upon details of transformer design, load cycle and ambient conditions.

Oxygen that has become dissolved in the oil is also a contributory factor in accelerating the rate of deterioration of the insulation and the oil, particularly in the presence of water and under conditions of high operating temperatures. The vacuum and heat treatment a transformer undergoes following manufacture and prior to testing and eventual service (see chapter 7) reduces the water content of the insulation and prepares it for impregnation with dry degassed oil. The amount of dissolved oxygen in the oil at this stage is negligible but can increase during the service life of the transformer owing to oxidation of the oil in the presence of copper and iron, water and other materials such as varnishes. Oxygen can also become dissolved in the oil if air-borne water vapour is drawn into the transformer with cyclic changes of the load.

The third factor affecting transformer life is the water content of the insulation and oil. The rate of thermal deterioration of paper is directly proportional to its water content. After vacuum and heat treatment, paper insulation may contain between 0.5 and 1 % residual water by mass. Practical experience and experiment show, however, that the water content of the insulation and of the oil increases during the life of the transformer, partly as a result of the natural deteriorating processes caused by temperature and oxidation but particularly as a result of the entry of moisture from the atmosphere.

The ingress of moisture to the oil and eventually, through migration, to the paper arises from the need to permit a transformer to breathe under cyclic loading. This, under the normally specified operating temperatures, can cause a 6% change in oil volume due to thermal expansion. To accommodate the increased oil volume, expansion chambers or conservators are built on or into the transformer either as separate vessels, as on most medium and large transformers, or as an air space within the main tank, as may be found on small transformers. This expansion space then exhausts either directly to the atmosphere or, more usually, indirectly through some form of breathing device. This limits the intake of moisture by extracting water vapour from the air entering the transformer during the cooling period of its load cycle. These devices are commonly known as dehydrating breathers. Some small transformers are constructed as totally sealed units where the oil expansion space is normally filled with an inert dry gas such as nitrogen (figure 10.17 (b)).

The type of breather most commonly found in service is shown in figure 10.18. It consists of a vessel containing a quantity of impregnated silica gel, usually cobalt chloride, over which the air entering the transformer passes and is dried, the moisture in the air being absorbed into the desiccant by reason of its sub-microscopic capillary structure[7]. In reverse, the air which is exhausted from the transformer during the heating period of the load cycle is vented to the atmosphere through an oil bath seal in the breather. The efficiency of this type of breather is dependent upon a number of factors[7], of which humidity and temperature of both desiccant and air, together with the rate of air flow, are the most significant. The rate of absorption of moisture by the desiccant rapidly reduces when its water content exceeds 23 % by mass[G2.16], at which stage it should be removed from the breather and should be reactivated by drying in an oven. The amount of moisture absorbed by the desiccant can be observed by the degree of colour change of the cobalt chloride impregnant.

A type of breather known as a freezer dehydrator or automatic insulation dryer[7] is illustrated in figure 10.19. It is based on the Peltier effect of thermoelectric modules to obtain low temperatures within an air duct so that moist air passing through the duct is dried continuously, the water vapour being collected as ice on the cold surfaces of the modules. If the polarity of the power supply to the modules is reversed, this increases their temperature and causes the accumulated ice to melt and drain to atmosphere. The defrost cycle of the breather is about 8 min, depending on the ambient conditions, and is repeated automatically by means of a time switch every 6 h. In addition to efficient drying of the incoming air, this type of breather can be fitted to the conservator to form a closed loop with the air in the conservator. By this means, thermosyphon action

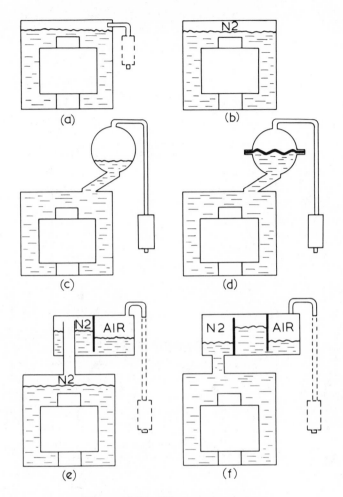

Figure 10.17   Oil expansion systems

of the air between the conservator and the cold air duct of the breather occurs, permitting continuous drying of the air within the conservator.

In addition to dehydrating breathers, a great deal of consideration has been given to evolving more efficient methods of oil protection[6]. A number of different systems are available which provide for the expansion of oil and also reduce the degree of possible atmospheric contamination. Some of these methods are illustrated in figure 10.17. Of those shown, the methods shown in figure 10.17 (a), (b) and (c) are generally applicable to countries where atmospheric conditions are not so extreme as to warrant the more sophisticated methods illustrated by figure 10.17 (d), (e) and (f). A further point arises if a gas-actuated protective (Buchholz) relay is to be fitted. This practice requires a separate oil-expansion vessel connected to the main tank by means of an oil-filled pipe. It thereby

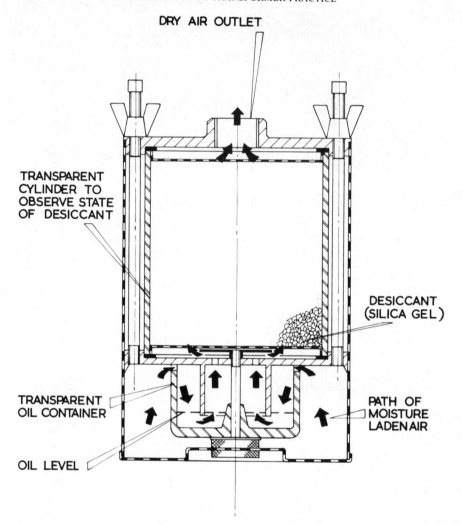

DRY AIR OUTLET

TRANSPARENT
CYLINDER TO
OBSERVE STATE
OF DESICCANT

DESICCANT
(SILICA GEL)

TRANSPARENT
OIL CONTAINER

PATH OF
MOISTURE
LADEN AIR

OIL LEVEL

Figure 10.18   Transformer breather

excludes any oil protection system employing an inert gas cushion similar to that shown in figure 10.17(e).

Finally, with regard to checking oil quality and carrying out oil maintenance programmes the details of the tests and procedures are considered essential[G2.23]. There are no hard and fast rules for the frequency of oil tests, but these should be made at such intervals as to permit the observation of any tendency of the oil to deteriorate. The principal recommended tests are to determine the electric strength, acidity and water content of the oil. A useful test is that of checking the resistivity of the oil, since this gives an indication of its overall condition[G1.18]. If we plot on a graph the results of such tests, it will enable any change in condition

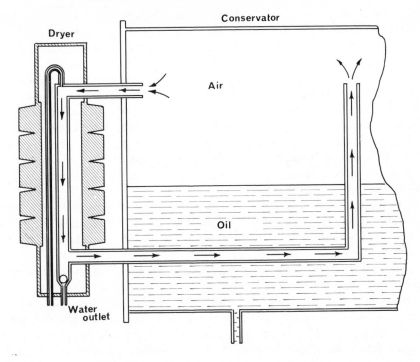

Figure 10.19 Drycol automatic insulation dryer

to be detected and remedial action to be taken.

If the oil has become contaminated to a degree which necessitates treatment, it may be replaced with new oil, or alternatively the oil may be purified by some regenerative treatment. This usually consists of filtering, drying and vacuum treatment and may be carried out on the transformer at site by using continuous circulation of the oil through the purifier. Otherwise, if suitable oil storage facilities are available, the oil may be passed through the purifier into the storage tanks and back again into the transformer as required.

Unless the oil has become contaminated by some internal fault or accidental ingress of moisture or as a result of prolonged operation at high temperature, every care should be taken to ensure that some action is taken to prevent the oil reaching a serious level of deterioration. Otherwise it is likely that many of the by-products of this deteriorating process will impregnate the insulation and will cause irreparable damage and permanent reduction of transformer life.

10.7.4 Bushings

The methods employed in bringing out the ends of the windings through the transformer tank for connection to the external system involve in nearly all cases the use of either bushings or cable boxes. An exception occurs with heavy-current low-voltage terminations, as used for furnace transformers, where the low-voltage busbar connections are made directly through the tank onto the windings. Where cable boxes are used, the connection from the winding is usually brought through the tank by means of a plain porcelain bushing into a chamber containing the cable ends. In practice, cable box terminations are found chiefly on distribution and medium-voltage transformers, though they are to be found also on some transmission transformers where direct connection to cables is desirable. Generally, however, where such connections are required, the practice adopted is to fit oil-to-air bushings in the transformer and to connect these by means of short overhead busbars to separately mounted cable sealing ends.

In general, there are two types of bushing: (1) the synthetic-resin-bonded paper bushing and (2) the oil-impregnated paper bushing. These are both of the conventional capacitor construction, in which the stress is graded throughout the bushing insulant by means of distributed electric screens contained within the body of the bushing. They are both porcelain clad. At 33 kV and above both types are oil self-contained.

The majority of bushings for 275 kV and above are of the oil-impregnated paper type, since this has a characteristically lower value of power factor and permittivity and in consequence greater thermal stability at these voltage levels than the resin-bonded paper type. Improvements in materials are, however, being sought constantly, and it is highly probable that a more satisfactory resin paper insulant with lower loss and power factor characteristics will be found eventually.

Thermal instability can occur in a capacitor bushing if the heat generated in the insulant due to its losses is not dissipated at a sufficiently high rate to limit the insulant temperature. Failure to achieve an adequate rate of heat dissipation results in increasing the undissipated part of the losses cumulatively until the accelerated process of thermal deterioration of the insulant causes breakdown. In addition to the inherent thermal conductivity and characteristics of power factor against temperature of the insulant, other related factors affecting thermal stability of the bushing include the ambient temperature, applied voltage and cooling-surface area of the insulant. The problem becomes more pronounced at the higher-voltage levels since the losses are a function of the square of the applied voltage. At voltages of 132 kV and below it is not so severe; within this range, the resin-bonded paper bushing finds almost universal application.

Bushings of 66 kV and above usually require the connection between the end of the winding and the line end of the bushing to be made by means of a flexible pull-through lead which extends from the winding to the bushing helmet connection. The oil end of the bushing can take one of two forms, and, depending upon which is used, the bushing is referred to as being of the re-entrant or conventional type. High-voltage bushings for 275 and 400 kV are of the re-entrant type as shown in figure 10.20(a) and (c). This method of construction provides a shorter oil-immersed end compared with a conventional bushing for the same voltage and hence requires less space within the tank. The conventional type of construction, as shown in figure 10.20(b), is common to 66 and 132 kV bushings, though in the latter instance the re-entrant type is sometimes used. Where the re-entrant type is

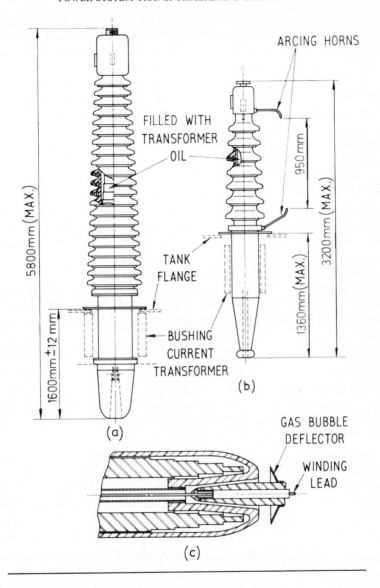

Figure 10.20   High-voltage oil-impregnated-paper capacitor-type bushings

employed, the flexible pull-through lead is fitted with a gas bubble deflector, as shown in figure 10.20(c), in order to prevent any gas released by the transformer being trapped within the bushing and thus not being detected by the gas-actuated protective (Buchholz) relay, if fitted. To prevent damage to the porcelain by a power arc resulting from an external flashover, arcing horns or co-ordinating rod gaps are usually fitted to all bushings (see, for example, figure 10.20(b)).

The bushings are subject to routine and type tests, depending upon whether the bushing is of new or established design. The routine tests are mandatory on all bushings and include amongst others oil leakage, voltage withstand, internal discharge, surge-voltage and power factor tests. Type tests are made on one complete bushing of each design selected from those which have previously passed the routine tests. The scheduled tests carried out include thermal stability, wet voltage withstand, surge-voltage, visible corona, temperature rise and oil-immersed voltage withstand test, in addition to confirmatory power factor and internal discharge tests. The sequence and methods of testing are outlined in the relevant specifications[5, G1.11, G2.3, G3.24].

In addition to factory tests, it may be necessary to carry out power factor and capacitance measurements on a bushing during its service life in order to check the condition of the insulant. Such tests are usually required to be made on site, and for this purpose all capacitor bushings for service at 33 kV and above are fitted with a tapping brought out to a terminal on the bushing flange. The connection also provides a convenient point for taking internal discharge measurements during high-voltage tests on the transformer.

The development of the present high-voltage bushing has taken place over a number of years and has incorporated new and improved manufacturing and design techniques which result from both research and the experience gained in service[8].

10.8   TRANSPORT AND SITE ASSEMBLY

10.8.1   General

Transport from factory to site is always required. Access to site may be limited by load restriction on bridges, culverts, etc., or by gauge restriction through bridges, tunnels or narrow roads. Although some relocation of site may be possible for transmission transformers, this is not so with generator transformers. In the latter case the site of the power station as a whole will be the dominant factor.

In considering the various types of transport which may be used and the effect of these and the various limits they impose, the overall design and cost of the system is the overriding factor. Thus it may be justifiable to use a relatively expensive type of transformer construction if this produces overall saving.

10.8.2   Transport

Restricted loading gauges on the railways and the fact that it is usually necessary to employ road transport for at least the last part of the route from the place of manufacture to a sub-station or power station have led to the UK practice of using either road transport for the whole journey or, sometimes, a combination of road and sea transport. Loads not exceeding 152 t gross may be transported by road without special authorisation; above this mass, permission is required, and routes need to be carefully selected. To minimise difficulties in route selection and to reduce road congestion, much use is made of sea transport, particularly for generator transformers and the larger transmission transformers. Access routes to and from suitable ports have been established for heavy-load vehicles, permitting in some cases loads up to 305 t nett.

In other territories, transport is frequently by rail, since roads are often insufficient to carry more than light loads, particularly in less-developed areas. Again, restricted loading gauges are a significant factor and may require special designs of transformer, the use of single-phase units or even some degree of site assembly. When transport into a country is by sea, cranes or other heavy lifting equipment are required either on shore or on the ship. It is often found that the choice of design is controlled by lifting facilities at the port of off-loading.

For road or rail transport, three basic methods exist.

(a) On a flat-bottomed low vehicle.

(b) On a vehicle with side girders supported at each end on bogies, the transformer hanging between the girders from appropriately positioned supports on the tank. Road vehicles of this type are available for loads up to 305 t, the largest of such vehicles weighing 75 t (see figure 10.21).

(c) Where the transformer tank itself forms the girder structure between the two bogies (figure 10.22).

Figure 10.21 Heavy-load road-transport vehicle with built-in girder structure

Method (a) is applicable to the smaller sizes of transformer and presents little difficulty. Methods (b) and (c) are used for the larger transformers. Method (c) is feasible for slightly higher loads than (b) owing to the saving of total mass by the omission of the vehicle side girders, despite some increase in the mass of the tank which must be stiffened to withstand the stresses involved.

When considering transport of heavy transformers, it should be remembered that the prime function of the tank is to contain the oil during the normal stresses encountered in service. Thus the tank may be of relatively light construction provided that means are found to minimise stresses imposed during transport, for

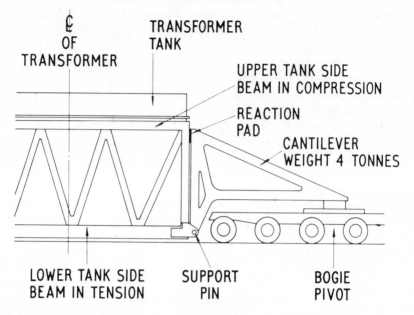

Figure 10.22   Transformer tank forming girder structure for road or rail transport

example by arranging for the core and windings to be supported directly from the vehicle rather than through the tank. One method of achieving this requires brackets to be built out from the ends of the core frames. These brackets project into overhangs on the tank ends; the mass of the core is thereby supported by two cross-beams lying across the vehicle towards the ends of the side girders. The tank is not involved, therefore, in transmitting the lifting force from the vehicle to the core and windings.

The method mentioned in (c) above involves extra stiffening of the tank sides so that they act in place of the side girders. The weight of the transformer is taken by pins through lugs at the bottom of the stiffened tank sides and corresponding lugs on the bogies. The resultant moment on the bogies is counteracted by forces transmitted through pressure pads towards the top of the tank side stiffening; thus the lower part of the tank side is in tension and the upper in compression.

In European countries and the USA this type of vehicle is made for rail use, whilst a road equivalent has been developed in Great Britain. The European transformers designed in this way use the spaces between the tank side stiffeners for control equipment, heat exchangers, etc.

Aluminium tanks are commonly used for large generator and transmission transformers. For example, many 400 kV, 570 and 600 MVA generator transformers for the Central Electricity Generating Board in the UK are of conventional three-phase construction and have aluminium tanks. Two stations have transformers with steel tanks. As illustrated in figure 10.23 their three-phase construction embodies the use of three identical 200 MVA single-phase transformers,

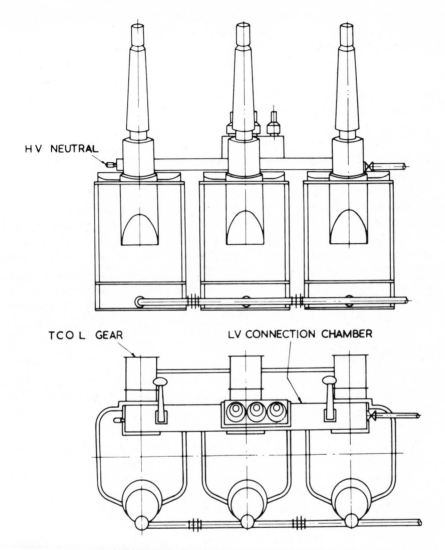

Figure 10.23   Three-phase bank composed of three identical single-phase transformers

mechanically and electrically coupled together on site to form a 600 MVA three-phase bank. The delta connections are made in a common trunking on which the low-voltage bushings are mounted. These units have a transport mass of under 152 t per phase, excluding oil.

To facilitate movement of transformers, two special methods of transport have been evolved. The first obviates the need for heavy lifting facilities at the ports by using a vessel designed on the roll-on – roll-off principle. Such a vessel requires minimum dock facilities. They are easily provided at many ports and form part of the installations at some estuarial and coastal power stations. The transporter, complete with transformer, is driven aboard, and the transformer is lowered into the hold for the voyage. Two such vessels are in commission.

The other special transport method is an application of the air-cushion vehicle principle. With a conventional heavy-load road vehicle, the limiting features are the axle load and the loading on a bridge when one bogie is on the span. By pressurising the space under the transformer, the axle and bogie loading can be reduced, thereby allowing heavier loads over established routes or the use of some routes previously found inadequate. The vehicle is similar to that shown in figure 10.21 but with air-retaining skirts fitted to the side and cross-girders and sealed to the transformer. The space under the transformer is pressurised from a separate compressor vehicle.

Extra mechanical stresses to those of normal service are associated with any method of transport. The design of both the internal and external parts of the transformer must be suitable to withstand these. Special consideration must, however, be given to extra stresses which may be imposed by a particular form of transport, for example shunting forces in rail transport or the rolling and pitching of sea transport. Cases are known of considerable damage which has ensued from disregard of these hazards. To check on conditions during a particular journey, especially when insurance liability may need to be proved, it may be advantageous to fit shock or roll recorders.

10.8.3 Site assembly

Where transport limitations either entirely prevent the use of a three-phase transformer or place severe restrictions on the design, alternative forms of construction must be sought. The design restrictions may be physical in that very high mechanical or electric stresses must be employed or monetary in that the combined cost of the transformer and its losses become uneconomic.

Two alternatives are available: (1) the use of single-phase transformers to form a three-phase bank or (2) the partial dismantling of a three-phase transformer after manufacture and its subsequent reassembly at site.

Compared with a normal three-phase transformer, the extra costs entailed in using single-phase transformers or in site assembly are of the same order but depend on the degree of dismantling required for the latter units. Either method can provide lower losses. The relative advantages of the two methods depend greatly upon the loss evaluation basis employed, the transport restrictions and the emphasis to be placed on the facility of replacing a damaged transformer or of removing to another site. If a site-assembled transformer is chosen, subsequent failure or removal may necessitate site dismantling prior to taking away.

Experience on transformers for use within Great Britain has shown that even 1000 MVA 400 to 275 kV auto-transformers are fairly readily transportable as three-phase units. It is principally for generator transformers that site-assembly techniques could be employed, since these transformers have double-winding and are, therefore, larger than auto-transformers. In addition, they tend to have higher values placed on their losses because they usually operate at full load.

However, British practice for generator transformers over 600 MVA has been to prefer single-phase construction from spares considerations.

Nevertheless, in order to gain experience with possible methods of site assembly, three such transformers have been constructed by British manufacturers for use on the 275 kV system of the Central Electricity Generating Board. Two other 275 kV transformers have been installed in the Cruachan underground power station in Scotland; these units were dismantled to a considerable extent because of very severe mass and dimensional restrictions on the access route.

Whatever method of site assembly is employed, a paramount requirement is that the integrity of the insulation either shall not be affected by the dismantling, transport and subsequent reassembly or shall be reinstated to its factory condition by processing after re-erection. Because it is not usually easy to make other than simple site tests, it is preferable to adopt, where possible, a construction which limits exposure of insulation to atmosphere, although the Cruachan transformers mentioned above were successfully subjected to surge-voltage tests on site after re-erection and reprocessing.

Apart from extensive dismantling or reassembly, two basic constructional methods exist for site-assembled transformers[9]. Both involve a division of the transformer into parts convenient for transport. In one method the transformer is split vertically into sections each comprising a part of the tank, a phase leg of the core, together with part of the yoke and a phase winding. The other method involves horizontal splitting of the transformer so that the windings are retained within a sealed centre section of the tank, whilst the core is dismantled as far as necessary to permit transport.

10.9 FUTURE DEVELOPMENTS

Power system transformers and inductors today are very efficient and reliable pieces of equipment. As systems continue to grow in size, voltage and complexity of design and operation, these characteristics of efficiency and reliability will continue to be demanded of the transformers and inductors concerned.

It is not anticipated that there will be much change in power system transformers or inductors in the foreseeable future. Core steel with improved loss characteristics will continue to be introduced, but, with similar fundamental characteristics of permeability, hysteresis and magnetostriction, the need to suppress third-harmonic effects and to reduce noise output will remain. Despite scarcity and rising costs, copper is likely to remain the usual material for conductors; the usual material for insulation continues to be paper, other cellulosic materials and transformer oil.

Ultimately, any further increase in the capacity of a transformer may be limited by the need to provide it with adequate short-circuit strength and to have available suitable testing stations.

It can be anticipated that site assembly will be advantageous or even necessary more frequently in the future. Several methods of achieving it have successfully been demonstrated. Their adoption, together with advanced forms of transportation, are likely to meet all future requirements.

The transformer or inductor is only one item in an electric power supply system, although a vital one. It is certain that transformer designers and production engineers continue to meet any future requirements of the system engineer.

ACKNOWLEDGEMENTS

The authors wish to express their thanks to GEC Power Transformers Limited for the photograph in figure 10.13 and the illustrations in figures 10.11 and 10.19, to Ferranti Limited for the photograph in figure 10.21 and to Messrs F. C. Pratt and A. C. Hall of the UK Central Electricity Generating Board for helpful information.

REFERENCES*

(Reference numbers preceded by the letter G are listed in section 1.14.)
1. Rippon, E. C., *Problems Peculiar to Large High-voltage Auto-transformers*, *CIGRE Rep.*, No. 140 (1960)
2. Wilkinson, K. J. R., and Carter, G. W., A method of measuring losses in reactors of low power factors, *BTH Act.*, September–October (1939) 2
3. Deutsch, F., Measuring the active power losses of large reactors, *Brown Boveri Rev.*, **47** (1960) 268
4. de Bourg, H., Jenkins, R. S., Slettenmark, I., Tengstrand, C. A., and Wester, C. E., *Calorimetric Loss Measurement on Alternators and Reactors*, *CIGRE Rep.*, No. 119 (1964)
5. British Electricity Supply Industry, *BEBS T2, Transformers and Reactors* (34 parts), The Electricity Council, London (1966)
6. Lutz, H., *Transformer Oil Preservation Systems and Associated Problems*, *CIGRE Rep.*, No. 134 (1960)
7. Brown, W. J., Kerr, H. W., Singer, D. E., and Walshe, L. C., *Accessories and Parts for Transformers*, *CIGRE Rep.*, No. 101 (1966)
8. Barker, J. H., Marlow, J. H., and Mellor E. J., *400 kV Transformer and Wall Bushings, Inst. Electr. Eng. Conf. Publ.*, No. 15, Part 1 (1965) 202
9. Haselfoot, A. J., Hartill, E., Kerr, H. W., Palmer, S., and White, E. L., *Design, Construction and Transport of Large Transformers Suitable for Reassembly on Site, CIGRE Rep.*, No. 106 (1964)

---

* See also the list of references in chapter 12.

11

Special Transformers

T. Kelsall*

11.1 DRY-TYPE TRANSFORMER

11.1.1 The case for dry-type transformers

In supplying electric power to large office blocks and flats, the supply authority would naturally prefer to install the distribution transformer in the building itself, so as to reduce the length of low-voltage cable runs, which would result in lower costs and fewer problems with voltage regulation. The consequence of fire in multi-storey buildings becomes more and more serious as the population density of the building increases. The installation of an oil-filled transformer in the building is held to increase the fire risk.

The transformer is not condemned as a source of fire outbreak, but a 500 kVA transformer (a reasonable rating for this type of installation) will contain about 0.9 m³ of oil. If during an outbreak of fire the tank were to split, this oil might add disastrously to the fire. To prevent this, oil-retaining walls often have to be built round the transformer, or the transformer has to be removed from the building altogether, the installation costs increasing in either case.

An air-insulated transformer needs no special precautions in this respect; although the transformer itself will be more expensive than an equivalent ON transformer, the overall cost of the installation can be appreciably cheaper.

11.1.2 Insulation and temperature

Insulation classification
The life of a transformer is determined by the life of its insulation which, apart from mechanical damage, is dependent on the temperature at which it is made to

* GEC Distribution Equipment Limited.

work. Insulation materials are classified according to their maximum safe working temperature [G1.9, G2.10].

Based on the specification for maximum working temperature and if we take into account the ambient temperature which may be expected, the permissible rated temperature rise [G1.7, G2.2, G3.2] for windings for the insulation classes recognised in transformer manufacture may be summarised as follows.

Insulation class	Winding temperature rise (°C)
A	60
E	75
B	80
H	125
C	150

In practice, few manufacturers differentiate between class H and class C. Most air-insulated transformers made in Great Britain are to class C specification.

The materials commonly associated with class C insulation comprise mica, porcelain, glass, quartz, treated glass fibre textiles and built-up mica with the use of silicone varnish as a bonding and impregnating medium.

Temperature

The winding temperature rises listed above represent the average temperature of the winding. Naturally, under working conditions thermal gradients exist within the coil. It is essential that the coil be so designed that the hottest spot in the coil structure does not exceed the critical temperature for the class of insulation. Cooling ducts are provided in the coil to enable the cooling medium to circulate within the coils and to remove the heat generated within them. It is found that a layer of stagnant fluid—sometimes called a boundary layer—is present over the surface to be cooled; thus heat has to be transmitted across this layer before the fluid flow can remove it. The thickness of the boundary layer is dependent, among other factors, on the fluid. For air, it is about 1 mm. The relatively narrow cooling ducts found satisfactory, when oil is the fluid, are totally inadequate for air ducts, and widths of 15 mm or more are commonly used. (For efficiency of the duct width, see figure 11.1.)

Electric stress

The electric strength of the insulation in an ON transformer is little different whether the transformer is cold or at its rated temperature. However, the electric strength of air is proportional to its density which is inversely proportional to absolute temperature. Therefore, at working temperature the electric strength of the air in a class C transformer is only about 60% of its value at room temperature.

When a voltage is applied across a mixed insulation, such as solid insulation in

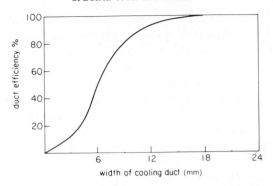

Figure 11.1 Variation of duct efficiency with width

series with an air space, the voltage distribution will be inversely proportional to the permittivity of the two materials. The relative permittivity of air is 1 and that of solid insulation about 5. For equal thickness of air and solid insulation in series the electric stress across the air will be five times as great as that across the solid insulation.

Suppose a gap of 25 mm is to withstand a voltage of 25 kV. The stress of $1\,\text{kV}\,\text{mm}^{-1}$ might be considered too high for air, and it might be proposed to 'strengthen' the insulation by inserting a barrier of 5 mm solid insulation in the air space. The insulation system can be considered as two parallel-plate capacitors in series, and the voltage appearing across each capacitor will be inversely proportional to the capacitances. Thus

$$V_a/V = 1/(1 + \varepsilon_{ra}\delta_s/\varepsilon_{rs}\delta_a) \qquad (11.1)$$

where V_a is the air gap voltage, V is the total voltage, $\varepsilon_{ra} = 1$ for air and $\varepsilon_{rs} = 5$ for the solid insulant are the relative permittivities, and $\delta_a = 20$ mm and $\delta_s = 5$ mm are the respective thicknesses. The evaluation of equation 11.1 gives $V_a = 23.8$ kV, which means an electric stress of $1.19\,\text{kV}\,\text{mm}^{-1}$. Insertion of the solid insulant has only increased the electric stress in the air gap. This is likely to break down, causing the stress on the solid insulant to increase to $5\,\text{kV}\,\text{mm}^{-1}$ at which it will also fail. Far from strengthening the insulation, the mixed insulation only reduces the safety margin.

The only safe approach is to ensure that the air space alone, irrespective of any additional solid insulation, is capable of withstanding the full voltage.

11.1.3 Construction

Cores

Cores are usually cut from cold-rolled grain-oriented steel, the phosphate coating put onto the steel during manufacture serving as interlaminar insulation (see chapter 4). An iron space factor of about 0.95 or better is achieved. Cores may be cut to have either a mitred or a lapped joint according to individual manufacturer's preference. After building, the cut edges are coated with a suitable heat-resistant paint or varnish to protect against rusting.

Although similar in construction to cores for ON transformers, cores for class C transformers are generally larger than for ON transformers of equivalent rating. Since clearances and cooling ducts are necessarily so much larger and since current densities usually lower than for air-cooled transformers, the core window opening must be larger than for an ON transformer of equivalent rating. The larger coil implies a higher reactance. To keep down the reactance value, the volts per turn is generally higher, and thus a larger core cross-section is used. The operating flux density is largely governed by the necessity to limit the temperature rise of the core, although low noise level requirements may also limit flux density.

Figure 11.2   Typical transformer with class C insulation

Coils

The simple helical coil is comparatively cheap to wind. It has a favourable surge-voltage distribution; it is generally adopted for ON distribution transformers. However, such a coil used for the high-voltage winding, with its great number of turns per layer, produces an appreciable electric stress across the interlayer insulation. The high-voltage winding of the class C transformer is usually subdivided into a number of individual coils of few turns per layer, stacked one above the other and connected in series (see figure 11.2). In this way the interlayer stress is greatly reduced. Unfortunately, this assembly has much less uniform surge-voltage distribution than the helical winding, and considerable care has to be taken to ensure that the coil will withstand the surge-voltage test requirements, particularly at the tapping sections.

Low-voltage coils are usually simple helical coils of two layers with a cooling duct between the two layers. The high-voltage stack is assembled concentrically over the low-voltage coil, again leaving an annular cooling duct. Spacing strips of silicone-impregnated glass-laminate board are disposed radially in the duct to space the high- and low-voltage coils apart.

The conductor covering is generally highly refined glass braid, whilst the major insulation is usually mica or woven glass laminate. Silicone-impregnated glass board is used for spacing strips and wedges. The whole coil assembly, after carefully drying out and shrinking, is vacuum impregnated with silicone varnish.

These insulating materials, particularly the silicone varnish, are expensive, and the cost of insulation in a class C transformer is a significant portion of the total cost.

On assembly on the core, the high-voltage coils are clamped between steel end rings at top and bottom, pressure being exerted by tensioning bolts from the frames.

Enclosure

The completed transformer may be intended for installation without any enclosure whatever. It is perhaps more usual to mount the transformer in a sheet steel ventilated housing fitted with detachable panels through which access is gained for inspection and adjustment of tapping links. In such housings the inlet and outlet ventilating grills should be as large as possible, and the vertical distance between them as great as can be obtained.

In more ambitious installations the housing may be designed to accommodate a high-voltage isolating breaker and low-voltage distribution fuses. A very compact packaged sub-station is formed.

If the transformer is to work in a corrosive atmosphere, then it may be enclosed in a completely hermetically sealed enclosure which it is usual to fill with an inert gas such as nitrogen under pressure. A pressure gauge is usually fitted to warn of loss of pressure. Such an arrangement naturally adds greatly to the expense.

11.1.4 Limitations

The dry-type transformer is essentially an indoor unit for use in relatively 'clean' conditions. Attempts to use it calls for increased costs in protection and maintenance.

Its surge strength is only about half that of an ON transformer. Thus dry-type transformers can only be used in electrically unexposed situations which are not subject to lightning strikes.

Because of the large electrical clearances required it would appear that the highest practical system voltage for a dry-type transformer is 15 kV.

11.2 BURIED TRANSFORMERS

11.2.1 General

In urban and suburban districts it is becoming increasingly difficult to obtain suitable sites for the installation of transformers that supply the distribution networks. One solution is to put transformers below ground when there is not room for them above ground.

Transformers have been buried direct in the ground for many years, but the practice was restricted to small transformers of about 100 kVA rating, as this was the largest transformer the losses of which could be dissipated direct through the soil. The actual thermal performance of such a transformer must always be conjectural, since the temperature rise depends on the thermal conductivity of the soil. This property can be expected to vary from place to place and from season to season according to the humidity of the soil. In very dry conditions the soil could even shrink away from the transformer tank, leaving an air film through which the transformer losses would have to be dissipated.

Modern interest in buried transformers has been directed to larger sizes, 500 kVA and upwards, and the design philosophy has been to dissipate the waste heat not through the surrounding soil but to the air above ground. Two distinct types have evolved according to whether the oil–air heat exchanger is located above or below ground.

11.2.2 Cooler above ground

Design

The transformer tank is buried direct in the ground, mounted on a concrete raft, and the radiator mounted above the transformer so that the bottom of the radiator is at ground level (see figure 11.3). The radiator carries an oil conservator onto which can be mounted fittings such as oil gauge, thermometer pocket and silica gel breather. Alternatively the radiator can be designed to permit within

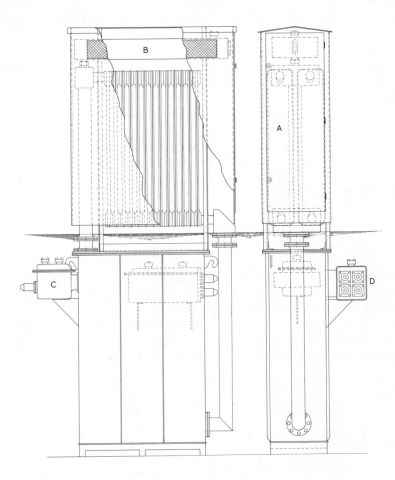

Figure 11.3   Buried transformer with above-ground cooler; A, access door to tap switch;
B, header tank; C, high-voltage cable box; D, low-voltage cable box

itself expansion of the oil, thus eliminating the conservator. The radiator housing
can be made to accommodate the ratio-adjusting switch and can also enclose a
standpipe into the transformer through which the tank can be emptied if
necessary.

On occasion, site conditions are such that the radiator cannot be located
immediately above the transformer but must be mounted some distance away.
This will entail a loss in cooling efficiency due to increased resistance to the
thermal syphon flow of the oil. It will necessitate either a slight reduction in
transformer output or the use of a larger radiator.

Advantages and disadvantages

The obvious advantage of this type of cooler is that the excavation is kept to a minimum. In certain situations this factor might be decisive. Because of the head of oil in the radiator all transformer tank joints are below oil, with a slight positive pressure, and water will not enter the transformer. The radiator bank is probably more efficient than in the buried transformer type. Against this must be set the very real danger of what happens if the radiator housing suffers physical damage, say, as the result of a traffic accident. Should this occur, not only will a large quantity of oil be lost, but the transformer must be taken out of service quickly before this too is damaged owing to overheating consequent on the loss of all its cooling. Following such an accident, it is highly desirable that the transformer and tank be cleaned and dried out to ensure that water and dirt have not entered the transformer through fractured pipework.

11.2.3   Cooler below ground

Design

In this type of transformer the radiator is carried directly from the tank in the conventional manner, except that it is usual to mount the entire radiator bank on one side of the tank to facilitate air ducting (see figure 11.4). The radiators are surrounded by a steel housing which serves to duct the air. The tank and cooler are buried direct into the ground, and a ventilating pillar is erected above ground

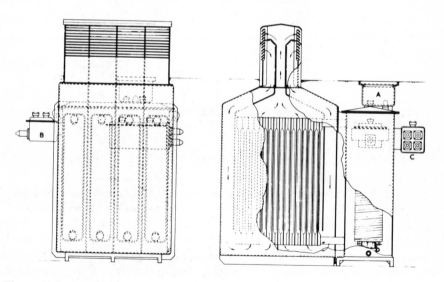

Figure 11.4   Buried transformer with below-ground cooler: A, access to tap switch and dip stick; B, high-voltage cable box; C, low-voltage cable box

over the cooler. Cold air enters the pillar at the sides and flows down into the bottom of the radiator housing. The air then flows over the radiator bank, back into the pillar and out to the air. Large ducts are essential, and care must be taken to ensure that hot air is not drawn back into the cooler. It is also essential to design the ventilating pillar to be independent of wind direction. A simple pillar with inlet on one side and outlet on the other would be of no use at all, since a strong wind on the outlet side could stop circulation completely and the transformer temperature would increase rapidly.

The pillar can also be designed to give access to a tap switch, oil gauge, thermometer, etc.

Advantages and disadvantages

Removal of the ventilating pillar, whilst causing a loss in radiator efficiency, is not immediately serious, and the transformer can function normally for several hours without injury whilst awaiting replacement of the pillar. There is therefore some advantage in designing the pillar so that in the event of an accident the pillar will break easily and cleanly away from the transformer so that the mounting suffers no damage. A new pillar can then be readily fitted, and the transformer need not come out of service.

Since there is no external head of oil, however, the transformer tank must be hermetically sealed to guard against ingress of water in the event of flooding. Of course, this construction demands an appreciably larger excavation.

It is not really practicable to mount the pillar remote from the radiator bank.

11.2.4   General construction of buried transformers

The core and windings of a buried transformer are as for a conventional ON-cooled design except that they might be designed to have lower losses to reduce the size of radiator. Cable boxes are designed for horizontal entry and are compound filled with all joints below compound level. Because of the high corrosive nature of most soils special protection of all buried steelwork is essential. It is usual to zinc spray and paint with an epoxy resin. On filling the excavation, fine sand is used against the tank wall to ensure that stones do not cut through the protective paint and do not expose bare metal to corrosion.

The design of the pillar or above-ground radiator housing should be such that it can be easily cleaned of street litter and moreover should be such that an accumulation of litter will not cut off air circulation. Aesthetic considerations will also have to be taken into account in designing pillars, and town and country planning authorities often have to be consulted over the appearance and dimensions of the pillar.

It would be possible to increase the kilovoltampere output from the transformer or alternatively to reduce the radiator size by circulating cooling air by a fan. There is, however, little interest in this approach, since the system reliability becomes dependent on the reliability of the fan.

11.2.5 Buried dry-type transformers

A conventional class C transformer can be buried in a vault, and a cooling pillar can be mounted above it to direct cold air to the bottom of the windings in a manner similar to the below-ground radiator type. However, the exhaust air is so much hotter owing to the higher operating temperature of the class C transformer that means have to be provided for mixing the hot air with cold air in the pillar before exhausting it so as not to inconvenience or endanger the public.

Such an arrangement does not have much to recommend it, since regular maintenance of the cooling ducts would be required and since there are obvious difficulties in using a dry-type transformer in what are essentially outdoor conditions.

11.2.6 Testing of buried transformers

No special problems are encountered in the routine testing of buried transformers, since they are to all intents and purposes conventionally designed transformers. The temperature rise test, however, does present problems. It is essential that the test simulates the site condition in that virtually all heat must be dissipated by the cooling pillar and heat loss from the tank itself must be prevented.

Different manufacturers will have their individual solutions which might range from actual burial of the transformer to the enclosure of the transformer in an insulated shell that effectively prevents heat transmission. The effectiveness of such a shell is easily checked by measuring its surface temperature and by seeing that there is no temperature difference between the shell surface and the air.

A well-designed buried transformer should be capable of operating in exactly the same manner as a conventional transformer as regards overloading, and its internal temperature differences should not be significantly different from those of a normal ON transformer of similar size. Owing to the blanketing effect of the earth fill it can, however, have an appreciably longer thermal time constant.

11.3 COAL-MINE TRANSFORMERS

11.3.1 Introduction

A transformer in a coal mine is in an area of high explosion hazard; therefore, it must be flame-proof. However, there is further consideration. Although they are flame proof, oil-filled transformers are at a disadvantage, largely because of the potential fire hazard of the oil they contain[G1.8, G2.5, G2.19]. For example, the British National Coal Board has issued its own specification[1] for dry-type flame-proof transformers for use underground. Amongst other things, this specification lists standard kilovoltampere ratings, primary and secondary voltages and impedances. It is usual to call for tappings on the high-voltage side at -5 and -10% of nominal voltage to cater for the voltage drop on the primary leads as the transformer is moved further and further away from the pit shaft as the coal face advances. The impedance is set at 4% as a compromise between too low an impedance which allows excessive short-circuit current in the event of a fault and too high an impedance which causes an excessive voltage drop.

11.3.2   Design and construction

General

Transformers are the three-phase type, supplied at one of three voltages: 2200, 2750 or 3300 V. The standard secondary voltage is either 565 or 1130 V. It is sometimes necessary to accommodate more than one primary or secondary voltage in the one transformer. This is done by tappings or by series-parallel arrangements on the windings. Such arrangements are, however, to be avoided wherever possible, since the space restrictions make it difficult to achieve a sound design.

Maximum dimensions are specified for height, width and length, and mass must be kept to a minimum. Transformers are usually fitted with wheels to accept standardised rail gauges. Each lifting lug must be capable of supporting the whole transformer. The core and windings are rigidly braced to the tank so as to withstand such handling conditions as lowering end on down a mine shaft.

The transformer is limited to a temperature rise of the windings to 150 °C as measured by change of resistance, but the tank temperature must not exceed 85 °C in an ambient of 25 °C, that is a rise of 60 °C.

Core and windings

Cores are of low-loss cold-rolled grain-oriented silicon steel, of similar design and construction to a normal dry-type transformer, with bolted yokes. It is not usual to have core bolts in the limbs of transformers of the ratings concerned. Because of the dimensional restrictions the design may appear to be misproportioned, tending to be long and low, with large core areas to achieve high voltage per turn. It is not generally found necessary or desirable to use the five-limb core construction which, though it saves height in much larger transformers than are considered here, does so at the expense of length and often mass.

Coils are constructed of class C or H materials as described in the section on dry-type transformers and are usually of rectangular shape to save space. The low-voltage coil is commonly a two-layer helical winding of multi-strip paper-covered conductor. High-voltage windings are disk type with external brazed connections. The coil assembly, after careful drying out, is fully impregnated with silicone varnish.

Tanks

Enclosures must be certified flame-proof by the appropriate testing authority.

The guiding principle for certification that they are flame-proof is that, if an explosion takes place within an enclosure filled with the appropriate gas, it should not be transmitted through the enclosure to cause a further explosion in a gas-filled atmosphere surrounding the enclosure. Thus a flame-proof joint is a joint between two faces of an enclosure, such as tank cover to tank flange, that will not permit a flame to pass from the enclosure to the atmosphere when an explosion occurs within the enclosure. It has been established that the critical factors are the length of the gap and the separation of the gap. Therefore, flame-proof joints require to be machined to achieve tight tolerances.

In addition to preventing the passage of flame, the tanks must also withstand the internal pressure resulting from an explosion.

The design of enclosure must. also be such as to preclude the phenomenon known as pressure piling that can take place if there are two or more compartments in restricted gaseous communication. Under these conditions, an explosion in one section of the tank can result in compression of the gas in another section, a subsequent explosion of this compressed gas giving rise to excessive pressure on the tank.

In addition to providing a flame-proof enclosure, the tank must also dissipate the heat generated in the core and windings. Various types of tank are in general use from cylindrical tanks with solid ends to conventional rectangular tanks. To dissipate the heat, most tanks require an extended surface area obtained either by a corrugated tank wall or by welding external fins to the tank wall. The design of fins or corrugations should be such that they do not collect dirt to such an extent that it would reduce the thermal efficiency of heat dissipation. Such dirt as does accumulate should be regularly brushed away.

Handhole covers in the tank are provided to give access to link boards for adjusting the voltage tappings. Usually one end is flanged for attachment to the relevant switchgear. All these joints and flanges are of course flame-proof.

11.4   WELDING TRANSFORMERS

11.4.1   Classification

Welding technology has made considerable and diverse advances, particularly since the end of the Second World War. With the development of new processes power sources have themselves become more sophisticated. It will be convenient to consider power sources under the two broad headings of resistance welding and arc welding.

In resistance welding the pieces to be welded together are placed between two suitably shaped electrodes made of copper alloy to give both good electric conduction and high resistance to wear. A current of hundreds or even thousands of amperes, depending on the joint to be welded, passes through the electrodes and the workpiece between them, raising the temperature at the interface of the joint locally to fusion value. At fusion temperature hydraulic pressure forces the two electrodes towards each other, exerting pressure on the joint; then the current is switched off to allow the joint to cool. The result is a spot weld between the two metals of the workpiece. The whole cycle, which is automatically controlled, is 2 or 3 s or even less. The cycles of current and pressure depend on the nature of the joint.

The distance between the spot welds can be reduced until the spots overlap to give a continuous weld. Difficulties can then be encountered if the current finds an easier path through a previously made joint instead of across the desired path, with the result that the weld is faulty.

In arc welding an arc is struck between an electrode and the workpiece such that the joint is brought to fusion temperature by the heat of the arc. In general, the weld is reinforced by the addition of filler metal into the molten weld pool. This is most commonly done by using a consumable rod as the electrode which melts away under the action of the arc and is carried into the weld pool to provide the filler metal.

The voltage required to initiate the arc is between about 60 and 100 V, the lower voltage being safer but causing more difficulty in striking. The voltage to maintain the arc depends on the type of electrode and the current used; it generally lies between about 20 and 30V. The current used is determined by the size and position of the weld that is being made. It varies from some 20 A for thin-gauge steel to 600 A for thick sections. Currents above 600 A are technically possible, but the weight of cables and electrode holder and the intense heat from the arc make it impracticable for manual welding. The electrode diameter is chosen to suit the current; it varies from 1.6 mm to 8 mm.

11.4.2 Duty cycle

In both resistance and arc welding, the process is essentially intermittent. In resistance welding current passes for 2 or 3 s and then is switched off whilst the next weld is set up. In arc welding an electrode rod may be used up after 2 to 3min, and a new rod must be fitted. Time is also needed for the removal of the solidified slag that is deposited over the weld and comes from the electrode coating. The ratio of actual welding time to total time is known as the duty cycle. The overall duty cycle of a resistance welding operation is typically 50 % and of an arc welding operation 36 % over a working day.

Welding equipments are designed to meet the rated temperature rise at a continuous current equivalent to the long-term heating effect of the intermittent load. Strictly, therefore, the continuous current is equal to the welding current multiplied by the square root of the duty cycle. In practice this is modified to take into account the heating time constant of the transformer and the general experience gained over many years in the design, manufacture and use of the equipment.

11.4.3   Resistance welding

Transformers to supply power for resistance welding[G2.13] are essentially short-time rated and are designed to give high currents of several thousand amperes at voltages of 5 to 10 V. Taps are normally provided on the primary winding to enable the secondary voltage to be adjusted. For a given electrode configuration and workpiece the welding current is determined by the secondary voltage. Thus the current control is achieved by the primary-winding tapping switches.

The construction generally takes the form of sandwich coils on a shell-type core. Water cooling is used on all but the smallest transformers. The secondary coils will normally be single-turn coils connected in parallel and interleaved with the disk-wound high-voltage coils. Low-voltage coils are made from solid copper bar or are cast in copper. Water cooling is usually achieved by brazing a copper tube to the copper coil which is cheaper in manufacture than making hollow-section coils. Because of the close proximity of high- and low-voltage coils resulting from the interleaved construction, the water-cooling tubes extract heat from the high-voltage coils as well as from the low-voltage coils, and it is possible to operate at current densities in excess of 5 A mm^{-2}. The resulting transformer is very compact, an important consideration when multi-head transformer units are required for high-speed welding in mass production lines.

In operation the electrodes are short circuited through the workpiece, and the current is limited by the impedance of the secondary circuit. The resulting pulses of high current cause mechanical forces to be exerted on the coils. The sandwich construction causes the forces on the inner coils to be self-cancelling. Only the outer coils need to be braced to withstand mechanical shock; they tend to bend outwards where they project beyond the core.

The design of the low-voltage busbars or cables taking current to the weld, whilst not strictly part of the transformer, needs to be considered with the transformer. At the currents which arise, the reactance voltage drop in a badly designed loop can be an appreciable percentage of the transformer output, with consequent reduction in welding current.

11.4.4   Arc welding

(For a general description of arc-welding plant equipment and accessories, see reference G2.8.)

Manual welding

For this familiar process, which uses stick electrodes, a drooping static characteristic is required. With a power source of this nature the variations in arc voltage that result from the inevitable variations in arc length have very little effect on the value of the current (see figure 11.5). The open-circuit voltage is substantially higher than the load or arc voltage, which facilitates striking the arc while the short-circuit current is only about 20 to 30% higher than the welding current. It limits the danger of burning through the plate to be welded.

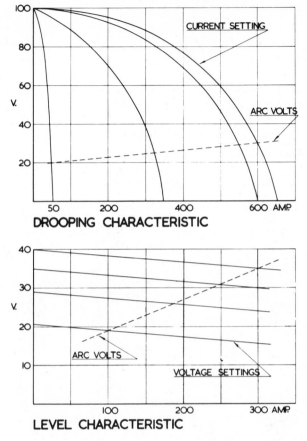

Figure 11.5   Power source: static characteristics

The difference between open-circuit voltage and load voltage is absorbed by the impedance of the welding circuit. It is usual to provide the bulk of this impedance by means of an added inductance. Apart from absorbing the voltage difference with minimum power loss the inductance also serves to produce a lagging power factor of approximately 0.35. This is a distinct advantage in welding, since at a current zero there is adequate voltage immediately available to restrike the arc. At unity power factor a voltage zero would coincide with current zero, and the arc would remain extinguished until the voltage had recovered sufficiently to restrike. This would have a deleterious effect on the weld quality.

By providing means of adjusting the inductance to any pre-determined value the welding current can be selected. Variation of the inductance (and thus current control) is achieved in many different ways. A tapped inductor may be used in conjunction with a selector switch: the more inductor turns in circuit, the higher is the reactance and the lower is the current. Alternatively, the air gap in the inductor might be varied by mechanical means, thus altering the inductance.

The added inductance is sometimes incorporated into the transformer itself, and the leakage reactance is varied by altering the air gap in magnetic shunts that bridge the core legs. Moving coils in the transformer leg may also be used to control the leakage reactance which is determined by the relative positions of primary and secondary windings.

In other systems the effect of variable reactance is obtained with a dc-controlled transductor by varying the magnitude of the controlling direct current.

Whenever a separate current-regulating inductor is used, the transformer may be of normal reactance (2 to 5%) because the bulk of the required reactance is built into the current regulator. Alternatively, the transformer may be designed to have a high reactance and the current regulator correspondingly less. Some of the more common systems are illustrated diagramatically in figures 11.6 and 11.7.

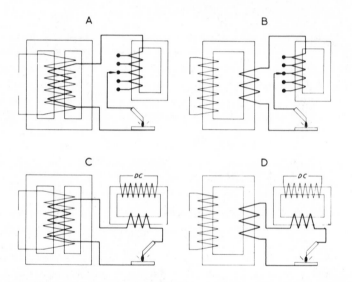

Figure 11.6   Transformer combined with current regulator: (a) low-reactance transformer plus tapped regulator; (b) high-reactance transformer plus tapped regulator; (c) low-reactance transformer plus transductor; (d) high-reactance transformer plus transductor

For single-operator equipments it is usual to combine inductor and transformer into one unit of whatever form it takes. For large welding shops it is usually more convenient and cheaper to provide a three-phase delta – star transformer which steps down the mains voltage to the required welding voltage (maximum 100 V line to neutral[G2.8]) and to utilise separate switched-tap inductors at each work station to control the welding current.

The low power factor so desirable in the welding equipment is often prohibitive in terms of supply. In such cases capacitors are connected across the primary terminals to improve the load power factor.

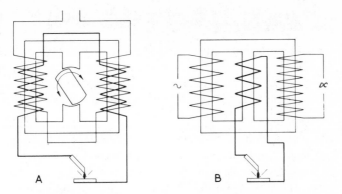

Figure 11.7   Variable-reactance transformer: (a) with adjustable air gap in magnetic
shunt; (b) with dc saturation of magnetic shunt

Automatic welding

Automatic processes involve the feeding into the arc of a continuous electrode
wire. Although some such processes use a drooping-characteristic power source
(see figure 11.5), the majority of applications require a flat characteristic,
sometimes known as a self-adjusting arc (see figure 11.5). The value of current is
determined by the rate at which wire is fed into the arc, a high feed rate requiring a
high current. Stability is achieved when the rate of feed is equal to the rate of
burning. If the wire feed is faster than the burn-off rate, the arc length decreases,
the arc voltage decreases, and the current rises to burn off wire more quickly.
Hence the system tends to be self-stabilising or self-adjusting.

The basic construction of such equipment consists of a transformer, usually
but not necessarily three-phase, with suitable voltage tappings over a range of
about 20 to 50 V, and a rectifier since these equipments are almost invariably dc
operating. The complexity and comparatively high cost of this type of equipment
arises mainly out of the auxiliaries and control circuits built into the equipment to
facilitate the particular process.

11.5   GAS-INSULATED TRANSFORMERS WITH CLASS A
INSULATION

11.5.1   General

The application of dry-type air-insulated transformers is limited to system
voltages of about 15 kV owing to the limitations of air as an insulating medium.
These transformers are also restricted to electrically non-exposed conditions
because of their low surge-voltage strength. Several gases are superior to air in
terms of electric strength, but, if they are to be used as an insulant in transformers,

several other aspects must also be considered. Some of the more important criteria are as follows.

(a)   Toxic effects. The gas must be non-injurious to human welfare.
(b)   Chemical effect. It must be inert to the materials used in the transformer.
(c)   Thermal effect. It must be stable over the temperature range applicable.
(d)   Economic effect. It must be reasonably cheap.

Sulphur hexafluoride (SF_6) is found to be the most suitable gas in the light of these criteria, with certain fluorocarbons also of interest. The growing interest in transformers of this type seems to be centred around SF_6.

The electric strength of SF_6 appears to be comparable with that of oil if the gas pressure is held at about $200\,kN\,m^{-2}$. Its surge-voltage strength is about half that of oil. There is, however, still a great deal to be learned about the properties of SF_6. Although its electric strength appears to be proportional to gas pressure in a uniform field, this does not appear to hold in a non-uniform field.

11.5.2   Thermal characteristics

Although SF_6 has a specific heat capacity about 75% that of air, its density is about five times greater. It is therefore able to absorb about three and a half times as much heat as air. Provided the heat absorbed by the gas can be given up to the atmosphere, the temperature of the transformer can be controlled by the pressure and velocity of the gas that flows over the coils and core.

The art of constructing a gas-insulated transformer lies in first circulating the gas through the windings and in then extracting the heat from the gas. To achieve the first part of this cycle it is usual to use forced circulation of the gas, pumps or blowers forcing the gas through ducts in the coils and core.

The second phase is rather more complicated. It is usually found necessary to use a two-stage heat exchange: the first stage is integral with the transformer; the second stage is mounted externally (see figure 11.8).

Stage 1 of the heat exchanger is an evaporator filled with the refrigerant Freon, over which the heated SF_6 flows after absorbing heat from the coils. Heat from the SF_6 is given up to the Freon in this stage, and the cooled SF_6 is recirculated through the coils.

The Freon contained in the first stage is evaporated in absorbing the heat and passes into the second stage which is an external condenser artificially cooled by fans or water or other available means. Here the Freon is condensed into liquid, and it returns to the evaporator stage. This action is self-motivated, the Freon system circulating by gravity/thermal head.

There are thus two separate closed heat exchange cycles. They are isolated from each other but depend upon one another.

11.5.3   Overload capacity

The time constant of a gas-filled transformer is noticeably less than that of an oil-

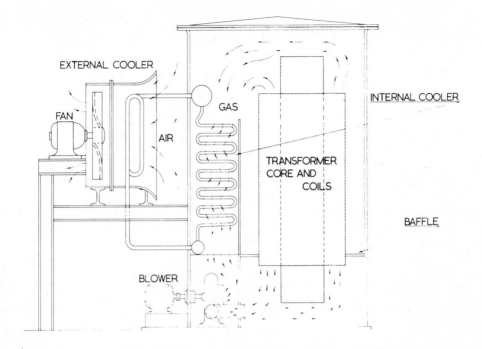

Figure 11.8   Two-stage heat exchanger

filled transformer. Thus operating temperatures are reached more quickly, and it would appear that the gas-filled transformer would have less overload capacity. However, there are indications that, provided moisture is rigorously excluded from the insulation, most materials can be run at slightly higher temperatures than in oil for the same predicted life. If this factor can be taken into account, then the overload capacity of a gas-filled transformer becomes comparable with that of an oil filled transformer.

11.5.4   Advantages and disadvantages

Primarily this type of transformer offers a solution when an oil-filled transformer is not acceptable, say, because of fire risk, and when the dry-type class C air-insulated transformer is technically not feasible because of either its size or its voltage rating.

Its surge-voltage strength is less than that of an ON transformer, although better than that of a class C transformer. Therefore, if the system is electrically exposed, an auxiliary protection, such as lightning arrestors, will be necessary.

Although lighter than an oil-filled transformer, it occupies just about the same physical space.

It is essentially a sealed unit and must be maintained in that condition during its life. Gas pressure must also be maintained.

It is entirely dependent on the continued operation of the fans or pumps used for circulating the gas and condensing the Freon.

11.5.5   Properties of sulphur hexafluoride compared with air and nitrogen

Property	SF_6	air	nitrogen
Density ($kg\,m^{-3}$)	6.2	1.20	1.17
Boiling point ($^{\circ}C$)	-62	-200	-196
Viscosity ($mPa\,s$)	0.018	0.0185	0.0175
Thermal conductivity ($Wm^{-1}K^{-1}$)	0.0033	0.0065	0.0065
Specific heat capacity ($J\,kg^{-1}K^{-1}$)	730	1000	960

SF_6 has been found to be non-toxic, but it is an asphyxiant. It has been widely used without health hazard for many years in industrial and medical X-ray transformers.

11.6   RECTIFIER TRANSFORMERS

11.6.1   General

The general principles governing the design of rectifier transformers are very similar to those employed in the design and manufacture of transformers for more normal applications. It will be understood that transformers for rectifier service may be oil cooled or air cooled, and the general engineering considerations apply equally to rectifier transformers as to other transformers. The differences in rectifier transformers arise because of the peculiarities which can be reflected onto the transformer through the presence in its secondary circuit of the rectifier.

In the past, the mercury-arc rectifier was widely used for power rectification. The susceptibility to 'backfire' of the rectifying element imposed frequent short circuits on the transformer. A prime consideration with transformers for use with mercury-arc rectifiers is, therefore, that they should be capable of withstanding repeated short circuits. Windings must be heavily braced and clamped to withstand this duty. There is something to be said for the use of foil-wound coils for this duty, since such coils can be readily designed to have virtually no resultant axial force from electromagnetic forces.

The widespread use of silicon rectifiers has largely superseded mercury-arc rectifiers. With this type of rectifier, 'backfire' does not occur, and the danger of short circuit on the transformer in normal operation is virtually eliminated.

The dc power largely determines the choice of pulse number or ripple frequency. It is well known that this number governs the harmonic content of primary current and secondary voltage. The higher the pulse number, the closer the ratio of the rms value of the fundamental of the primary current is to its total rms value and the lower is the inherent distortion of the secondary voltage. An increase in pulse number, however, increases the cost of the transformer and the number of rectifier elements.

The discontinuous anode currents can result in a secondary winding that is much larger than the primary winding, owing to the extra heating effect of the anode current form.

With the above remarks in mind, the transformer parameters are largely determined by the rectifier configuration that is supplied. Some of the more usual circuits are outlined briefly below.

11.6.2 Three-phase single-way connection

From figure 11.9(a), the secondary current contains a dc component which is not reflected in the primary. One-third of the mean dc current in the load must pass through each winding. The primary must be delta connected. With a star-connected primary the current in an active phase must return to the line through the other two phases; since there are no secondary ampere – turns on the inactive phase, the circuit would be unbalanced.

11.6.3 Three-phase single-way interstar connection

To eliminate the effect of static magnetisation of the core, the secondary winding may be connected in inter-star, so that each rectifier arm is now supplied by two phases (see figure 11.9(b)). The mean rating is higher than the simple star connection since the secondary winding kilovoltamperes are now $2/3^{1/2}$ times as great as for a star winding. The primary can be connected in either star or delta.

11.6.4 Three-phase double-way connection

The circuit in figure 11.9(c) is one of the most useful rectifier connections. It is very flexible where the transformer is concerned. Primary and secondary can be connected star or delta as convenient. There is no dc component in the transformer, and primary and secondary ampere – turns are balanced.

11.6.5 Six-phase single-way connection

The circuit in figure 11.9(d) is an extension of the circuit in figure 11.9(a). Each secondary is wound to have a voltage of twice the dc output, and a tap is brought out at the centre of each winding. All three centre taps are connected together to provide a neutral point which is connected to one side of the rectifier. Each of the resulting six arms carries current for one-sixth of a period. Static magnetisation of the core is eliminated, but the equivalent kilovoltamperes value is higher than for the connection in figure 11.9.

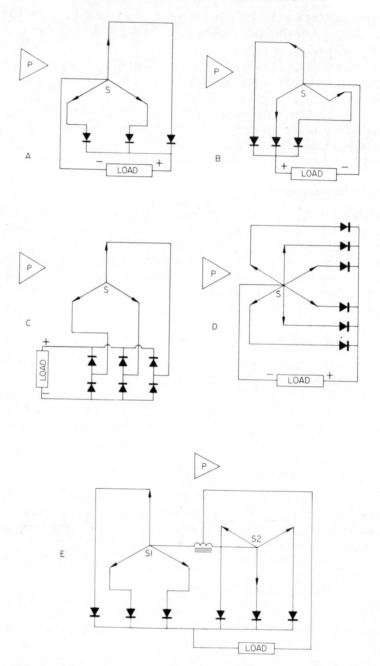

Figure 11.9 Rectifier circuit arrangements: (a) three-phase single-way connection; (b) three-phase single-way interstar connection; (c) three-phase double-way connection; (d) six-phase single-way connection; (e) three-phase single-way parallel-compound connection

11.6.6 Three-phase single-way parallel-compound star connection

Figure 11.9(e) shows the combination of two circuits like figure 11.9(a), connected in parallel with one of the transformer stars turned through π rad with respect to the other to give a phase displacement of $\pi/3$ rad between the six arms. The combination is in pulse number equivalent to a six-phase single-way star connection.

In order to maintain independent operation of the two component circuits, their transformer neutrals are interconnected through an interphase inductor which absorbs the instantaneous voltage difference between the star points. This has the effect that the time of operation of each rectifier element is doubled to $2\pi/3$ rad in comparison with the time of $\pi/3$ in the circuit in figure 11.9(d).

ACKNOWLEDGEMENT

The author expresses his thanks to GEC Distribution Equipment Limited for permission to publish the material in this chapter.

REFERENCE*

1. The British National Coal Board, *NCB 297, Dry-type Flame-proof Transformers 150–300 kVA* (1962)

12

Transformers in Distribution Systems

L. Lawson*

12.1 TRANSFORMER SELECTION—INTRODUCTION

The customer's specification (see section 3.2) meets constraints set by the supply system, the load to be supplied, the environmental aspects and the need to provide a reasonably secure supply by the most economic means.

12.2 TRANSFORMER SELECTION—CONSTRAINTS OF THE SUPPLY SYSTEM

12.2.1 Choice of phasor group

Where the transformer is inherently used for one-way traffic of load and hence does not need neutral earthing facilities on the higher-voltage winding, a delta–star combination is preferred, since the delta winding and tap changer will carry lower current and since third harmonics are shunted out. By careful choice paralleling is possible with various combinations of transformation (see figure 12.1).

The star–star transformer without tertiary delta has been found suitable in distribution networks not closely associated with generating points. With modern designs the zero-sequence impedance is so low that some users find it necessary to use resistance in the neutral to limit earth fault currents to avoid unacceptably high voltage gradients near the supply point (see figure 12.2).

In general the choice of phasor group is determined by the existing system.

---

* North-Western Electricity Board.

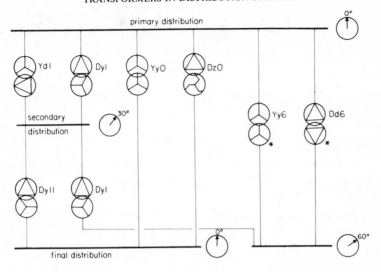

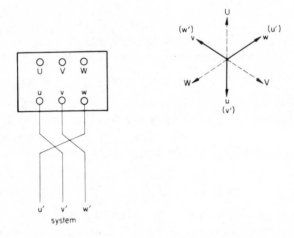

Figure 12.1   Paralleling of transformers (the notation is that of reference G1.12, and the last digits referred to clock face indicate phase difference):*, with connections to lower-voltage winding made non-standard to convert transformer π rad shift to system π/3 rad shift (see lower diagrams)

12.2.2   Choice of voltage ratio, tap range and tap step

For primary distribution transformers it is necessary to allow for an input voltage which may vary between limits above and below the nominal voltage, with an output required at or above nominal voltage. The impedance on any one tap will be specified to limit the fault current to a specified level when delivered by the

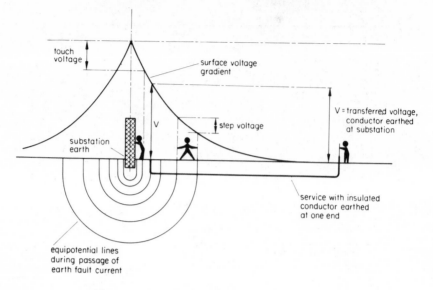

Figure 12.2   Effect of high earth fault currents

maximum number of transformers in parallel, supplied from a system with a specified maximum fault level; a manufacturing tolerance must be allowed.

The transformer regulation results in the expression

$$(V_i/n)^2 = V_0^2 + I_0^2(X_n/m + X_i/n^2)^2 + 2V_0I_0(X_n/m + X_i/n^2) \qquad (12.1)$$

where I_f is the fault current, I_0 the load current, m the number of transformers in parallel, n the transformer ratio on particular tap, V_i the input voltage, V_0 the output voltage, X_i the input reactance of the high-voltage system and X_n the reactance of transformer on a particular tap. The fault current on the output busbars would be

$$I_f = nV_0/(X_i + n^2X_n/m) \qquad (12.2)$$

From these equations it is possible to plot curves of X_n/m against n for various conditions which satisfy the fault limitation requirements. The maximum and minimum steps to satisfy the voltage requirement can then be calculated.

There is a practical and economic limit on the number of taps and hence on the minimum step. A step value of 1.5% is not noticeable on networks with a voltage variation at the consumer terminals of not more than $\pm 6\%$; this value includes all the system voltage variation.

12.2.3   Choice of impedance

The main advantage of inherent impedance is in short-circuit limitation. A further use is in limiting circulating currents when transformers in parallel are on different tappings. With one transformer out of step for n transformers in parallel, directly coupled to the same high- and low-voltage busbars and each with the impedance Z per phase, the circulating current is

$$I_c = 0.01 \times \text{per cent tap step} \times \text{phase voltage} \times (n-1)/nZ \qquad (12.3)$$

12.2.4 Mode of operation

As a general definition distribution transformers are transformers with a loading only determined by the load to be supplied. They may operate at voltages up to 275 kV. Table 12.1 shows an example for a decision taken in respect of the security of supplies in a highly developed area.

TABLE 12.1

Load	One fault		Two faults		Mode
(MVA)	Load lost	Restoration time	Load lost	Restoration time	
up to P_a	all	repair time	—	—	single
P_a to P_b	all	switching time	—	—	switched duplicate
P_b to P_c	all	manned control of switching	—	—	switched duplicate
P_c to P_d	nil	—	all	repair	duplicate
P_d to P_e	nil	—	all	part switching, part repair	multiple
over P_e	nil	—	part	repair time	multiple

The levels of load P_a to P_e are determined by the economics of supply. This includes the concept of firm supply, which means that a system is firm if at the time of highest loading it can accept the loss of the largest single unit due to any one failure without losing the supply. The simplest case for transformers is that of duplication. Each transformer normally carries a maximum of 50% of the load but must be capable of carrying the total load. Installations of three transformers provide up to 67% utilisation (one-third sub-station load on each transformer under normal conditions, each transformer capable of one-half sub-station load) but at higher capital cost. The output fault level of the parallel group may be above that of the switchgear standard adopted. To avoid increased switchgear costs while minimising copper losses, automatic switching may be utilised, where the lower-voltage busbars are normally run in sections and are coupled automatically on loss of any one transformer.

The above are installations concentrated within one site with little, if any, interconnection between sites. An alternative method is to utilise one transformer on each site, with the lower-voltage network interconnected. Theoretically this gives a high possible utilisation, but in practice the maximum is about 80%. This alternative method requires heavier interconnection between sub-stations with less concentration of cable around each sub-station. It is, therefore, economically difficult to change from one method to the other in an existing situation.

Selection of the mode of operation is determined by the overall system economics.

Industrial transformers

Industrial transformers may supply a mixed load or a single process. In both cases the load cycle is readily predicted.

12.3   TRANSFORMER SELECTION—LOADS SUPPLIED FROM DISTRIBUTION TRANSFORMERS

12.3.1   Load

Distribution transformer loads depend upon the consumer installation and habits, climate and degree of development of the country and can be split down into the following.

(1) Lighting.
(2) Space heating.
(3) Air conditioning (cooling).
(4) Cooking.
(5) Water heating.
(6) Small domestic power.
(7) Industrial (motive power, process heating, electrolysis).
(8) Traction.

Lighting has a predictable time of day and time of year variation, with superimposed variation due to heavy cloud or fog. Space heating is largely dependent on weather, the peak occurring at low ambient temperature, particularly with high winds. In a temperate climate an unusually high peak develops after 10 to 14 days of continuous freezing conditions owing to loss of heat stored in the fabric of buildings, frozen coal stocks, inability of distributors to meet peak demands for oil, gas pressure drop and the use of additional electric heaters to prevent water pipes from freezing as internal temperatures drop. In cold climates this additional peak does not occur, since this weather is normal, since heat insulation is better, and since the population is better prepared for low temperature.

Air conditioning is again dependent on weather and is sensitive to increased temperature and humidity, giving a summer peak. Other loads listed are reasonably predictable.

For load estimation it is usual to reduce measured loads to standard conditions to assess annual trends or to use the estimated load assessed from annual trend to assess peak loads. For example, in a temperate zone the peak load is

$$P_m = P_{ms} + K_3(K_2 - \theta_{amb}) + K_4H + K_5(E_p - E_f)^2/K_6 + K_7v_W \quad (12.4)$$

where P_{ms} is the peak load expected under standard conditions, K_2 to K_7 constants which depend on day of week, seasons and type of load, θ_{amb} the ambient temperature, H the humidity, E_p the potential zenith solar illuminance, E_f the forecast illuminance and v_W the wind velocity. K_3 has been measured at $2\% \, ^{\circ}C^{-1}$ in one temperate situation. For summer peak, the second term would become $K_3(\theta_{amb} - K_2)$.

12.3.2 Load of final distribution transformers

The loading may vary widely in its daily pattern. Some examples are given in figure 12.3 taken from actual situations in a temperate zone. Figure 12.3(a) shows a domestic load showing storage heaters on off-peak tariff, switching on at a predetermined time and tailing off as individual thermostats operate. An unusually high lunchtime load is due to the prevalence of young married women who work mornings only in this particular area. Figure 12.3(b) shows an old commercial area with a high proportion of storage heating, a mid-day boost being more obvious on Sundays. The 0600 peak is due to office and shop cleaners. Figure 12.3(c) shows a mix of one- and two-shift industrial work.

When a number of individual final distribution sub-station loads are summated at the next higher distribution level, the result approximates towards a total system daily load curve such as that in figure 12.3(d). Such system load curves depend on climate, on state of development, on national habits and, if the system is large, on the existence of different time zones. This electrical summation may possibly be used as the average shape for final distribution transformers. The annual curve of daily peak demands will also differ with climate and type of load (see figure 12.4).

12.3.3 Optimum transformer rating

The basic system design to secure the optimum economic conditions must take account of the cost of the system to supply the transformer, the cable type and cost for distribution, the load density, the maximum permissible voltage drop at the consumer terminals, out-of-balance conditions and how far practical situations and design may vary from the theoretical maximum. Various methods are available[1e], but a typical method, with all these factors inserted could result in a simplified formula

$$\text{load/sub-station} = \text{constant} \, (\text{kW}10^{-4} \, \text{m}^{-2})^{1/2}$$

In 1973, in British conditions, this formula indicated sub-station ratings of 500 to 1000 kVA per sub-station but, with overhead distribution and lower load densities, a much lower result is to be expected.

The cabling and switchgear should normally cater for the ultimate size.

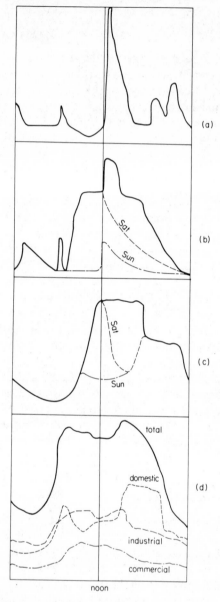

Figure 12.3   Examples of daily load curves

However, it may be economic to install a smaller rating of transformer initially, changing to a higher rating as load develops.

12.3.4   Loading of transformers

A transformer is a thermally loaded device because the hottest spot of the

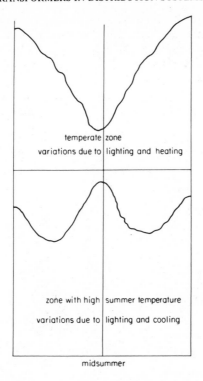

temperate zone

variations due to lighting and heating

zone with high summer temperature

variations due to lighting and cooling

midsummer

Figure 12.4   Examples of annual curve of daily maximum demands

insulation is the limiting factor. An excessive temperature shortens its life. With paper insulation, the paper becomes brittle and reaches the final stage when a through-fault current would literally shake off the insulation.

The winding temperature depends on the load, on the efficiency of the cooling system, on the coolant temperature and on the time from the application of load. The transformer does not have one fixed rating but, for convenience, is given a nominal rating formed by the product of rated voltage, rated current and phase factor, with the principal tap in use, to give a specified temperature rise under steady-state conditions. The intended temperature rise is designed to give negligible loss of life with specified coolant temperatures.

Those systems which have peak loads during colder weather can utilise a greater temperature rise without exceeding the maximum hot-spot temperature. This situation has led to the development of transformers which are designed for running in pairs, with natural cooling. With loss of one transformer throughout a peak load season, demand can be met by the remaining transformer by using forced cooling without excessive loss of life on the assumption that peak load will not occur above a specified ambient temperature[2].

In Great Britain, with an average winter temperature range of $+5$ to $-5\,°C$ and a system temperature sensitivity of load of about $2\%\,°C^{-1}$ drop, the increased temperature rise (under forced cooling conditions) due to increased load is approximately equal to the fall in ambient temperature. The temperature rise under natural cooling conditions is far greater for the same marginal increase of loading. While this does not affect paired transformers, it is important to note that a group of three transformers require auxiliary cooling supplies which are individually firm since loss of a common cooling supply may result in cascade tripping (see figure 12.5).

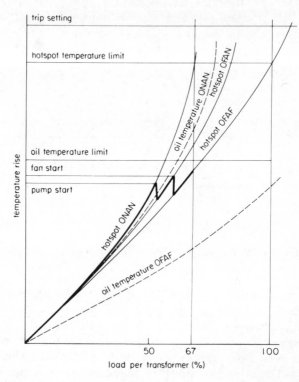

Figure 12.5   Requirement for firm cooling supplies to peak-rated transformers run in groups of three

These transformers are co-ordinated with the preferred current ratings of switchgear and have been described by various manufacturers as an integrated system transformer, a continuous emergency-rated transformer, a temperature-loaded transformer or a peak-rated transformer. The rating is continuous, with some loss of life, since at these loadings the thermal time constant is usually much shorter than the duration of heavy load.

This principle was used in the past by the use of loading tables which ascribed a cylic overload capacity[G1.20, G2.23, G3.26]. With a more realistic approach to loss of life and with regard to load cycle, load growth during life and temperature sensitivity of load, it is possible to exceed the values given in such tables, which of necessity refer to the general case. For a specific case, where manufacturers have performed a temperature rise test, it is possible for detailed calculation to be made (see chapter 7).

It is necessary to consider the ratings of associated cabling, switchgear and tap changer in association with transformer peak rating. This is particularly the case where switchgear is indoors and does not have the same benefit from reduced ambient temperature.

12.4 TRANSFORMER SELECTION—ENVIRONMENT

Noise may be of concern in urban areas; hence specifications impose a limit on sound output. This is treated in depth in chapter 8.

The thermal effects of housings are covered in section 12.7.

12.5 TRANSFORMER SELECTION—DESIGN PHILOSOPHY

It is becoming usual to leave the designer more freedom by specifying only the main parameters such as ratio tap range, tap step, phasor group, impedance, nominal load rating(s) and a capitalisation figure for losses and by leaving flux density and other design parameters to the designer. As a precaution against overfluxing, that is excessive flux density, noise (and harmonic output on star–star transformers) may be limited by specification.

12.6 ECONOMIC ASPECTS OF TRANSFORMER SELECTION AT THE PRE-SPECIFICATION STAGE

12.6.1 General

The total cost of a transformer is in two parts: (1) the initial purchase costs, together with maintenance, and (2) the loss costs throughout life. Maintenance costs vary little, but from low cost–high loss to high cost–low loss there is a multiplicity of designs with an optimum overall cost.

The cost of losses depends on tariff, on daily and annual load curve and on load growth over the life of the transformer[3,4,5,6].

12.6.2 Summation of future costs

The most accurate system of summation of future costs is by the method of discounted cash flow which places most significance on the early years when tariffs and loads can be more accurately assessed. Thereby the effects both of currency inflation and of indeterminate life span are also minimised. In brief a nett present sum of costs is estimated by theoretically setting aside a sum for each estimated future cost; this sum, if invested at compound interest at current rates of return on capital, would produce by the due date the sum required. The more distant the date, the less is set aside and the less the error that may be involved by poor estimation or changing conditions.

12.6.3 Tariffs

Losses are costed against the cost of production or purchase, and hence for other than generation systems at the bulk supply tariff of purchase. Tariffs are usually of two basic parts: a demand charge plus an energy charge. The latter may be subdivided into several rates, usually by time of day. Fuel price surcharges and other additions may be made, so that the final tariff reflects the cost of capital investment to generate and distribute peak load, and the cost of generated energy with the mix of generators of various merit factor which must be employed.

Further complexity may arise where a number of distribution undertakings pay a demand charge based on the proportionate demand of each at a time of system peak met by a common generating undertaking. This can only be decided in retrospect.

12.6.4 Cost of losses

List of symbols

Symbol	Meaning	Unit
C_f	annual cost of fixed losses	U_m year^{-1}
C_{fd}	component of annual cost of fixed losses due to demand charge	U_m year^{-1}
C_{fe}	component of annual cost of fixed losses due to energy charge	U_m year^{-1}
C_p	maximum demand charge	U_m year^{-1}
C_v	annual cost of variable loss	U_m year^{-1}
C_{vd}	component annual cost of variable loss due to demand charge	U_m year^{-1}
C_{ve}	component of annual cost of variable loss due to energy charge	U_m kWh^{-1}
C_{w1}	energy charge at first rate for x h year^{-1}	U_m kWh^{-1}
C_{w2}	energy charge at second rate for y h year^{-1}	U_m kWh^{-1}
C_{w3}	energy charge at third rate for z h year^{-1}	U_m kWh^{-1}
D	diversity of peak demand (see below)	ratio
I	load current	A
I_p	peak load current	A
I_r	rated current	A

Symbol	Meaning	Unit
I_{sp}	current at time of system chargeable peak	A
K_b	overfluxing constant	ratio
K_s	loss load factor	ratio
P_f	fixed loss at rating	kW
P_{fe}	iron loss including overfluxing component	kW
P_v	variable loss at rated load	kW
U_m	monetary unit	U_m
x, y, z	see C_{w1}, C_{w2}, C_{w3}	h

Fixed loss

Iron loss in a transformer will vary with flux density and hence with applied voltage. The final distribution transformers with off-load tap selection normally operate within narrow limits of applied voltage; the major portion of the allowable variation of voltage at the consumers' terminals is utilised in the lower-voltage system. A typical example is where total voltage variation at the consumers' terminals is limited to $\pm 6\%$; this would usually be made up as follows.

Control dead band	2%	on the assumption that not all will be
High-voltage system	3%	maximum at same time, in any
Transformer	1%	one situation total not to
Low-voltage system	5%	exceed 12%
Consumers' service line	2%	

The resultant change in transformer loss between full load and no load may be neglected.

A system including a transformer with on-load tap changer is designed to operate over a range, as described in sub-section 12.2.2, where highest output voltage is required at peak load with lowest input voltage. This results in a higher flux density and hence increased iron loss, typically of about $+16\%$. For practical purposes it will be assumed that the variation follows a square law in relation to load.

If load is defined as a function of I, then iron loss P_{Fe} will be increased by an amount dependent on $f(I^2)$. Thus

$$P_{Fe} = P_f\{1 + K_b f(I^2)/I_r^2\} \tag{12.5}$$

If we apply the effects of a tariff, with time diversity D (which is the peak load divided by load at the time of system chargeable peak demand), the cost due to demand charge is

$$C_{fd} = P_f C_p + P_f \frac{C_p}{D} K_b \frac{f(I^2)}{I_r^2} \tag{12.6}$$

where the second term is the overfluxing component. The cost due to energy charge is

$$C_{fe} = P_f(xC_{w1} + yC_{w2} + zC_{w3}) + \frac{K_b P_f}{I_r^2}\{C_{w1}\int_0^x f(I^2)\,dt$$

$$+ C_{w2}\int_0^y f(I^2)\,dt + C_{w3}\int_0^z f(I^2)dt\} \qquad (12.7)$$

where the three components of the second term reflect the overfluxing component at the times of the varying tariff rates, $x + y + z$ being the hours in a year.

It is shown in sub-section 12.6.6 that the ratio of mean square value of load to the square of rated load may be represented by a loss load factor K_s. The total cost of fixed losses as the sum of equations 12.6 and 12.7 simplifies to

$$C_f = P_f\left\{C_p\left(1 + \frac{K_b K_s C_{w1}}{D}\right) + xC_{w1} + yC_{w2} + zC_{w3} + K_b K_s C_{w1}\right\} \qquad (12.8)$$

Variable loss

Variable loss consists mainly of resistive loss in the windings. Stray loss, that is the load-related losses other than resistive loss, consisting of eddy current losses, etc., varies with current somewhere between a linear and a square law relationship. However, tolerable accuracy may be obtained by assuming that total variable loss varies as the square of load current for other than the largest transformers.

At the time of system chargeable peak, the load current will have a time diversity. The ratio of load peak to load at the time of system peak will be

$$D = I_p/I_{sp} \qquad (12.9)$$

The chargeable demand is therefore that of losses* due to I_p/D and will be

$$C_{vd} = P_v C_p(I_p/DI_r)^2 \qquad (12.10)$$

The energy charge will be of the form

$$C_{ve} = \frac{P_v}{I_r^2}\left\{C_{w1}\int_0^x f(I^2)\,dt + C_{w2}\int_0^y f(I^2)\,dt + C_{w3}\int_0^z f(I^2)\,dt\right\} \qquad (12.11)$$

where $\int f(I^2)\,dt$ represents the integral of load squared during the chargeable period.

12.6.5 Measurement of factors

Tariff and rated loss quantities are known. The elements to be measured relate to load and are I_m, D and $f(I^2)dt$.

---

* Losses of the peak-rated transformers (see sub-section 12.3.4) are usually stated relative to the *peak* rating. Since transformer peak load is normally only half this rating (with paired transformers), a factor of 2 must be introduced into the denominator of peak load, resulting in the stated variable loss being divided by four. Similarly with three transformers a divisor of $(3/2)^2$ is required.

I_m is obtainable from a simple maximum-demand indicator. Where continuous demand indications are recorded, D is readily available by reading demand at time of system peak. For final distribution it will be adequate to measure demand at the time of day and day of week of anticipated system peak and to relate this to maximum demand during that week, provided there are no unusual disturbing factors of weather, public holidays, network transfers, etc.

$\int f(I^2)\,dt$ can be measured by squared amperes hour meters switched to correspond with tariff time rates. Such meters are available for measurement of machine losses or for estimation of transformer loss when metering is on the opposite winding of a transformer to the agreed supply terminals. This instrumentation may be applied on a sampling basis as a feedback check on the accuracy of estimating methods.

As a cross-check, in the average case

$$\text{source load factor} = DK_p \frac{\text{load} + \text{system losses}}{\text{load}} \tag{12.12}$$

where K_p is defined in sub-section 12.6.6.

12.6.6 Estimation of loss costs

List of further symbols

Symbol	Meaning	Unit
a	proportion of annual energy assumed at peak demand	ratio
b	proportion of annual energy assumed at mean demand	ratio
K_p	annual load factor	ratio
P	power	kW
P_{av}	mean power	kW
P_m	peak power	kW
t	time ($T = 1$ year $= 8760$ h)	h

Owing to the complexity of distribution load pattern in a general supply network it is usual to simplify by accepting an approximation to a load curve consisting of a proportion of load taken at peak load rate with the remainder at mean rate. This is illustrated in figure 12.6, where a and b are proportions of the total energy $P_{av} T$. This method gives a result based on load factor, which may be measured as total energy divided by maximum demand.

If we define the annual load factor K_p as the ratio of mean load P_{av} to peak load P_m, we obtain

$$K_p = P_{av}/P_m = (1/P_m T) \int_0^T P\,dt \tag{12.13}$$

The definition of loss load factor K_s as the ratio of mean load of losses to peak load of losses gives

$$K_s = (1/P_m^2 T) \int_0^T f(P^2)\,dt \tag{12.14}$$

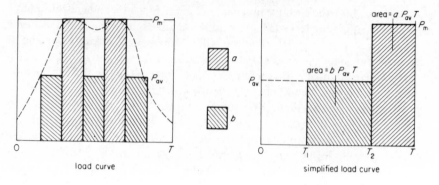

Figure 12.6   Transformer annual load curve

If we use the simplified load curve (figure 12.6) with the total energy area $W = TP_{av}$, the energy ratio $b = (T_2 - T_1)P_{av}/W$ and the ratio $a = (T - T_2)P_m/W = (T - T_2)P_m/TP_{av}$, the integration of equation 12.14, with equation 12.13, results in

$$K_s = b(P_{av}/P_m)^2 + aP_{av}/P_m = aK_p + bK_p^2 \qquad (12.15)$$

The proportions a and b may be adjusted to allow for time of day energy rates. They are estimated by experience confirmed by sample studies as described in sub-section 12.6.5, if we note equation 12.13.

The energy cost of variable loss may be stated as

$$C_{ve} = P_v K_s C_{w1}(I_p/I_r)^2 \qquad (12.16)$$

An example of estimation of loss load factor is given in appendix A12.1.

12.6.7   Effect of load growth—changing transformers

List of further symbols

Symbol	Meaning	Unit
C	any of the costs arising annually	U_m
C_a	annual cost of capital	U_m
C_c	total amount cost of changing a transformer	U_m
C_1	recovered value of an existing transformer	U_m
C_2	capital cost of a replacement transformer	U_m
C_3	cost of changing a transformer	U_m
g	annual growth rate (100g is percentage growth)	ratio
I_m	maximum annual peak load	A
I_0	initial annual peak load	A

Symbol	Meaning	Unit
m	multiplier of cost to change to annual charge	ratio
n	rate of return on capital ($100n$ is percentage return)	ratio
q	number of years	years
Q	number of years to pay off costs of a change	years

Load, except to a fixed process unit, will change from year to year. In general networks the trend is a compound growth rate of the form

$$I = I_0(1+g)^q \qquad (12.17)$$

This leads either to change to a higher rated transformer or to reinforcement by a sub-station elsewhere.

In the general case of final distribution radially from a single transformer the rating may be selected from the preferred number series with steps of rating approximately equal to the fifth root of ten, for example 315, 500, 800, 1250 kVA.

$$(1+g)^q = 10^{0.2} \qquad (12.18)$$

from which

$$q = 0.2/\log(1+q) \qquad (12.19)$$

The life between changes is indicated by table 12.2. The individual transformer therefore starts life with $10^{-0.2} = 0.63$ of full load, rising with compound rate of growth to full load. Full load in this context is that load at which it is economic to change a transformer to the next higher rating.

TABLE 12.2

100g	1	2	3	4	5	6	7	8	9	10
q	46	23	15	11	9	8	7	6	5	5

The costs of actually changing a transformer (civil works, transport, cabling, jointing) should be paid off well within the life before the next change becomes due. The payback period Q may be selected from table 12.3. The annual charges of interest and repayment will be

$$mC_3 = \frac{(1+n)^Q n}{(1+n)^Q - 1} C_3 \qquad (12.20)$$

where C_3 is the cost of changing, excluding the capital costs of transformers.

The capital value of a recovered transformer in a small industrial system may be resale or even scrap value, but in a large system, where the transformer is almost

TABLE 12.3

Q	2	3	4	5	6	7	8
m	0.576	0.402	0.315	0.264	0.230	0.205	0.187

certain to be reused, thereby avoiding the purchase of a new transformer, it may be given a 'notional' value equal to the current purchase price of a new transformer. When reused it should again be charged at notional value.

The purchase cost of a new transformer is spread over its life using the constant charge method of calculation. Since the life may confidently be expected to be 40 years, the annual cost C_a becomes

$$C_a = \frac{(1+n)^{40} n}{(1+n)^{40} - 1} C \approx nC \qquad (12.21)$$

For example, if the interest rate is 10 %, the expression becomes $0.102C$. The total annual charge C_c of changing a transformer, therefore, becomes during the initial period

$$C_c = (C_2 - C_1)n + C_3 m \qquad (12.22)$$

The costs of supplying transformer losses at varying loads will be as shown in figure 12.7. Curve A relates to a smaller transformer than curve B. The costs of zero load are equal to the cost of fixed loss of the transformer. The vertical intercept CD indicates an annual amount available to pay for change to a larger rating or EF to pay for change to a lower rating. The economic decision may be stated in the following way.

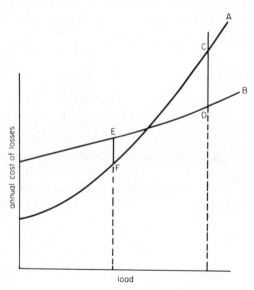

Figure 12.7 Costs of supplying transformer losses at varying loads

If by changing a transformer the losses will be reduced such that the saving in annual cost will be greater than the return required on the capital employed, that change should take place provided there is no foreseeable reversal of the trend of load growth.

Allowance may be made for load growth and optimum loading found for changing transformers[5], but, if we take a pessimistic view and a decision on facts as they exist, the following expression may be derived from equation 12.22 with equations 12.8, 12.10 and 12.11 or 12.16

$$C_c \leq (C_{f1} - C_{f2}) + (C_{vd1} - C_{vd2}) + (C_{ve1} - C_{ve2}) \qquad (12.23)$$

All quantities may be measured or closely estimated. The widest variation between sites is possibly the cost of changing a transformer, since this may vary from bare connections with bare bushings, as the cheapest example, to compound-filled cable boxes, additional cable per phase required and difficult access, as the most expensive. It is, therefore, convenient to modify calculations to provide a value of C_3 as the justifiable cost of change.

Where the exercise is repetitive, it is convenient to provide tables (recalculated after every tariff change) and a standard format for site calculations. In a large enough system a computer programme is practical.

Factors leading to change to larger rating are as follows.

(a) Higher load.
(b) Higher load factor.
(c) Flatter load curve.
(d) Low diversity of demand.
(e) Slower rate of growth (longer to pay back).
(f) Lower costs of change.
(g) Lower interest rate.
(h) Existing transformers with high losses.

It is possible that, owing to change of process, an industrial load will fall with little change of revival. It may be economic to change to a lower rating of transformer. The same format and tables will apply.

It can be seen that for the converse case of a specific method of connection there will be an optimum loading at which to change a transformer. This is not necessarily at or above the full-load rating. Lower-loss transformers will be economically loaded to a higher proportion of their rating before a change can be justified.

The maximum economic load is essential information in the other comparisons given in appendix A12.2.

12.6.8 Selection of initial rating of transformer on a given site

In this case, the method must be one of estimation. The basis of comparison used lessens the effect of future changes of tariffs, interest rates, equipment prices and errors in estimation of future growths by discounting future costs to the present day. Theoretically, the amount at the present day which, invested at compound interest, will meet the future cost is used in comparison studies. This is the standard technique known as a discounted cash flow study.

The problem resolves to comparing different combinations of transformers, by using the annual loss costs, and the economic load at which a change is made as developed in sub-section 12.6.7. Such a study is amenable to computer treatment, but, since it is at most an annual determination of policy, it is not beyond the bounds of a modern desk calculator.

A typical study is shown in appendix A12.3, showing the worth *under the stated conditions* of adopting a programme of changing transformers to meet load growth and of using low-loss transformers.

12.6.9 Selection of transformers for purchase

Specifications can only assess the costs of losses at one point in history and for one particular load pattern. Owing to the time unavoidably taken in revising specifications, it follows that specified loss values are not the most economic. Given that transformers are available with lower losses at a premium price, the overall cost should be considered. It is useful to reduce the cost of losses to two capitalisation factors.

The simplest case is the industrial process with known load characteristics. In the general case it is necessary to estimate the loss load factor, as in sub-section 12.6.6, and to estimate future load annual changes. In general, the annual maximum demand on a transformer will grow to a maximum economic value (see sub-section 12.6.7) and then will fall to a lower value when it is removed and installed elsewhere to relieve the next lower size. As described in sub-section 12.6.7 for the preferred rating series, this will be in a fixed ratio to the maximum. The annual maximum demands will therefore follow a saw-tooth pattern.

The cost of losses throughout the life will be the sum of the costs of the individual years, discounted to the present day. For the fixed loss this may be obtained precisely as the sum of the series

$$\sum_{s=1}^{q} \frac{1}{(1+n)^s} = \frac{1-(1-n)^{-q}}{n} \tag{12.24}$$

multiplied by the constant cost in each year.

For the variable loss, the loss in each individual year must be multiplied by the term of the above series. While it is possible to write an estimated load pattern calculated for each year of life and summated, the overall accuracy is relatively low. It is usual to assess the mean square value, to treat this as a constant annual charge and to summate as for the fixed loss.

The load growth is usually of the form given in equation 12.17, with I starting at I_0 rising to a maximum I_m and then restarting at I_0, forming a saw-tooth. With the symbols listed in sub-section 12.6.7 the mean curve square value I_{ms}^2 is

$$I_{ms}^2 = I_0^2 \int_0^Q \frac{1}{Q}(1+g)^{2q}\mathrm{d}q = \frac{I_0^2(1+g)^{2Q}-1}{Q} \frac{1}{2\ln(1+g)} = \frac{I^2-I_0^2}{2\ln(I_m/I_0)} \tag{12.25}$$

in which $I_m = I_0(1+g)^Q$, from equation 12.17, where I_m is the maximum current before a change of transformer. This value is independent of growth rate.*

The complete expression for the fixed-loss capitalisation factor K_{1f} per kilowatt is, therefore, with equations 12.8 and 12.24,

$$K_{1f} = \frac{1-(1-n)^{-q}}{n} \frac{C_f}{P_f} \tag{12.26}$$

and that for the variable-loss capitalisation factor K_{1v} per kilowatt is, with equations 12.10, 12.16 and 12.24,

$$K_{1v} = \frac{1-(1-n)^{-q}}{n} I_{ms}^2 \frac{C_{vd}+C_{ve}}{I_p^2} \tag{12.27}$$

A worked example is shown in appendix A12.4.

When reinforcement is based on the addition of another sub-station, one method is to assume that reinforcement stages are firstly to relieve two fully loaded substations by one addition, next to relieve these three when fully loaded by a further addition and then to consider the four resulting as two separate pairs (see table 12.4).

TABLE 12.4

Stage	1		2		3	
Network	○	○	○	○	○	○
					○	○
	⊗		○	⊗	⊗	
Load before	100%		100%		100%	
Load after	67%		75%		67%	

This makes I_0 alternately 67 and 75% of I_m; the mean square value of annual maximum demand becomes 0.723, which may be used in equations 12.25 and 12.26 for calculation of capitalisation factors. This idealised method is equivalent to a transformer utilisation of 85% (42.5% of peak rating on each of a pair of transformers).

It should be noted that, where voltage variation is reasonably small ($\pm 2\%$) such as on the input of final distribution transformers, the overfluxing factor K_s may be ignored by making it equal to zero in equation 12.8.

* From equation 12.25 the rms value may be obtained. With ideal utilisation of transformers this should be equal to the sum of maximum demands divided by the installed capacity, that is equal to the utilisation factor.

12.6.10 Reduced voltage drop in transformers

The change to a larger rating of transformer will reduce the voltage drop in meeting a given load. For a supply authority this will mean an increase in demand at peak periods. Energy sales are only marginally affected on supplies to cooking, water heating and storage-type heating but will increase to lighting, direct-acting heating and iron loss of transformers and machines.

The improvement in energy sales will be largely self-cancelling with the increased demand charge and has been ignored, particularly since indications are that the economic loading of transformers should not involve excessive voltage drops.

12.6.11 Capital rationing

The economic judgment must be based on the cost of capital. This is not always the same as interest rate, since an undertaking may have capital restricted by political or economic pressure. The cost of capital is then equivalent to the rate of return of competing projects.

12.6.12 Relations to other methods of using factors

It is common in the USA to utilise rather different factors, which may be equated as follows, with the symbols given in the sub-sections 12.6.4 and 12.6.6

(system investment/kilovoltampere peak load) × (responsibility factor) × (discount factor) × (carrying charge rate) = C_p/D

annual load factor = K_p
annual loss factor = K_s
cost of energy to utility = C_w

where

$$8760C_w = xC_{w1} + yC_{w2} + zC_{w3}$$

In the situation with smaller final distribution transformers transformer regulation is of greater import where the load is of a voltage-sensitive form, that is direct-acting devices (see sub-section 12.6.10).

Parts of Continental Europe use the principle of working hours per annum, that is the hours at full load to incur the losses actually incurred in a full year with the actual load

$$K_s = \frac{\text{Working hours per annum}}{8760}$$

Also

$$\text{Simultaneity factor} = 1/D$$

Other factors are defined by common methods and should be readily transcribed.

12.6.13 System losses

The transformer losses themselves present a load to the supply system, causing additional costs of line and transformer losses in that part of the supply system from the point at which energy is metered up to the terminals of the transformer in question. It is, therefore, normal practice to increase the calculated loss costs of that transformer by a fixed factor of about 2 to 4%, depending on the supply system.

12.7 TEMPERATURE EFFECTS OF ENCLOSURE OF AIR-COOLED TRANSFORMERS

List of symbols used

Symbol	Meaning	Unit
A	wall or roof area	m^2
A_v	ventilator area	m^2
h_e	effective height between inlet and outlet	m
h_i	inlet height	m
h_o	outlet height	m
h_t	height to top of transformer coolers	m
K_v	head loss factor of ventilator	
P_c	power dissipated by convection	W
P_d	power dissipated through walls	W
P_t	total transformer loss	W
p	fan pressure head	$N m^{-2}$
V	fan discharge	$m s^{-1}$
θ_i	temperature at inlet	°C
θ_o	temperature at outlet	°C
$\Delta\theta_e$	temperature rise at outlet	°C
$\Delta\theta_w$	temperature difference through wall	°C
Λ	wall thermal conductance	$W m^{-2} °C^{-1}$

12.7.1 Foreword

The contents of this section, excluding the paragraphs on extreme cold and humidity, are based on reference 7 and are in metricated and greatly abbreviated form. This reference should be consulted for full details and discussions of the effects of sub-station layout, calculations of ventilation and temperature conditions, the effects of ambient and varied loading conditions, etc. The author is grateful to the Electrical Research Association for their permission to publish these brief extracts and to Mr M. R. Dickson for his personal help and advice.

12.7.2   General principles

Obviously a transformer which eventually transfers the energy of its losses to the air is most efficient without any enclosure. For various reasons it may be necessary to enclose a transformer, thus reducing the efficiency of cooling and therefore reducing the rating of the transformer for the same maximum temperature rise. The majority of larger transformers are outdoors, or with at least the oil–air heat exchanger outdoors with no reduction of rating, but there are many of the smaller ground-mounted transformers which are placed indoors and for which the following remarks apply.

Some heat will be dissipated by conduction through wall and roof. This proportion is easily calculated by using the temperature difference through the wall, the thermal conductance, and the wall thickness. The temperature difference will depend on height and efficiency of the convective cooling, if we note that ventilation is essential, except for a transformer which is very small compared with the wall area. This dissipated loss is

$$P_d = \Lambda A \Delta\theta_w \tag{12.28}$$

Some representative values of Λ are given in table 12.5. It is to be noted that the value is not linear with thickness owing to boundary layers of static air. In general the floor and below ground portions of walls may be neglected since the low temperature rise and low thermal conductances (due to length of path) cause only a very small proportion of the total loss.

TABLE 12.5

Material	Thickness (mm)	Λ $(W m^{-2} {}^\circ C^{-1})$
brick	114	2.8
	228	1.9
concrete	75	3.8
	150	3.2
	225	2.8
breeze block	114	1.8
	228	1.5
tiles	standard	4.0
slates	standard	4.4
glass	any	4.4
metal	any	4.5
average ground	large	0.54

Figure 12.8 indicates the air convection currents that arise, carrying the loss energy from the transformer to the enclosing structure and to the ventilating air. Figure 12.9 shows the temperature rises. The shaded area indicates the convection energy available to circulate the air, indicating that, where ventilators are not

placed at the extreme top and bottom of the walls, the reduction in heights of the outlet is more detrimental than a similar increase in height of the inlet.

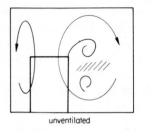
unventilated

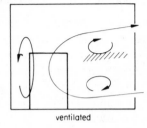
ventilated

Figure 12.8 Air convection currents

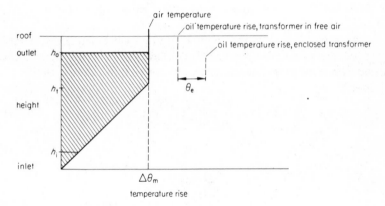

Figure 12.9 Temperature gradients

12.7.3 Ventilator position

Except as noted later under the effects of wind, for given ventilator heights and sizes the orientation does not greatly affect the main temperature distribution.

Any major obstruction near a vent where the speed of air flow is highest will form an obstruction to flow and should be avoided, leaving the transformer in a direct line of sight from the ventilators; however, if we place the transformer directly in the cold air stream which persists for 1 to 1.5 m from the inlet, this may reduce the overall temperature rise by up to 15 %. The best results are with the inlet at one-third to one-half the height of the transformer coolers. If we place the transformer too close to a wall, this may, however, restrict the air flow, leading to 3 to 4 °C excess over the temperatures that result if the spacing of the transformer from the walls is not less than 0.5 m.

Roof vents should be applied with care since, if above the transformer, the heated air may tend to be removed without mixing with the general air flow, which means that less air, but at higher temperature, is circulated and that the transformer temperature may be higher.

12.7.4 Calculation of naturally convective flow

Essentially, the calculation is an iterative method with the following steps: (1) assume a reduced rating, (2) calculate the transformer temperature rise at this rating, the permissible air temperature rise for the transformer hot-spot limit (allowing for jet effects direct on the transformer, etc.), (3) calculate the wall and roof loss and (4) arrive at the loss to be dissipated by convection. The effective height between inlet and outlet is then applied to find the effective area of ventilators required.

If the calculations are repeated for a second reduced rating and the outdoor rating is used as a third point with infinite ventilator area, this enables the construction of a curve of $1/A_v^2$ against P_t which may be used to read off either the reduced rating with a particular ventilator size or the ventilator size to be fitted for a particular required rating.

Full details of the calculation method are given in reference 7. Metricated versions of the essential formulae are appended here for convenience of conversion when utilising that reference. The convection loss is, with the symbols given above,

$$P_c = 14.85 A_v \left\{ \frac{h_e \Delta \theta_e}{(\theta_e + 273 + \Delta \theta_e)^{1.5}} \right\}^{1/2} 10^{-3} \qquad (12.29)$$

$$h_e = h_o - \tfrac{1}{2} h_t - \tfrac{1}{2} h_i^2 / h_t \qquad (12.30)$$

$$\frac{1}{A_v} = \frac{K_{v1}}{A_{v1}^2} + \frac{K_{v2}}{A_{v2}^2} + \cdots \qquad (12.31)$$

Representative values of K_v are given in table 12.6, or reference may be made to textbooks on ventilation.

TABLE 12.6

Change of section	K_v
sharp-edged orifice	2
perforated screen	1.5–2
plain louvres	1.5–2
grilles	2.2–3.5
stalled fan	10–12
windmill fan	3–4
square-edged entry to duct	0.5
pipe projecting into space	0.8
streamline flare	0
exit from duct to large space	1
right-angle bend	1
duct length (for each duct length of fifty times diameter of duct)	1

12.7.5 Forced ventilation

The air circulation will be substantially independent of air temperatures with higher air velocities and consequent turbulence. Almost all the air, except the cold jet stream from the inlet, will be at almost the same temperature. It is important to avoid the possibility of an airstream which leaves the transformer in a static air zone. So, unless the transformer can be sited directly in the inlet flow, it is wise to install louvres or baffles to break up the incoming jet of cold air to secure adequate mixing and circulation. It is preferable to arrange that, without the fan, the natural air flow is a maximum using the principles stated earlier.

The usual choice of fan is at the inlet where maintenance and physical arrangement is usually easiest. This also leads to slight pressurisation, which, in dusty situations, permits the use of an air intake filter as a ready means of reducing dust ingress. There should be no additional unblown inlets.

Propeller fans are preferred. Larger diameters with slower speeds are quieter and offer less resistance to convective air flow when switched off, particularly if the bearings are free enough to permit the fan to 'windmill'. However, other choices may be necessary with long ducting runs where higher pressure heads are necessary.

12.7.6 Calculation of forced ventilation flow

Calculation is easier since the flow of air is now controlled. Once the flow rate has been calculated, the next larger standard fan is chosen. If we assume the transformer is not directly in the incoming cold jetstream, the metricated formulae of reference 7 become for the fan discharge

$$V = 0.84P_{t}(\theta_{o} - \theta_{i})^{-1} \times 10^{-3} \qquad (12.32)$$

at a pressure head

$$p = k(V/A_{v})^{2} \qquad (12.33)$$

where $k = 0.58$ if p is in newtons per square metre or $k = 0.06$ if p is in millimetres of water head.

12.7.7 Underground sub-stations

The principles are no different from those already covered, except that the inlet must be by a shaft reaching near the base of the chamber; as for a chimney this shaft should be thermally insulated from the main air space to reduce pre-heating which would reduce air circulation. Precautions should be taken to avoid the entry of rubbish and rainwater through ventilators.

12.7.8 Environmental effects

Altitudes

Although the lower air density reduces the coolant effects at higher altitudes, the ambient temperature is generally also reduced to more than compensate. In general, altitude need not be considered.

Wind

Wind is critical in its effect on natural air circulation, generating pressures when blowing onto a ventilator, and it may generate high- or low-pressure zones due to surrounding structures. Ground drag will result in differing pressures on ventilators at different heights (see figures 12.10 and 12.11).

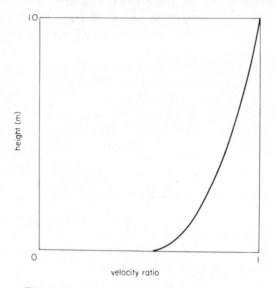

Figure 12.10   Effect of ground drag on wind

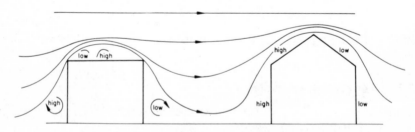

Figure 12.11   Typical aerodynamic pressure effects

In general, if inlet and outlet vent pairs are on the same vertical face of a structure, little difficulty will be met. In certain cases an outlet baffle may be necessary. Roof ventilators require special care.

Solar heating

Solar heating is of minimal effect with heavy structures in temperature zones, but, where required, an allowance of additional heat input should be made (for the area of roof and zenith facing wall) of $800\,\mathrm{W\,m^{-2}}$ in low latitudes and of $400\,\mathrm{W\,m^{-2}}$ in temperate latitudes, with a mean effect (within heavy structures) of 10% of these values.

Extreme cold

Extreme cold increases the viscosity of cooling oil to the point where radiator circulation is inhibited while the core and coils overheat. In these extreme cases it may be necessary to provide tank heaters when loads are insufficient to maintain the oil temperature above the critical point[8].

Humidity

Humidity is not normally a problem to loaded transformers. Cyclically loaded transformers will breathe, and, if there is no dehydrating breather, an oil-immersed transformer feeding a daytime load may breathe in moist air at dusk. However, free-breathing oil-immersed transformers are confined to lower voltages where moisture content of oil below saturation does not usually give rise to trouble. Dry-type transformers are usually confined to indoor situations provided with other heating. If other heating may be off for extended periods when the transformer is not loaded, conditioning heaters may be advisable to prevent the possibility of condensation.

Sealed type and dehydrating-breather type transformers are not affected.

Where transformer losses are utilised to keep dry the remainder of a single-chamber sub-station, the onset of a warm wet mass of air at times of low structure temperature and low load will lead to condensation. Where this combination of circumstances can occur other means of either conditioning or equipment, which can withstand condensation, must be supplied.

12.8 TRANSFORMER FAULTS AND PROTECTION

12.8.1 Transformer faults

Faults which may arise may be classified as follows.

(1) Phase-to-phase faults on high-voltage windings or connections.
(2) Earth faults on high-voltage windings or connections.
(3) Phase-to-phase faults on low-voltage windings or connections.
(4) Earth faults on low-voltage windings or connections.
(5) Phase-to-phase faults on tap-change gear.
(6) Earth faults on tap-change gear.
(7) Interturn faults in main tank.
(8) Interturn faults in tap-change compartment.
(9) Core faults.
(10) Low oil.

(11) Overload.

(12) Tap-change mechanism.

The causes of these faults may be insulation deterioration, faulty manufacture or maintenance, or excess applied voltage.

Over-voltages may be either due to lightning or to switching surges. Switching surges are usually no problem on systems of 33 kV or below, oil-filled circuit breakers act as a damping resistor during clearance, and current chopping is unlikely. The majority of lightning surges are due to induced rather than due to direct strokes. For small pole-mounted transformers the normal equipment includes a duplex horn gap which will not be self-clearing, but owing to the wider use of auto-reclosing this is not a great disadvantage.

For larger transformers a cable connection is some protection, but surge protection is fitted at the point of connection of cable and overhead line. With direct connection to an overhead line lightning arresters (surge diverters) should be provided at the transformer terminals; an average wavefront of $2\,kA\,\mu s^{-1}$ would cause a 60 kV drop in 30 m of connection, thereby reducing the effectiveness of the voltage surge by this amount. The arrester earths should also be bonded direct to the transformers for the same reason. At higher voltages arresters become expensive and co-ordinating rod gaps are used, mounted as above. These should not be confused with rod gaps provided for bushing protection.

Transformers connected direct to systems which are all underground may not require surge protection.

12.8.2 Fault protection

With good design phase-to-phase winding faults are unlikely. It is therefore the practice, at other than the highest voltage, to omit overall differential protection and to provide high-speed earth fault protection for each winding, relying on back-up over-current protection to clear the less likely phase-to-phase faults.

Interturn faults give little over-current externally and are not detected by overall or earth protection. To provide for these cases the surge of oil created by the sudden heat at the fault position is used to operate a pressure-actuated protective relay (Buchholz relay). Oil surge protection can operate on release of air pockets due to heating on load; it is preferable to precede commissioning by vacuum treatment to avoid this nuisance tripping on all but small transformers.

On final distribution transformers an interturn fault may occur at mid-winding, particularly if tappings exist. The resulting exceptionally heavy current in the shorted single turn, together with the loss of magnetic balance, causes an electromagnetic hammering effect which can rapidly wreck a winding, giving an appearance of a failure due to a through short circuit.

Core faults create a local hot spot which decomposes the oil into gas at a relatively slow rate. The gas may be collected in a gas-actuated relay (Buchholz relay) to give an alarm.

Low oil will operate the surge float.

The measure of over-load is winding temperature. A suitable reliable device is fitted to read the temperature accurately under all conditions. It is used for forced-cooling control, alarm and finally a trip. A most unusual case of overheating is due to ferroresonance; cases have been reported of such events[9].

Any faults that occur in the tap-change compartment of a delta or line-end tap changer are readily detected. Intertapping faults or faults on a neutral-end tap changer are not so easily seen by electric protective devices; where oil surge protective devices are provided, care must be taken that tapping operations at peak load do not cause any activation.

12.8.3 Action following relay operation

With any relay operation involving solid insulation it is most unlikely that reclosure will be successful. Unsuccessful reclosure leads to further damage and destruction of evidence of the cause. Action is therefore confined to determine whether solid insulation is not involved and successful reclosure is likely and to deduce the type of fault. The initial action is to consider relay indications (see table 12.7) which may require gas analysis. Any subsequent actions may include more sophisticated methods[10, 11].

12.9 MAINTENANCE

Maintenance will be related to the cost and complexity of equipment. The cost of failure of a pole-mounted transformer may not justify any maintenance during life. However, large generator or transmission transformers merit the other extreme of expenditure. It is important to ensure that even at this level the unit is not overmaintained, since each maintenance operation has its own risk of introducing a cause of failure. Each undertaking has usually performed the exercise of extending maintenance periods until fault rates increase and then of retracing one step. The periods adopted will depend on environment, on loading and on mode of use. Table 12.8 should therefore be considered in the light of local circumstance.

In most cases the procedure is to inspect the possible causes of failure and, only if necessary, to carry on to further work. As with all maintenance schemes there must be a system of reporting defects found, with a feedback to the originator to maintain confidence in the scheme.

APPENDIX A12.1 EXAMPLE OF ESTIMATION OF LOSS LOAD FACTOR

The network examined was of mixed domestic, small commercial and small industrial loads in a temperate zone. A rule-of-thumb first approximation of loss

TABLE 12.7 *Relay indications and consequent actions*

Relay operation	Indication	Further action	Possible cause	Reclose
current operated	fault in zone	insulation test; eliminate cable; check gas (or dissolved gas)	most internal faults	check further
oil surge	no gas	were pumps starting?	pump surge	yes
	air	vacuum treatment	entrained air released by heating or vibration	yes
	oil breakdown	was there an external voltage surge?	flashover from bare connection	yes
	solid insulation breakdown	see further testing	if no current-operated relay operation— interturn fault, earth fault near neutral or mid-delta	no
gas alarm (switch out as soon as possible)	air oil breakdown	as above switch in with minimum time settings if no more gas—was there a through fault? if more gas, switch out, drop oil level, open inspection hatch, energise at low voltage with maximum fire precautions including CO_2 above oil; trace bubbles–if on core top this may be removable cause.	as above overheated connection— site action may be possible	yes —
			core fault	—
	solid insulation breakdown		developing fault between turns or near neutral[a]	no
winding temperature (no gas)		check load	over-load	yes, after reducing load
		check cooler isolation valves	valve wrongly closed	
		check operation of forced cooling	auxiliary failure	yes
		check temperature rise by other means	WTI failure	yes
		if temperature high check oil acidity	sludge	yes, but take further action

[a] Includes fault between conductors of adjacent turns of multi-start winding with voltage of one turn, impedance to ends of winding and back giving unusually limited current.

TABLE 12.8

Item	Period	First stage	Second stage, if necessary
small pole-mounted ground-mounted up to 15 kV	6 years	visual inspection oil test and visual check	paint oil treatment, paint
above 15 kV	2 years	oil test and visual check	oil treatment, paint
dehydrating breathers	monthly	check colour of silica gel	change breather element
forced cooling	1 year	run up	fan lubrication, contactor contacts, control relay check
control relays	4 years	contact and pivot check, functional check	
oil surge relay	4 years	functional test by injected pressure wave	
OIP bushings	4 years	modern types, no maintenance necessary; older types, oil change	
on-load tap-changer mechanism:			
(a) in air	2 years	inspect, lubricate	
(b) in oil	4 years	inspect	
current-breaking contacts:			
(a) inductor type	2 years or 10 000 operations	inspect contacts filter oil and test	
(b) high-speed resistor	2 years 4 years or 20 000 operations	oil test inspect contacts, filter oil and test	
selector contacts	2 years	oil test	
where selectors in separate chamber	8 years	inspect contacts	

load factor would give $K_s = 0.3$. Detailed enquiry revealed a high off-peak heating load; the revised estimate was $K_s = 0.25$.

If we take load curves at various dates during the year and manually integrate the square of the load (see equation 12.14), this resulted in $K_s = 0.22$, where $K_p = 0.4$, $a = 0.25$, $b = 0.75$ (see equation 12.15).

Using an square-ampere hour meter, a meter reading of 17151.2 per phase was obtained, the meter being connected to the secondary of a 1600 to 5 current transformer. The annual peak loading was $I_m = 950$ A. From the load curve 10% of losses occur at lower energy rate. Equating the right-hand sides of equations 12.11 and 12.16, we obtain for K_s, with $C_{w3} = 0$, $I_p = I_m$ and t varying from x to T in the second integral,

$$K_s = \frac{C_{w1} \int_0^X f(I^2)dt \times C_{w2} \int_X^T f(I^2)dt}{C_{w1} I_m^2 T}$$

Therefore, with $C_{w1} = 0.00305\,U_m\,\mathrm{kWh}^{-1}$, $C_{w2} = 0.00360\,U_m\,\mathrm{kWh}^{-1}$ and the values stated above, K_s is evaluated as

$$K_s = \frac{(0.00305 \times 0.9 + 0.00360 \times 0.1) \times 17151.2 \times (1600/5)^2}{0.00305 \times 950^2 \times 8760} = 0.218$$

The use of a meter (or preferably meters switched to correspond to different tariff times) provides a result better than the best estimate without the laborious work involved, and this eliminates the errors that are always possible with estimation.

APPENDIX A12.2 EXAMPLE OF DETERMINATION OF JUSTIFIABLE COSTS OF CHANGING A TRANSFORMER

In monetary units U_m with tariff and load factors as shown in table A12.1, the relative costs of transformers supplying the same load are as follows.

TABLE A12.1

	500	750	1000
Existing available standard loss transformers (kVA)	500	750	1000
Transformer price multiplied by interest charge	80	110	130
Fixed loss cost per annum	50.67	69.85	87.07
Variable loss cost per annum	245.97	151.39	106.01
Total charges per annum	376.64	331.24	323.08
Justifiable cost of change		171.96	30.9
(5 year pay-back period)			202.86

Existing maximum load 600 kVA, well within present loading limits for a 500 kVA transformer. However, it would be worthwhile to change an existing 500 kVA transformer to 1000 kVA if the costs of changing are 200 U_m or less.
The actual existing transformer is of 750 kVA rating; change to 1000 kVA cannot be justified if costs of changing are 30 U_m or more.

$C_p = 18\,U_m\,\mathrm{kW}^{-1}$ $C_{w1} = 0.00305\,U_m\,\mathrm{kWh}^{-1}$ $C_{w2} = 0.00360\,U_m\,\mathrm{kWh}^{-1}$ $C_{w3} = 0.525\,U_m\,\mathrm{kWh}^{-1}$
$I_m = 800\,\mathrm{A}$ $D = 1.1$ $K_s = 0.318$

It is entirely feasible to prepare sets of tables for ready use in the field or to use computer methods. Both methods have been examined.

APPENDIX A12.3 EXAMPLE OF DETERMINATION OF INITIAL RATING

One example is given in full in table A12.2; the comparison is as follows.

Standard loss (typical)

Case A	315 to 500 to 800 kVA	$1580.66\,U_m$
Case B	500 to 800 kVA	$1707.15\,U_m$
Case C	800 kVA	$2185.66\,U_m$

Low losses (typical)

Case D	315 to 500 to 800 kVA	$1462.38\,U_m$
Case E	500 to 800 kVA	$1562.94\,U_m$
Case F	800 kVA	$1912.89\,U_m$

This is based on a starting load of 200 A per phase taken up to a point where all methods require the same-size transformer.

Under the conditions stated low-loss transformers would be used, starting with a 315 kVA rating.

APPENDIX A12.4 EXAMPLE OF LOSS CAPITALISATION

Tariff

$C_p = 18\,U_m\,kW^{-1}$

$C_{w1} = 0.00390\,U_m\,kWh^{-1}$ for 5590 h (day)
$C_{w2} = 0.00335\,U_m\,kWh^{-1}$ for 2720 h (night)
$C_{w3} = 0.00535\,U_m\,kWh^{-1}$ for 450 h (peak)

System

$D = 1.35 \qquad K_p = 0.4 \qquad a = 0.2 \qquad b = 0.8$

Mode of use

Install at 63 %; change at 100 % (80 % utilisation factor).

Financial

The discounting rate is 10 %, and the transformer life is 40 years.

Fixed-loss capitalisation

With equations 12.6, 12.7 and 12.24

$$C_{fc} = (C_{fd} + C_{fe}) \sum_{s=1}^{q} 1/(1+n)^s$$
$$= 51.32 \times 9.779 = 501.86\,U_m\,kW^{-1}$$

TABLE A12.2   *Example of determination of initial rating (case A)*

Year	Growth	Discount	Load (A)	(kVA)	Cost (U$_m$)	Fixed loss	Variable demand	Loss energy	Total	Discounted total (U$_m$)
0	1	1	200	315	659	—			659	659
1	1.05	0.909	210			35.99	11.85	12.30	60.14	54.67
2	1.10	0.826	220			35.99	13.00	13.50	62.49	51.62
3	1.16	0.751	232			35.99	14.34	14.88	65.21	48.97
4	1.22	0.683	244			35.99	15.87	16.46	68.32	46.66
5	1.28	0.621	256			35.99	17.47	18.13	71.59	44.46
6	1.34	0.564	268			35.99	19.30	20.03	75.32	42.48
7	1.41	0.513	282			35.99	21.22	22.02	79.23	40.64
8	1.48	0.467	296			35.99	23.55	24.43	83.97	39.21
9	1.55	0.424	310			35.99	25.83	26.80	88.62	37.57
10	1.63	0.385	326			35.99	28.56	29.64	94.19	36.26
11	1.71	0.350	342			35.99	31.43	32.62	100.04	35.01
12	1.80	0.319	360			35.99	34.64	35.94	106.57	34.00
13	1.89	0.290	378			35.99	38.20	39.64	113.83	33.01
14	1.98	0.263	396			35.99	42.14	43.73	121.86	32.05
15	2.08	0.239	416	500	311	52.95	26.35	27.35	417.65	99.82
16	2.18	0.218	436			52.95	28.95	30.04	111.94	24.40
17	2.29	0.198	458			52.95	31.94	33.15	118.04	23.37
18	2.41	0.180	482			52.95	35.23	36.56	124.74	22.45
19	2.53	0.164	506			52.95	38.84	40.30	132.09	21.66
20	2.65	0.149	530			52.95	42.78	44.39	140.12	20.88
21	2.79	0.135	558			52.95	47.42	49.20	149.57	20.19
22	2.93	0.123	586			52.95	52.12	54.08	159.15	19.58
23	3.07	0.112	614			52.95	57.41	59.58	169.94	19.03
24	3.22	0.102	644			52.95	63.35	65.74	182.04	18.57
25	3.39	0.092	678			52.95	69.80	72.43	195.18	17.95
26		0.084		800	442					37.13
										1580.66

Based on $K_s = 0.3$, $g = 0.05$, $n = 0.1$.
Other quantities as appendix A12.2; transformer prices as in table A12.1; change cost, 100 U$_m$; change transformer at $I_m = 100\%$ of rated value.

Variable-loss capitalisation

With equations 12.10, 12.16, 12.25 and 12.24

$$C_{vc} = (C_{vd} + C_{ve})I_{m}^2 \sum_{s=1}^{q} 1/(1 + n)^s$$

$$= 16.983 \times 0.654 \times 9.779 = 108.53 \, U_m \, kW^{-1}$$

ACKNOWLEDGEMENT

The author wishes to thank the present and previous chief engineers of the North-Western Electricity Board for permission to publish and for facilities granted. Opinions expressed and methods quoted are not necessarily policies of the North-Western Electricity Board.

REFERENCES*

(Reference numbers preceded by the letter G are listed in section 1.14.)

1. a. British Electricity Supply Industry, *ESI 35–1, Distribution Transformers (from 16 kVA to 1000 kVA)* (1971)
   b. British Electricity Supply Industry, *ESI 35–2, Continuous Emergency Rated System Transformers (24 MVA to 11 kV delta – star or star – star connected), Issue 2* (1971)
   c. British Electricity Supply Industry, *ESI–3, Cable Boxes*
   d. British Electricity Supply Industry, *BEBS T2, Transformers and Reactors* (34 parts)
   e. British Electricity Supply Industry, *Design of Underground Distribution Systems*

(All these are obtainable from the Electricity Council, 30 Millbank, London, Great Britain)

2. Davis, P. A., Design philosophy of the integrated system transformer, *Electr. Rev.*, **179** (1966) 494
3. Szwander, W., Valuation and capitalisation of transformer losses, *Inst. Electr. Eng.*, **92**, Part II (1945) 125
4. Berry, R. N., Economics of high-voltage transmission, *J. Inst. Electr. Eng.*, **94**, Part II (1947) 573
5. de Vlieger, H. C., and de Keyzer, A. J. M., Technical and economic considerations about distribution transformers, *UNIPDE Pap.*, No. 1.8, October (1972)
6. Chang, N. E., Dynamic analysis of economic loading of distribution transformers, *IEEE Power Eng. Sect. Winter Meet. Pap.*, No. C73 002-1 January (1973)
7. Dickson, M. R., *Ventilation of Transformer Sub-stations, Electr. Res. Assoc. Rep.*, No. 5096 (1964)
8. Eastgate, C., Sub-zero operation of oil-immersed transformers, *Electr. Rev.*, **181** (1967) 648
9. Dolan, E. J., Gillies, D. A., and Kimbark, E. W., Ferroresonance in a transformer, *IEEE Trans. Power Appar. Syst.*, **91** 3 (1972–3)
10. Waters, M., Farr, J. C., Stalewski, A., and Whitaker, J. D., *Short-circuit Testing and Detection and Location of Damage, CIGRE Rep.*, No. 12–05 (1968)
11. Hall, A. C., and Parratt, P. G., *Experience with Low-voltage Impulse Testing of Power Transformers, Inst. Electr. Eng. Conf. Publ.*, No. 94 (1973)

---

* See also references G1.2, G1.6, G1.7, G1.10, G1.12, G1.14, G1.17, G1.18, G1.21, G1.22, G2.2, G2.17, G2.24 and G3.2.

BIBLIOGRAPHY

Bolton, D. J., *Electrical Engineering Economics*, Chapman and Hall, London, 2nd edn (1936)

Brownsey, C. M., 'The Problem of Noise with Particular Reference to Transformers', *Mem, No.* 3, Central Electricity Research Laboratories (1956)

Franklin, E. B., 'Distribution of Water in Transformer Insulation', *Electr. Times*, **147**, 787 and 839 (1965)

Ellis, R. M., *Transformer Vibration Isolation Materials* Central Electricity Research Laboratories (Leatherhead) Report No. RD/1/R1511 (1968)

Norris, E. T., 'Loading of Power Transformers', *Proc. Inst. Electr. Eng.*, **114**, 428 (1967)

Sealey, W. C., Paralleling Load-tap-changing Transformers, *Allis Chalmers Rev.*, third quarter (1954)

British Electricity Supply Regulations, British Electricity Supply Industry, H.M.S.O., London (1937)

Guide to Transformer Noise Measurement, BEAMA Ltd Pub. No. 227, (1968)

Index